# The Lit Interior

William J. Fielder PE, m. IESNA

With contributions by
Frederick H. Jones PhD

OXFORD   AUCKLAND   BOSTON   JOHANNESBURG   MELBOURNE   NEW DELHI

Architectural Press
An imprint of Butterworth-Heinemann
Linacre House, Jordan Hill, Oxford OX2 8DP
225 Wildwood Avenue, Woburn, MA 01801-2041
A division of Reed Educational and Professional Publishing Ltd

℞ A member of the Reed Elsevier plc group

First published 2001

**British Library Cataloguing in Publication Data**
Fielder, William J.
   The lit interior
   1. Interior lighting   2. Lighting, architectural and decorative
   I. Title   II. Jones, Frederick H (Frederick Hicks), 1944–
   747.9'2

**Library of Congress Cataloguing in Publication Data**
Fielder, William J.
   The lit interior/William J. Fielder; with contributions by Frederick H. Jones.
      p. cm.
   Includes index.
   ISBN 0-7506-4890-2
   1. Interior lighting.   2. Electric lighting.   3. Lighting, architectural and
   decorative.   I. Jones, Frederick H. (Frederick Hicks), 1944–
   TH7703.F54 2001
   621.32'2–dc21                                           2001053540

ISBN 0 7506 4890 2

For information on all Butterworth-Heinemann
publications visit our website at www.bh.com

Composition by Scribe Design, Gillingham, Kent, UK
Printed and bound in Great Britain by Biddles Ltd, *www.biddles.co.uk*

FOR EVERY VOLUME THAT WE PUBLISH, BUTTERWORTH-HEINEMANN
WILL PAY FOR BTCV TO PLANT AND CARE FOR A TREE.

# Contents

# Preface

This book is intended as a design guide for those individuals in the fields of electrical engineering, architecture, and interior design who will one day design lighting systems for others to build.

The book is organized so that an individual with little or no training in lighting design will become familiar with the basic principles and psychology behind good lighting before design procedures are addressed. Discussions on the process of vision and the properties of light set the stage for exploring the various tools at the designer's disposal for creating and manipulating light to provide a desired effect in an architectural space.

The reader is then led through the conceptual design process, which entails the use of manufacturer's offerings, codes and guidelines for space lighting, as well as calculation methods to predict the performance of a design. The conceptual design is rounded out by exploring methods for powering and controlling a lighting system.

A realistic design problem is begun early in the journey, and is completed, bit-by-bit, as each new concept is explored and applied to the design. Documentation of the design is the final stage of the process, which culminates in a finished set of plans and detailed specifications for the project. A final segment of the book, called 'The Second Time Around', is devoted to retrofitting existing inefficient lighting systems with new, energy-efficient components to improve light quality and reduce the energy consumption of older systems.

Extensive use of the Internet is used throughout the design process. Instructions for downloading and using manufacturer's data, calculation engines, and other tools are included in the text and put to use in the exercises. In the interest of continuity, Internet information for this book is almost exclusively that of Lithonia Lighting Co., a lighting equipment manufacturer. Other manufacturers have similar information available, and the reader is encouraged to search the internet for other favorite sources of information.

William J. Fielder
South Carolina, USA

# 1 The design medium

The art and science of lighting design is just that, and more: a little artistic flair; some scientific knowledge; and last but not least, a healthy helping of psychology. While every well-done lighting design is attractive, and most provide adequate illumination for the task at hand, the superior design goes the extra mile: it takes into account the effect of the lighted environment on the eye and mind of the human observer. This psychology of the environment is always at play in the relationship of people and architecture, and it can be molded dramatically with effective lighting.

Light can be thought of as a 'building material' much like steel or concrete. Although structural components are needed to enclose a space, the space has no existence for an individual until it is seen and registered in the conscious mind. Light defines space, reveals texture, shows form, indicates scale, separates functions. Good lighting makes a building look and work the way the architect intended at all hours of day and night. It contributes to the character and effective functioning of the space by creating the desired attitude in the mind of the occupant. Change the lighting and the world around us changes.

The actual way the eye–mind combination evaluates light is a complex, dynamic process, which could fill volumes the size of this one. There are, however, some basic principles which bear consideration in the design of lighting systems. In this chapter we will consider both the process of vision, and the effect that light has on our perception of the lighted architecture. You should come away with a better understanding of both the physical and psychological aspects of a lighted environment.

## The process of vision

The process of vision can be roughly compared to the operation of a radio or television receiver: there is an antenna, the eye, tuned to

WAVELENGTH IN NANOMETRES

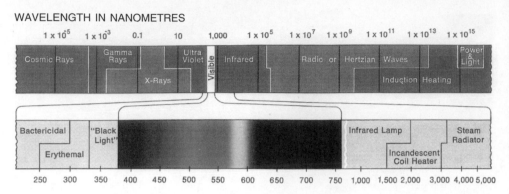

Figure 1.1. The electromagnetic spectrum (source: *Philips Lighting Handbook*).

a specific portion of the electromagnetic spectrum; there's a cable, the optic nerve, connecting the antenna to the decoding device; and then there's the decoding device, the brain, which processes the received information. The eye is tuned to that portion of the electromagnetic spectrum with wavelengths between 380 and 780 nanometers (1 nanometer = $10^{-9}$ m = 1 thousand millionth of a meter) known as the *visible portion* of the spectrum. Figure 1.1 shows the electromagnetic spectrum, with the visible portion expanded.

As you can see, the visible part of the spectrum covers the wavelengths from ultraviolet, which is commonly associated with skin damage from the sun, to the infrared, which is associated with the heat felt from the sun. This points out the fact that *the shorter the wavelength, the higher the energy* in electromagnetic radiation.

The 'visible' section of the electromagnetic spectrum (see Figure 1.1), when seen simultaneously, appears as white light, such as bright sunlight at noon on a clear day. When white light strikes an object, part of it is reflected, and part is absorbed. For example, a ball which is seen as blue is, in fact, reflecting the blue wavelengths and absorbing all the others.

Our eyes are sensitive to all the wavelengths within the visible spectrum. However, as stated before, they act as 'antennas' to receive reflected light and, like antennas, they are tuned to a specific frequency. In the case of the eye, that frequency lies approximately at the center of the visible spectrum, and has a wavelength of 550 nanometers. This means that the sensitivity of the average eye peaks in the yellow–green portion of the spectrum, and falls off sharply as the limits of the spectrum are approached. Figure 1.2 shows this as a bell curve of eye response relative to light wavelength.

Our eyes not only have to respond to a wide range of wavelengths, but they also must automatically adjust to a constantly varying light intensity. To see how this is accomplished,

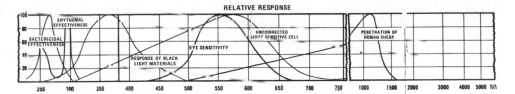

Figure 1.2. Color response of the eye (source: *Philips Lighting Handbook*).

let's take a look at the components of the eye, as shown in Figure 1.3. The 'front end' of the eye acts much like a camera to regulate the incoming light and focus it on the retina. This 'front end' is made up of the *cornea*, the clear outer layer of the eye, and the *pupil*, an opening whose size is constantly being adjusted by the *iris* to compensate for varying light intensity, and the *lens*, which uses the *ciliary muscle* to change its shape to focus the light on a special part of the *retina*, called the *fovea*. The retina contains from 75 to 150 million *rods* and about 7 million *cones*, which make up the actual antennae tuned to the visible spectrum. The rods and cones convert light energy into neural signals that are transmitted to the brain through the optic nerve.

Rods cannot detect lines, points, or colors. They can only detect light and dark tones in an image. Rods are highly sensitive, and they can distinguish outlines of objects in almost complete darkness. Cones are even more sensitive – they detect the lines and points of an image, such as the words you are now reading. Cones also detect

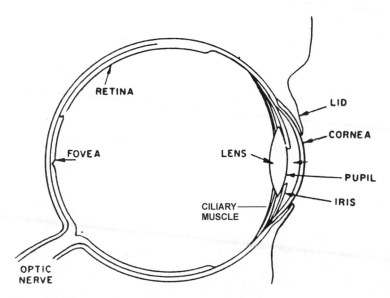

Figure 1.3. Components of the eye (source: F.H. Jones).

color, and there are three types of cones present in the eye: one that is sensitive to the blue–violet end of the spectrum; one sensitive to the yellow–green, or middle of the spectrum; and one sensitive to the red end of the spectrum.

The fovea contains only cones, and provides the optimal reception in brighter light conditions. Muscles controlling the eye work in conjunction with the ciliary muscles controlling the lens to keep the viewed object focused on the fovea. That's why you are moving your eyes while reading this page.

In higher light levels, the cones are the main receptors of light, and the response of the eye to the varying wavelengths of light is as shown in Figure 1.2. In a very low level of light, the cones cease to function, and the sensitivity peak of the eye shifts toward the light with the higher energy wavelengths at the blue end of the spectrum. This is known as the *blue shift*, or *Purkinje effect*, and it is the reason that, under very dim ambient light, the eye will perceive blue light as inordinately bright. This is why police cars in the US have switched from red to blue emergency lights.

As we get older, the components of the eye begin to deteriorate. The ciliary muscles get weaker, the lens loses elasticity, and our ability to focus, particularly on close objects, becomes less. The lens itself yellows with age, which affects color vision, particularly the differentiation between blues and greens. The lens also becomes thicker and less transparent, which results in light scattering and 'night blindness', or extreme sensitivity to glare. The pupil gets smaller, which reduces the overall amount of light which reaches the retina. The result of all this deterioration is that older people need more illumination, larger print, and more contrast in order to see clearly – and to function comfortably.

Now that we know something about the eye, let's take a look at some of the mechanics of those light rays which are constantly bombarding our rods and cones.

## Light mechanics

Light travels in a straight line until it strikes a surface. It is then modified by either *transmission, refraction, reflection*, or *absorption*. Figure 1.4 illustrates each of these light modifiers.

Light can also be modified by polarization, diffraction, or interference by other light rays, but these play a very small part in lighting design. For now, let's concentrate on the 'big four', and see how they affect light rays.

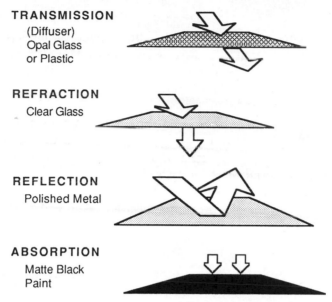

**TRANSMISSION**
(Diffuser)
Opal Glass
or Plastic

**REFRACTION**
Clear Glass

**REFLECTION**
Polished Metal

**ABSORPTION**
Matte Black
Paint

Figure 1.4. Types of light modification (source: F.H. Jones).

## 1. Transmission

There are three general categories of transmission: *Direct transmission* occurs when light strikes transparent material which can be seen through. These materials absorb almost none of the light in its passage through the material, and do not alter the direction of the light ray. *Spread transmission* occurs with translucent materials in which the light passing through the material emerges in a wider angle than the incident beam, but the general direction of the beam remains the same. *Diffuse transmission* occurs with semi-opaque materials such as opal glass, and the light passing through the material is scattered in all directions. These materials absorb some of the light, and the emerging rays are of less intensity than the transmitted rays. Figure 1.5 illustrates the types of transmission.

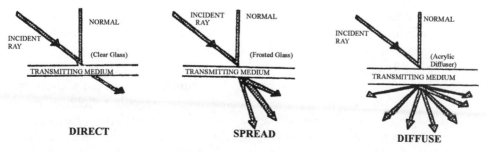

Figure 1.5. Types of light transmission (source: F.H. Jones).

## 2. Refraction

Refraction occurs when a beam of light is 'bent' as it passes from air to a medium of higher density, or vice versa. This occurs because the speed of the light is slightly lower in the medium of higher density. Two commonly used refractive devices are *prisms* and *lenses*. A prism is made of transparent material which has non-parallel sides. A large prism slows down the various wavelengths of light by different amounts and can be used to divide the light ray into its color components; smaller prisms are used in lighting fixtures to lower brightness or to redirect light into useful zones. Lenses are used to cause parallel light rays to converge or diverge, focusing or spreading the light, as desired. Figure 1.6 illustrates some refractive devices.

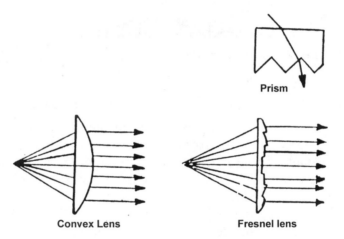

Figure 1.6. Refractive devices (source: F.H. Jones).

## 3. Reflection

Reflection occurs when light strikes a shiny opaque surface, or any shiny surface at an angle. Reflection can be classified in three general categories: *specular reflection, spread reflection* and *diffuse reflection*. Specular reflection occurs when light strikes a highly polished or mirror surface. The ray of light is reflected, or bounced off the surface at an angle equal to that at which it arrives. Very little of the light is absorbed, and almost all of the incident light leaves the surface at the reflected angle. Spread reflection occurs when a ray of light strikes a polished but granular surface. The reflected rays are spread in diverging angles, due to reflection from the facets of the granular surface. Diffuse reflection occurs when

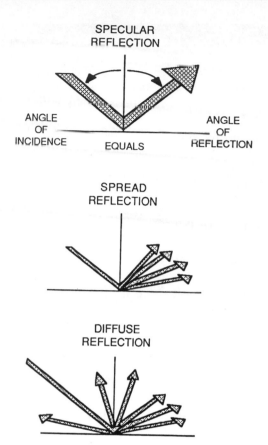

Figure 1.7. Specular, spread, and diffuse reflection (source: F.H. Jones).

the ray of light strikes a reflective opaque but non-polished surface, such as flat white paint. Figure 1.7 shows the types of reflection.

## 4. Absorption

Absorption occurs when the object struck by the light ray retains the energy of the ray in the form of heat. If you remember the blue ball example, the ball reflects only the blue wavelengths of the incident light, and absorbs all of the others. If the ball were in the sunlight, this energy absorption would heat the ball up. Some surfaces, like flat black paint, absorb nearly all of the incident light rays. These surfaces, such as those of a solar collector panel, tend to get very hot when placed in the sunlight.

With these principles in mind, you can predict how the light itself will behave when used with the various control devices. Now let's look at some of the factors of light which affect the way we see.

## Physical factors

In addition to color, the four factors which determine the visibility of an object are: *size, contrast, luminance*, and *time*. Of the four, luminance, or brightness, or the strength of the light falling on the rods and cones, is the underlying dominant factor. Let's look at these factors in more detail.

### 1. Size

Size is considered because the larger or nearer an object, the easier it is to see. A larger object, of course, reflects more total light, and offers a stronger stimulation of the rods and cones. Also, as we will see in a moment, light adheres to the inverse square law. This means that the strength of the reflected light decreases as the square of the distance between the object and the eye. In other words, the closer the object, the stronger the reflected light.

### 2. Contrast

Contrast is simply the difference in brightness of an object and its background. Distinct contrast allows the brain to differentiate easily between areas of strong and mild visual stimulation. For example, black words on white paper are read easily, but gray lettering on a slightly lighter gray paper is much harder to interpret.

### 3. Luminance

Luminance, simply put, is the brightness of an object, or the strength of the light reflected from it. The greater the luminance, the stronger the visual stimulation, and the easier the object is to see.

### 4. Time

Time refers to how long it takes to see an object clearly. Under the best conditions, it takes slightly less than one-sixteenth of a second for the eye to register an image. In a dim setting, it takes longer. This is especially important where motion is involved, such as in night driving.

Obviously, the luminance of an object, or the quantity of light reflected from it, determines the level of visual stimulation the object provides. Now it is time to look more closely at the mechanics of light quantity, and also to investigate another factor that influences visual acuity, namely, light quality.

## Light quantity

In evaluating light quantity, it will be helpful to examine the afore-mentioned inverse square law, and some of the nomenclature that is used to describe the features of light. Succinctly put, the inverse square law as applied to lighting states that: 'the luminance of an object is directly proportional to the light output of the illuminating source, and inversely proportional to the square of the distance between the source and the object'. At the risk of losing a few of you to the geometry, let's look at Figure 1.8, which graphically illustrates the inverse square law.

Light output from a source is normally expressed in *candlepower*, and light output in a given direction is expressed in *candelas*. The density of light flux radiating from the source is expressed in *lumens*, and the luminance, or light reflected from an object is expressed in *footcandles*. Footcandles has units of lumens per square foot. Figure 1.8 shows a point source of uniform candle-power, having 100 candela in all directions. If we approximate light propagation in a solid angle of 1 steradian and go out a distance of 1 foot from the source, we see that the angle circumscribes an area of 1 square foot. The glossary defines a *lumen* as the flux density generated within 1 steradian by a point source of 1 candela. We have 100 candela in the source of Figure 1.8, so the flux density will be 100 lumens. Using the lumens per square foot definition, we see that the luminance at 1 foot will be 100/1, or 100 footcandles (100 fc). If we go out 2 feet from the source, we see that the area circumscribed by the steradian envelope is now 2 squared, or 4 square feet. Similarly, at 3 feet, the area is 9 square

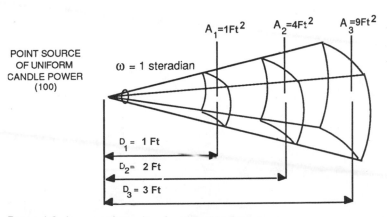

Figure 1.8. Inverse square law (source: F.H. Jones).

feet. Corresponding luminances are 100/4, or 25 fc, and 100/9, or 11 fc, respectively. For the mathematicians among you, this relationship can be expressed mathematically as $I = L/D^2$, where $I$ is illumination in footcandles, $L$ is the luminance of the source in lumens, and $D^2$ is the square of the distance in feet from the source to the point under examination.

The inverse square law works pretty well in predicting the illumination on a surface from a point source directly above the surface, but what happens when we want to predict the effects of a source that is at an angle to the surface under consideration? We can use an old static mechanics trick and expand the inverse square law to take care of the angle by breaking the angle down into its two components, one parallel to and one normal to the surface, and then discarding the parallel component. Figure 1.9a illustrates this graphically.

Now, if you've ever had a statics course, you'll remember that a force applied to a Point P on a beam at an angle ω from the normal is treated this way, and that the downward component of the force is equal to the total force times the cosine of ω. If you've never had statics, no matter, it still works that way. Taking luminance $L$ as the 'force' of the light, and using the inverse square law, we can say that the illumination $I$ on a point from a source that is at an angle $X$ from being directly above the point, and at a distance $D$ from the point is: $I = L \cos X/D^2$. This is called the *cosine law of incidence*. To get some idea of what this means, look at the light sources above you and all around you. All of these contribute to

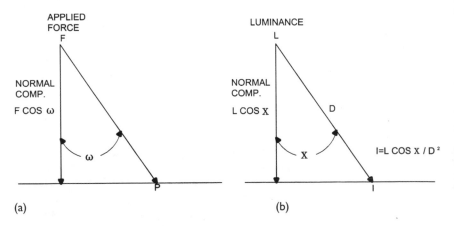

Figure 1.9. Two-component force vector.

the total illumination falling on your desktop. If you have a good calculator, and about a month of free time, you can calculate exactly what that illumination is, using this equation.

Fortunately, we don't have to get bogged down in extensive, tedious calculations of this sort. As we will see in a later chapter, there are plenty of good computer programs out there to perform these calculations for you. It is, however, helpful to know the logic behind the calculations, so that you will be able differentiate between valid output and computer-generated gibberish.

So there you have the factors involved in the quantity of light that illuminates a chosen area. To review in a nutshell, these are the *strength in candlepower* of the illuminating sources; the *distance* those sources are from the area; and the *angle* those sources are from the normal to the surface of the area. Adequate quantity of light, however, doesn't always insure good visibility. The quality of the light is often as important as the quantity.

## Light quality

What do we mean by light quality, and what are the factors which contribute to 'good' or 'bad' quality illumination? Simply put, good quality illumination is that which provides a high level of visual comfort, and allows us to view tasks clearly and easily. This affects our psyche in a positive way. On the other hand, poor visual comfort illumination irritates us. The four most important factors affecting visual comfort are *glare, brightness ratio, diffusion*, and *color rendition*. Let's look now at each of these factors in greater detail.

### Glare

We've all experienced glare in our everyday lives: bright lighting fixtures located in your field of view, or sunlight coming through a window. This is known as *discomfort glare*, and the degree of discomfort inflicted depends on the number, size, position, and luminance of the glare sources. In interior lighting design, we are primarily concerned with discomfort glare from windows and overhead lighting fixtures. Other forms of glare are *disability glare* and *veiling reflections*. Disability glare obliterates task contrast, and scatters the light within your eye to the point that visibility is reduced to zero. A common example is glare from a glossy magazine page that makes it impossible to read the page. Veiling reflections, such as lighting fixture 'images' on your computer monitor, make it hard to see what is on the screen. The severity of glare in any form is primarily dependent on two factors: the *brightness* and *position* of the source.

## Brightness ratio

The brightness ratio is the brightness contrast between the task and the background. This affects the amount of work our eyes have to do in order for us to perform the task. For example, a high brightness task in a low brightness surrounding forces the eye to continually adjust from one light level to the other. Conversely, a low brightness task surrounded by a bright background tends to obscure contrast, and the eye tends to be attracted away from the task. Obviously, a balance between task and background brightness is desirable for effective viewing.

## Diffusion

Contrary to the above factors, which affect viewing negatively, diffusion generally improves visual comfort. Diffusion results from light arriving at the task from many different directions. A highly diffuse lighting system will produce no penumbra, or sharply defined shadows. Diffuse lighting is desirable in office areas where computers are in use, in school classrooms, and in library reading areas. Diffuse lighting is accomplished through the use of many low brightness fixtures, or through the use of indirect lighting, where the light is reflected from diffuse surfaces, such as a white ceiling, before reaching the task.

## Color rendition

Color of light affects the 'mood', or emotional aspects of a space. It also affects the accuracy with which we perform tasks. We've all, at one time or another, purchased a garment under artificial lighting, only to have it change color when we got it out into the sunlight. That happens because the artificial light source does not contain the full visible spectrum of colors, as does the sunlight. As noted before with the blue ball example, we see only those colors which are reflected from a surface. Obviously, those tasks that involve color discrimination should be lighted by a source that contains as much of the visible spectrum as possible. In other situations, the mood of the environment can be altered by the use of 'warm' or 'cool' colors, high in reds, or blues, respectively.

Unlike light quantity, light quality is subjective in nature, and is not easy to calculate by mathematical formulae. The lighting industry has, however, come up with several methods that the designer can use to evaluate the relative quality of lighting systems. The first of these is *equivalent sphere illumination* (ESI). This is a complicated method of relating illumination of a task on a surface within the design space to that of a task on a surface in the center of a sphere that is equally illuminated throughout. The logic being that the lighted sphere will provide the optimum illumination, and that the space should be designed to match the footcandle requirement of the sphere as closely as possible. For example, if a task requiring 100 fc in the design space was put into the sphere, and the lighting level was adjusted to provide the same task visibility, and that level was 60 fc, then the ESI would be 60 fc. Equivalent sphere illumination takes into account room geometry and reflectance, fixture characteristics, and viewer position. Needless to say, only fixture manufacturers with big computers attempt ESI calculations. Another comparison type system evaluator is the *relative visual performance* (RVP) factor, which is expressed in percentages. The RVP represents the percentage likelihood that a standardized visual task can be performed within the designed lighting system. Age of the viewer, luminance and contrast are all included in RVP calculations. When comparing systems, the one with the higher RVP will provide better light quality. Also expressed in percentages is the *visual comfort probability* (VCP), which is the percent of viewers positioned in a specific location, viewing in a specific direction, who would find the lighting system acceptable in terms of discomfort glare. Visual comfort probability takes into account room geometry and reflectances, fixture number, type and luminance. As with RVP, the higher the VCP, the better the light quality of the installation.

The lighting industry has done a yeoman's job of trying to quantify the factors of light quality so that the above evaluators may be calculated numerically. There are so many non-direct factors involved, however, that these calculations are best left to fixture manufacturers with plenty of time and people, and large computers. Most fixture manufacturers publish some sort of visual comfort data for their fixtures. A good lighting designer is aware of the causes of visual discomfort, and develops an innate 'feel' for which fixtures will perform well where, rather than trying to rely solely on numbers and calculations to provide good light quality.

## The psychology of lighting

A seasoned lighting designer can visualize how a given lighting system will look and perform within a space. He also can predict

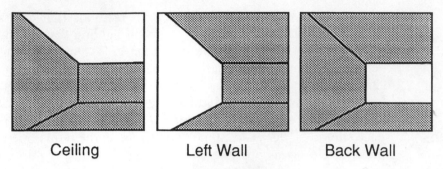

| Ceiling | Left Wall | Back Wall |

Figure 1.10. Planes of brightness (source: F.H. Jones).

how an observer will react to the system. This insight is gained with experience, of course, but certain basic relationships of light and space and the psyche are always present, and are worth mentioning. The first is the location of the plane of brightness, or the brightest surface in the space. Figure 1.10 illustrates some different planes of brightness.

A ceiling left in shadow creates a secure, intimate, and relaxing 'cave' environment suitable for lounges and casual dining. High brightness on the ceiling creates the bright, efficient, working atmosphere desirable for offices, classrooms, and kitchens. Brightness on the vertical planes draws attention to the walls and expands the space visually, and is appropriate for art galleries, merchandising, and lobbies. Such facilities often also use variations of light intensity on the walls to accentuate a desired feature.

Variations of light intensity form areas of light and shadow, which are desirable if you are trying to create a 'mood' environment, rather than an evenly illuminated workplace. The interplay of light and shadow add variety to a space, and provide visual relief to an otherwise monotonous environment. Scallops on a wall from downlights, shadows on the ceiling from uplights, or highlights from accent lighting create areas of visual interest, and can draw attention to a desired area or object. The designer must be careful not to overdo it, though, because too many lighting effects in one space have roughly the same visual effect that too many sidebars, colors, and font styles do to a magazine page: the original design intent is obscured or obliterated.

It is always best to work with the architect from the outset of a project to get in tune with the flavor or mood that he or she is trying to create in a space. Architectural features can be modeled through the use of shadows, as can objects within the space. A three-dimensional object lighted directly from in front will appear

Figure 1.11. Effects of shadows (source: F.H. Jones).

flat, but when lighted from an angle, will assume depth and round-
ness. We all remember the 'Frankenstein flash' from photography
class, where the hand-held flash is placed beneath the chin of the
subject. The resulting photo looks like a Boris Karloff publicity
shot. This happens because of the striking contrast between the
brightly lit and deeply shadowed facial features. The same effect
can be achieved in architectural spaces through the use of uplights,
downlights, and lighting from the side. Figure 1.11 illustrates
modeling through the use of shadows.

Areas of brightness can also be used to create mood, or accent
architectural features. Small pinpoints of brightness from tiny lamps
or reflections add sparkle and glitter to a space, which enhances
the gaiety, elegance, and festivity of the space. You can try this
yourself at home with a string of clear Christmas mini-lights. Wrap
that ficus tree in the corner with the lights, and plug them in. Note
the overall effect. Or string the lights around an architectural
feature, such as an arched doorway. You'll get the idea. The watch
word here is the same as with light and shadow, don't overdo it –
and keep some background light. Without sufficient background

Figure 1.12. Sparkle and glitter (source: F.H. Jones).

lighting to soften the contrast, glitter can become glare. The lighting is used to accent the architecture – not overpower it. Figure 1.12 shows an effective use of sparkle and glitter.

## Summary

We now know the nature of light, how the eye responds to light, and how light behaves when subjected to various control mediums. We've also explored methods for quantifying the intensity of light, and some methods for assessing the visual comfort of a lighting system. All of this knowledge will come into play in every lighting system that you design.

We have also looked at some of the artistic aspects of lighting design. Although these will probably not be a majority of your total design effort (unless you specialize in artistic lighting design), this is where you can use your creativity to create your own unique lighted environment.

In the next chapter we will explore some of the tools that you have at your disposal for creating a lighting system. But first, let's see how much you've retained from this chapter. . .

## Exercises

1. What is the wavelength range of the visible portion of the electromagnetic spectrum?
2. Which light rays contain the most energy?
   a. Ultraviolet
   b. Infrared
   c. Blue–Green
3. When we see an object, we are actually seeing
   a. Light striking the object
   b. Light reflected from the object
   c. Light being diffracted by the object
4. We see colors by using the eye's
   a. Rods
   b. Cones
   c. Iris
5. What are four things that happen to our eyes as we age?
6. Name four ways that a light ray can be modified.
7. What is the purpose of a lens?
8. What are the three types of reflection?
9. What are the most important factors determining the visibility of an object?
10. In 31 words or less, parrot back the definition of the inverse square law as it applies to lighting.
11. Problem: A point source of 200 lumens, located directly above a surface, and 2 feet from it, will produce how many footcandles on the surface?
12. What law would we invoke to calculate illumination by a source at an angle from the normal to the lighted surface?
13. What is considered 'good quality' illumination?
14. What is disability glare? Why do we call it that?

15. What determines brightness ratio?
16. Where would diffuse lighting be used?
17. What determines the perceived color of an object?
18. What does a visual comfort probability (VCP) of 70 tell us about a lighting system?
19. Which ceiling brightness would we try to obtain in an upscale dining space?
    a. High
    b. Medium
    c. Low
20. True or false: To add spice to a space, we always try to use as many lighting effects as the budget will allow.
21. What methods would you use to achieve effective shadows?

# 2 The design tools

Just as an artist has paints and canvas, and a sculptor has chisels and marble, a lighting designer has certain tools with which to create a lighted environment. There is a wide variety of lighting sources, or lamps, available to provide the desired light intensity and color, and there is an almost unlimited choice of lighting fixtures (called *luminaires* from here on out) to shape and place the light in just the right pattern, and there is a multitude of control systems to make those luminaires behave as the designer intends. Like the artist, whose canvas has finite limits, the lighting designer has energy, task, and code requirements that establish the boundaries of the design. To aid in meeting these requirements, the designer has several calculation techniques, both manual and computerized, available to help predict the adequacy and long-term performance of the design.

As with any endeavor, the lighting designer will develop skill and technique with experience. Until that experience is gained, there are a number of 'tried and true' universal lighting techniques that may be used to solve a variety of design problems. Some of these techniques are included in this chapter.

Understanding this powerful arsenal of tools will enable the designer to specify a superior lighting system for any space with confidence, be it a warehouse or a specialty retail shop.

Let's now take a closer look at these tools.

## Lamps – the light source

### Lamp theory

Artificial light, meaning something other than sunlight, may be produced either chemically, mechanically, or electrically. Chemical light is used for very special applications, such as the emergency wands that you see in sporting goods stores; mechanical (flame –

gas lamps) is primarily used to create atmosphere. By far, the greatest producer of usable artificial light is electricity. The electrically driven light source is what we will examine here.

The two most important methods by which we can produce light from electricity are *incandescence* and *photoluminescence*. These are big words used to describe the processes by which the lamps that we use every day produce light. Let's see how each of these processes works.

## Incandescence

*Incandescence is the visible radiation produced by a body at a high temperature.* In incandescent lamps, the temperature is created by passing an electric current through a wire filament which has a finite resistance to current flow. The result is called Joule heating, and is proportional to the resistance of the filament, and to the square of the current flowing. As the filament gets hot, it glows, first red, and then white. The white light produced by a glowing filament contains all of the colors of the visible spectrum to some degree, and therefore produces a *continuous spectrum* of light. In incandescence this spectrum is higher in intensity toward the longer wavelengths, or the 'red' end of the visible spectrum, but the actual intensities of the various wavelengths depend upon the temperature of the filament. Figure 2.1 shows the spectral energy distribution of a tungsten filament at 3000 K.

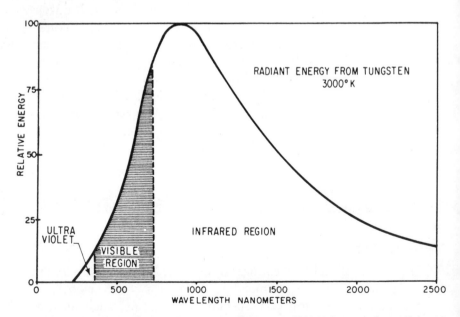

Figure 2.1. Spectral distribution of a tungsten filament at 3000 K (source: Osram/Sylvania).

## Photoluminescence

Photoluminescence is what happens when *neutral gas atoms collide with electrons in an electric arc discharge*. Some of the energy of the collision is released as visible radiation, and some as ultraviolet (UV) radiation with wavelengths below the visible spectrum. The energy is released at specific wavelengths which are dependent on the chemical makeup of the gas. The resulting spectral distribution is a discontinuous series of 'spikes', rather than the smooth distribution curve of the incandescent source. This is illustrated in Figure 2.2, which is the spectral energy distribution for a mercury vapor lamp.

Figures 2.1 and 2.2 are what is called *spectral energy distribution curves*. They show what wavelengths, and hence what colors of light are produced by the source under examination. These curves are a good indicator of the color-rendering abilities of the source.

Remember the blue ball example from Chapter 1? OK, let's assume that the color of the ball is about 440 nanometers in wavelength.

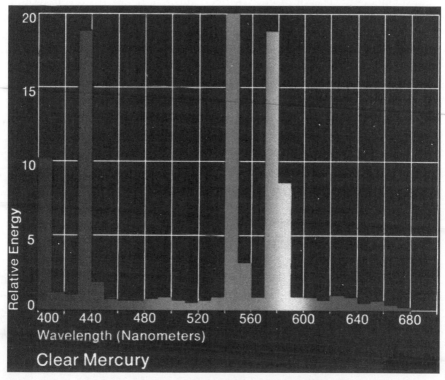

Figure 2.2. Spectral distribution of a clear mercury vapor lamp (source: Westinghouse).

What would the ball look like if it were illuminated by the mercury vapor source in Figure 2.2? Since the ball's color is contained in the spectrum of the source, the ball will reflect the 440 nanometer wavelength light, and appear . . . bright blue. What would a red ball of 700 nanometers wavelength look like? Right – brownish-black! Since very little 650 nanometers wavelength light is in the mercury vapor spectrum, the ball will absorb almost everything. Now you know why meat markets don't use mercury vapor lamps.

## Color temperature

It is common to speak of 'cool' colors such as blue and green, and 'warm' colors, such as red and orange. These colors, of course, are the result of the spectral distribution of the light source illuminating them. Light sources, then, can be called cool or warm, or in-between, dependent upon their spectral distribution. Lighting designers find it useful to be able to assign temperature numbers to define the degree of coolness or warmth of a source.

In correlating color to temperature, it is helpful to think of the old time blacksmith heating a piece of iron. As the iron started to heat up, it would glow a deep red. As it got hotter, it would become bright red, and finally, white hot. The easiest way to describe the color of the glowing metal was to give its temperature, because any two glowing pieces of the same metal having the same temperature would always have the same color.

The tungsten of an incandescent lamp filament behaves in the same way, and the color of its emitted light is always directly related to its temperature in Kelvin degrees. The whiter the light, the higher the temperature. The color temperature of a household incandescent lamp, for example, is a little less than 3000 K. Strictly speaking, color temperature applies only to incandescent sources and to natural sources such as the sun and the sky.

When it comes to photoluminescent sources such as fluorescent and gaseous discharge lamps, the correct term to use is *correlated color temperature*. When a fluorescent lamp is said to have a correlated color temperature of 3000 K, that means that its color looks more like that of a 3000 K incandescent source than that of an incandescent source of any other temperature. This does not mean that it will illuminate colored objects the same way a 3000 K incandescent source will, however. Figure 2.3 shows the spectral energy distribution curves for several popular fluorescent lamps.

As shown in Figure 2.3, fluorescent lamps do not have the smooth continuous output curve of an incandescent lamp, so its color temperature is only an apparent one. Figure 2.3 illustrates

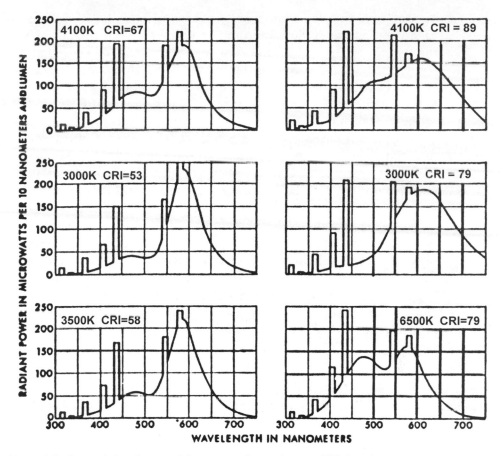

Figure 2.3. Spectral distribution of fluorescent lamps (source: F.H. Jones).

that two 4100 K fluorescent lamps can have very different spectral distributions. In fact, with the right mix of phosphors, it would be possible to make a fluorescent lamp that had most of its output in only two narrow peaks of light, one blue and one yellow, and balance the colors so that the lamp had a correlated color temperature of around 3000 K. The color performance of the lamp would be very poor, however. A red object would appear brown under this lamp, while it would appear normal under a 3000 K fluorescent source with an spectral distribution containing reds.

## Color rendering index

Lamp manufacturers, as well as lighting designers, find it useful to be able to compare how well the different sources render colors, so in 1965 an international panel of experts, the Commission Internationale de la Eclairage (CIE) devised an extremely complicated method of testing to assign a *color rendering index* (CRI) number to each lamp on the market. Fortunately, doing the mathematics falls to the manufacturer, and all the designer has to remember is that the higher the CRI, on a scale of 0–100, the better that lamp will render colors. Figure 2.3 lists the CRI for each lamp shown. It is seen that lamps of the same color temperature can have very different CRI ratings, depending on the colors contained in its spectral distribution.

In general, the higher the CRI, the better a fluorescent or discharge source compares with a natural source at the *same* correlated color temperature. It has been said that correlated color temperature is what the lamp is trying to be, and its color rendering index shows how well it is succeeding.

Lamp manufacturers have devoted large amounts of time and money to developing exotic phosphor coatings for the inside of photoluminescent lamps in order to improve CRI. As a result, some fluorescent lamps with a CRI of over 90 are readily available.

Let's take a closer look now at the three major types of light sources, or lamps, which you, as a lighting designer, will use. These are *incandescent, fluorescent,* and *high intensity discharge (HID)* lamps.

## Lamps – major types

### Incandescent lamps

*Standard incandescent lamps*

The standard incandescent lamp is the oldest electric lighting technology still available today. Although relatively short lived (750–3000 hours), and fairly inefficient (6–24 lumens per watt), incandescent lamps are widely used in residential and other applications. This is primarily because they are cheap, easy to obtain, work in inexpensive fixtures, have good color rendition, and perform well in low ambient temperatures. Standard incandescent lamps are still constructed today much the same way they were when they were originally introduced: two lead-in wires are attached to a metal screw base, and are insulated from each other by a glass stem. A tungsten filament is attached between the two lead-in wires, and a glass envelope, called the bulb, surrounds the filament structure and is attached to the base. The bulb is evacuated, then filled with an inert gas to prevent the filament from burning up. The envelope may be frosted to diffuse the light emitted from the filament. Figure 2.4 illustrates the construction of the incandescent lamp.

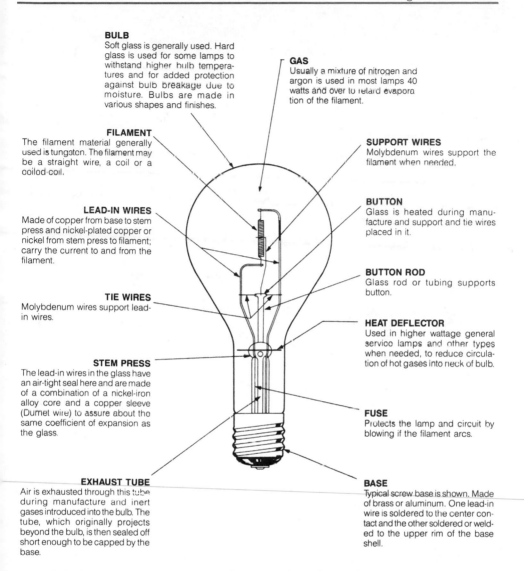

**BULB**
Soft glass is generally used. Hard glass is used for some lamps to withstand higher bulb temperatures and for added protection against bulb breakage due to moisture. Bulbs are made in various shapes and finishes.

**GAS**
Usually a mixture of nitrogen and argon is used in most lamps 40 watts and over to retard evaporation of the filament.

**FILAMENT**
The filament material generally used is tungsten. The filament may be a straight wire, a coil or a coiled-coil.

**SUPPORT WIRES**
Molybdenum wires support the filament when needed.

**LEAD-IN WIRES**
Made of copper from base to stem press and nickel-plated copper or nickel from stem press to filament; carry the current to and from the filament.

**BUTTON**
Glass is heated during manufacture and support and tie wires placed in it.

**TIE WIRES**
Molybdenum wires support lead-in wires.

**BUTTON ROD**
Glass rod or tubing supports button.

**HEAT DEFLECTOR**
Used in higher wattage general service lamps and other types when needed, to reduce circulation of hot gases into neck of bulb.

**STEM PRESS**
The lead-in wires in the glass have an air-tight seal here and are made of a combination of a nickel-iron alloy core and a copper sleeve (Dumet wire) to assure about the same coefficient of expansion as the glass.

**FUSE**
Protects the lamp and circuit by blowing if the filament arcs.

**EXHAUST TUBE**
Air is exhausted through this tube during manufacture and inert gases introduced into the bulb. The tube, which originally projects beyond the bulb, is then sealed off short enough to be capped by the base.

**BASE**
Typical screw base is shown. Made of brass or aluminum. One lead-in wire is soldered to the center contact and the other soldered or welded to the upper rim of the base shell.

Figure 2.4. Incandescent lamp construction (source: Osram/Sylvania).

In operation, electric current is passed through the filament, which heats up, and produces visible light. The amount of current flowing in the filament determines the brightness and the color of the lamp, from dull red to bright white. An incandescent lamp is easily dimmed by simply reducing the current through the lamp. This is easy to accomplish using a rheostat, or variable resistor, in the lamp circuit.

As you can see from Figure 2.1, the light produced by a tungsten filament is strongest in the long wavelengths, or red region, of the visible spectrum. Incandescent sources can therefore be considered

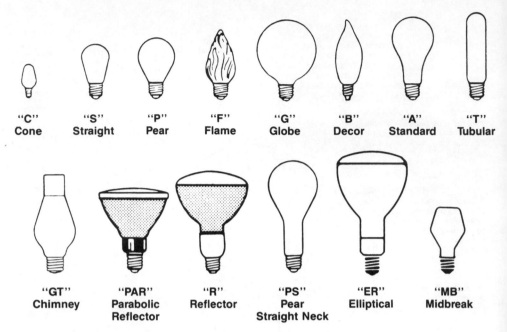

Figure 2.5. Incandescent lamp shapes (source: Osram/Sylvania).

'warm' sources, because reflection from an object will be strongest in the red region.

Standard incandescent lamps are manufactured in a large variety of shapes and sizes. Figure 2.5 shows some of the more common ones.

Each shape of incandescent lamp is assigned a letter, or letters, and each size is assigned a number. The letters loosely describe the lamp, and the number tells the diameter of the lamp in eighths of an inch. For example, the A-19 lamp in Figure 2.5 is of Arbitrary (or Apple) shape, and is 19/8, or 2 3/8 in. (6 cm) in diameter. Some combinations describe the function of the lamp as well as the size, like the PAR-38, which has a *p*arabolic *a*luminized *r*eflector coating inside the glass envelope, and can be bought in either flood or spot light configuration. It is 38/8, or 4 3/4 in. (12 cm) in diameter.

In the years since the invention of the incandescent lamp, lamp manufacturers have done extensive R & D to improve the life and the lumen output of the lamps. This work has been directed toward both the envelope and the gas which fills the envelope.

### Tungsten halogen lamps
The tungsten halogen lamp is one result of this effort. The halogen fill gas prevents the lamp envelope from darkening as

the lamp ages by carrying vaporized tungsten back to the filament, so more light is produced, and the life of the lamp is extended. The envelope itself is small, and made of quartz, instead of glass like regular incandescent lamps. This combination allows the tungsten halogen lamp to operate at a much higher temperature than a standard incandescent lamp, resulting in a higher efficacy, or lumen per watt, output. Since the lamp is hotter than a standard incandescent, the light output is whiter, and the color rendition is more evenly balanced than the red-rich output of the standard lamp. Tungsten halogen lamps operate with a color temperature of 3000–3300 K and have a CRI of 100. Two types of tungsten halogen lamps are available: the line voltage; and the low voltage lamp, and each type is available in a variety of shapes and sizes.

### Infrared reflecting lamps

One version of the tungsten halogen lamp, the infrared (IR) reflecting lamp, uses a dichroic coating that reflects infrared back into the lamp, and passes visible light as output. This serves two functions: it reduces the lamp's energy wastage to heat; and it helps to heat the filament to a higher temperature. A variation of this coating is used in the low voltage MR-16 lamp to produce a 'cool' brilliant white light beam. This lamp gets extremely hot, and cannot be used in an indoor luminaire without protective shielding.

### Incandescent lamp benefits

Some of the more prominent benefits of incandescent lamps are:

1. inexpensive;
2. high CRI (good color rendition);
3. operates in inexpensive fixtures;
4. easily dimmed;
5. instant on–off;
6. insensitive to ambient temperature;
7. available with a variety of built-in reflectors;
8. available in many wattage and colors.

### Incandescent lamp drawbacks

Some of the drawbacks of incandescent lamps are:

1. short life (750–3000 hours);
2. poor over-voltage tolerance;
3. low lumen output per input watt (efficacy);
4. high heat generation (90% of input wattage goes to heat).

*Incandescent lamp uses*

Incandescent lamps are generally best suited for use in:

1. dimmable systems;
2. accent and specialty retail lighting;
3. outdoor convenience and decorative lighting systems;
4. mood lighting.

## Fluorescent lamps

Fluorescent lamps are the most commonly used commercial light source in the US, and perhaps the world. In the US, about 75% of the commercial space is illuminated by a fluorescent source. This is because fluorescent lamps are relatively cheap, long lived, have a high light output to watts input ratio (efficacy), and are designed for use in fixtures which fit into the architectural schemes of commercial structures.

Let's look at how a fluorescent lamp is built, and see how it operates.

Fluorescent lamps are built using a tubular glass envelope coated on the inside with a mix of phosphors. Inert gas and a small amount of mercury is introduced into the tube to provide the atoms for photoluminescence. The tube is slightly pressurized, and the ends of the tube are capped with electrodes, which contain a cathode to generate an arc.

In operation, when the arc is struck between the electrodes at each end of the tube, the mercury vaporizes, and the electric arc colliding with the atoms of mercury vapor produces UV light along the length of the tube. This high-energy light strikes the phosphor coating, and imparts energy to it, which causes the phosphor to fluoresce, or produce light. The chemical makeup of the phosphor compound determines the color of the light produced.

If this process were allowed to go unchecked, the current flow in the arc would continue to increase until the lamp overheated and destroyed itself. To regulate the current flow, a ballast transformer must be used in conjunction with the lamp. The ballast transformer also serves to start the arc in instant start lamps, and to provide coil voltage in rapid start lamps. Since it must generate higher voltages, the ballast transformer used with instant start lamps is larger than that used with rapid start lamps. A larger ballast is also used for the high output (HO) and very high output (VHO) lamps, which are constructed to allow a larger current flow in the arc.

Fluorescent lamps are available in correlated color temperatures of 2700 K, which approximates incandescent lighting, all the way up to 6.500 K, which approximates daylight. The lower temperature lamps, up to 3400 K, are said to be 'warm' lamps; the mid-range lamps, from 3500 to 4000 K, are considered 'natural', neither warm nor cool; and lamps having temperatures of 4100 K and above, are called 'cool' in color. The lamps are available in color rendering indices (CRI) from 57 to above 90. In general, the higher the CRI, the more expensive the lamp.

The two most popular types of fluorescent lamps, the rapid start and the instant start, utilize different methods of arc starting.

### Rapid start fluorescent lamps

The rapid start lamp uses a cathode consisting of a coiled tungsten wire coated with an emission material. When voltage is applied across the coil, it heats up, and the coating emits electrons. These electrons produce an arc in the inert gas. This type of lamp necessarily has two pins in each electrode to power the coil.

### Instant start fluorescent lamps

The instant start lamp uses a ballast transformer to boost the voltage up to a sufficiently high level to strike the arc directly. This only requires a single pin in each electrode, although many lamp manufacturers use the two-pin configuration, and connect the pins together. Figure 2.6 shows the components of the rapid start fluorescent lamp.

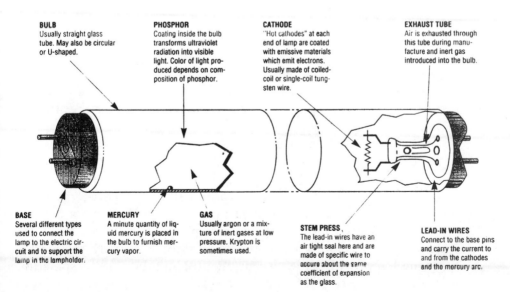

**BULB**
Usually straight glass tube. May also be circular or U-shaped.

**PHOSPHOR**
Coating inside the bulb transforms ultraviolet radiation into visible light. Color of light produced depends on composition of phosphor.

**CATHODE**
"Hot cathodes" at each end of lamp are coated with emissive materials which emit electrons. Usually made of coiled-coil or single-coil tungsten wire.

**EXHAUST TUBE**
Air is exhausted through this tube during manufacture and inert gas introduced into the bulb.

**BASE**
Several different types used to connect the lamp to the electric circuit and to support the lamp in the lampholder.

**MERCURY**
A minute quantity of liquid mercury is placed in the bulb to furnish mercury vapor.

**GAS**
Usually argon or a mixture of inert gases at low pressure. Krypton is sometimes used.

**STEM PRESS**
The lead-in wires have an air tight seal here and are made of specific wire to assure about the same coefficient of expansion as the glass.

**LEAD-IN WIRES**
Connect to the base pins and carry the current to and from the cathodes and the mercury arc.

Figure 2.6. Fluorescent lamp construction (source: Osram/Sylvania).

### High output and very high output fluorescent lamps

High output (HO) and very high output (VHO) fluorescent lamps are constructed in the same manner as standard fluorescent lamps, except that heavier components are used to allow higher than normal current flow within the arc. As a result of this higher current flow, the arc–electron collisions are more violent, and the lamp produces more lumens than a standard lamp. The cost of this extra output is the increased energy required to operate the lamp. This cost is often justified if higher mounting heights are required by the application.

### Compact fluorescent lamps

Compact fluorescent lamps have the same components, and operate the same way that the large tubular lamps do. The glass envelopes of the compact lamps are usually bent into a 'U' shape to offer more surface area in a small space. There are compact fluorescent lamps available that will serve as low-energy substitutes for incandescent lamps in all applications up to about 150 W. The general rule of thumb is that 1 input watt of compact fluorescent energy produces as much light output as 4 input watts of incandescent energy.

### Other fluorescent lamps

There are also fluorescent lamps available which have *no* electrodes, but instead control the arc through inductive coupling. The Osram–Sylvania 'Icetron'™ is one such lamp. Since there are no electrodes to deteriorate, these lamps have an extremely long life – about 100 000 hours on average. These lamps are relatively expensive, and are used primarily where lamp replacement is difficult, for example, in a theater with 40 foot (12 m) high ceilings.

Fluorescent lamps are described in the manufacturer's literature in much the same way that incandescent lamps are: the first letter in their designator is either 'F', for fluorescent, or 'CF' for compact fluorescent. The second letter in the designator is used to describe manufacturer's information about the lamp. For example, 'FB' would describe a U-shaped, or 'bent' tube. One lamp manufacturer, Osram–Sylvania calls their T8 energy saving lamps 'Octron', so the designator would be 'FO' for a Sylvania T8 lamp. Absence of a letter in this position indicates a standard lamp. The first number in the designator is usually the wattage of the lamp, with a few exceptions: the number '48' is sometimes used to indicate a 4 ft, or 48 in. (1.2 m), long tube, and '96' is used the same way for an 8 ft (2.4 m) long tube. Following the wattage number is the shape letter descriptor, usually 'T', for tubular. Following the shape descriptor is a number which is the diameter of the tube in eighths of an inch, and that completes the basic lamp designator. For example, an F32T8 lamp would be a fluorescent 32 W tube, 1 in. (2.5 cm) in diameter, understood to be standard, or 4 ft (1.2 m) long. Following the basic designator, and separated from it by a forward slash (/) are CRI and color descriptors, as well as manufacturers' specialty descriptors. For example, F32T8/841 denotes the above 32 W lamp, with a CRI of 80 or above, and 4100 K in color. High output and very high output lamps are designated with the suffix HO or VHO, respectively.

### Fluorescent lamp advantages
Some of the advantages of fluorescent lighting are:

1. long life (20 000 hours average);
2. low cost;
3. high lumen to input watt ratio (F32T8/841 has efficacy of about 8 times that of standard incandescent);
4. available in a wide range of sizes and colors;
5. available with high CRI ratings;
6. low heat generation.

### Fluorescent lamp drawbacks
Some of the drawbacks associated with fluorescent lighting are:

1. temperature sensitive. Output drops drastically in low temperatures. Lamps used outdoors require special low temperature ballasts;
2. require expensive dimming ballast for dimming;
3. lamps contain mercury, which is classified as an environmental hazard, and can present difficulties in disposal of burned out lamps;
4. can produce stroboscopic effect around rotating machinery, since the arc turns off and on at twice the frequency of the incoming power with magnetic ballasts.

### Fluorescent lamp uses
Fluorescent lamps should be the first choice for:

1. office space ambient lighting;
2. large retail space ambient lighting;
3. interior common space lighting;
4. compact fluorescent lamps should be used in recessed downlighting and accent lighting for energy savings.

## High intensity discharge (HID) lamps

High intensity discharge (HID) lamps operate on the photolumines-cence principle like fluorescent lamps. That is, they require an electric arc passing through, and colliding with gas atoms to produce light. However, unlike fluorescent lamps, which produce 90% of their light from phosphors excited by the generated light, most of the light produced by HID lamps is the arc-generated light itself.

This requires that the HID lamps operate at a higher tempera-ture, with a much higher intensity arc, and under a much higher pressure than fluorescent lamps. To do this, the HID lamp has an inner pressurized tube, called the arc tube, where the high intensity arc takes place, and an outer envelope to protect the arc tube. Like the fluorescent tube, this outer envelope can be coated with phosphors to improve the CRI of the lamp. The space between the outer envelope and the arc tube is evacuated to a high degree to maintain a constant temperature in the arc tube. High intensity discharge lamps utilize a starting electrode to ionize the gas mixture and start the arc, and a coated filament coil electrode in each end of the arc tube to maintain the arc. Like the fluorescent lamp, the HID lamp must be used in conjunction with a ballast to provide starting voltage, and to regulate the current flow once the arc has started. The ballast must be closely matched to the lamp for the system to operate properly. High intensity discharge ballasts are sensitive to momentary drops in input voltage, and for some ballasts, a drop of greater than 10% will extinguish the lamp. Figure 2.7 shows the construction of typical HID lamps.

The high intensity arc in an HID lamp will not strike until the lamp comes up to operating temperature, so there is an average warm-up time of 2–6 minutes from the time the switch is thrown until the lamp reaches full output. Conversely, if the lamp is turned off for any reason, the gasses inside the arc tube will be too hot to re-ionize immediately, and there will be a cool-down period of 5–15 minutes before the arc can be restruck. Table 2.1 is a comparison chart which shows the warm-up and restrike times for the various HID sources.

Despite these drawbacks, HID lamps are among the most efficient and long-lived lamps in use today. They are widely used to light large, high ceiling spaces, such as warehouses or gymnasi-ums. Included in the HID family of lamps are *mercury vapor, metal halide*, and *high pressure sodium*. A close cousin, the low pressure sodium lamp, will also be included in the HID group even though it operates at a lower arc tube pressure than the others.

Let's look now at the characteristics of each of these lamps.

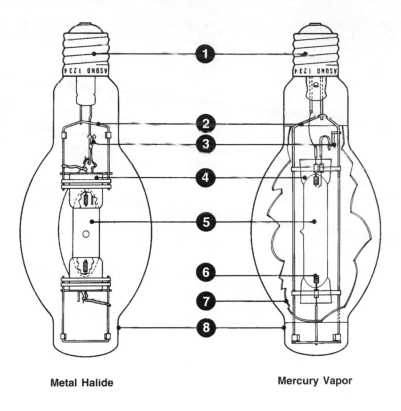

**Metal Halide**          **Mercury Vapor**

1. Base

2. Supports

3. Starting Resistor

4. Arc Tube Seal

5. Arc Tube

6. Electrode

7. Phosphor Coating

8. Outer Envelope

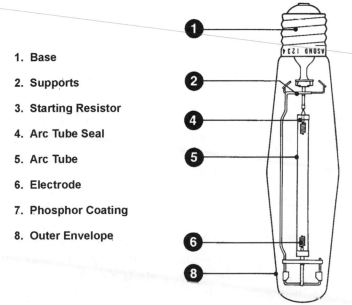

**High Pressure Sodium**

Figure 2.7. High intensity discharge (HID) lamp construction (source: Philips Lighting).

**Table 2.1.** High intensity discharge warm-up and restrike times.

| Light source | Warm-up time | Restrike time |
|---|---|---|
| Mercury vapor | 5–7 min | 3–6 min |
| Metal halide | 2–5 min | 10–20 min |
| Pulse start metal halide | 2–3 min | 3–4 min |
| High pressure sodium | 3–4 min | 1–2 min |
| Low pressure sodium | 7–10 min | 3–12 s |

*Mercury vapor lamps*

Mercury vapor lamps produce the color spectrum of mercury, which appears as a series of 'spikes' on the spectral distribution charts. The distribution for a clear mercury vapor lamp was shown in Figure 2.2.

Looking back at Figure 2.2, it can be seen that a large portion of the distribution is in the UV region, which is harmful to both the eyes and the skin. To counteract this, the outer envelope of the mercury vapor lamp is coated with a UV inhibitor. Even with the inhibitor, the clear mercury vapor lamp still produces light with a correlated color temperature from 5600 to 6400 K. In addition, there are a lot of colors that are not included in the spectrum of the lamp. This leaves the lamp with a CRI of less than 20, which offers very poor color rendering, particularly in the yellow–red region. To correct this, manufacturers have formulated phosphor coatings for the outer envelope which fluoresce in those colors that the mercury gas lacks. Two of the resulting lamps are the 'warm' mercury vapor lamp with correlated color of about 3300 K, and a 'white' lamp at about 4100 K. These lamps have a CRI of 50 or more, which is not great, but is a big color improvement over the clear lamp.

Mercury vapor lamps are the least efficient of all the HID lamps, having a lumens per input watt ratio (efficacy) lower than some of the fluorescent lamps. Mercury vapor lamps are, however, long lived, having an average life in excess of 24 000 hours. Mercury vapor lamps come in a number of shapes and sizes, including reflector lamps. The ballasts must be closely matched with the lamp in a mercury vapor system.

In the US, The American National Standards Institute (ANSI) has assigned a manufacturer's designator for mercury vapor lamps of 'H', taken from the chemical designation Hg for mercury, which (to throw in a little more useless trivia) comes from the Greek word 'hydrargyrum'. Therefore, the first letter in a mercury vapor lamp designator is H. The following numbers indicate the ballast required for use with the lamp; the two letters following the numbers indicate the physical characteristics of the lamp (i.e., size, shape, etc.). This is separated by a dash from another number, which is the wattage of the lamp. Letters following a forward slash after the wattage indicate manufacturer specialty features. Thus, an H38AV-100/DX lamp is a 100 W mercury vapor lamp requiring a type 38 ballast, having size and shape code AV and a Deluxe white phosphor coating.

Some of the beneficial features of mercury vapor lamps are:

1. long life (24 000+ hours);
2. blue–green output flattering to plant color;
3. available in a variety of sizes and shapes;
4. least expensive HID source.

Some of the drawbacks to mercury vapor lamps are:

1. long warm up time to full output (3–5 minutes);
2. poor CRI, even with phosphor coating;
3. inefficient – lowest lumen-per-watt ratio of all the HID lamps;
4. produces UV radiation which can cause skin and eye burn if the outer envelope is broken – must be shielded for indoor use.

Mercury vapor lamps are best suited for landscape or atrium lighting, where the color and long life are an asset.

### Metal halide lamps

Metal halide lamps are a very efficient source of 'white' light, and are available in wide wattage range. The efficacy, or lumen output per watt input ratio for metal halide is 3–4 times that of mercury vapor, and exceeds that of fluorescent above 400 W. Metal halide also has a relatively high color rendering index, with an average CRI of about 70, and some lamps are available with a CRI above 80. Metal halide lamps come in sizes from 50 to 2000 W, with correlated color temperatures of 2900–6000 K, and in a variety of shapes.

The construction of the metal halide lamp is very similar to that of the mercury vapor lamp. The major difference between the two is the gas which fills the quartz arc tube. Where the mercury vapor lamp contains a mixture of argon gases, with mercury for the arc, metal halide lamps contain argon, mercury, and several different iodide compounds. It is the iodides which produce light of a superior spectral distribution, and give the metal halide light a high CRI. Metal halide lamps are available with a variety of phosphor coatings for the UV inhibiting outer tube to provide a desired correlated color temperature. A second type of metal halide lamp, the pulse start lamp, uses the same iodides and coatings, but instead of using a filament coil, the arc is started using a high-energy pulse generated by an igniter, similar to a high pressure sodium lamp. In addition to using an igniter circuit, the pulse start lamp differs from the standard metal halide in several other ways:

1. an improved arc tube seal allows the lamp to run hotter with higher fill gas pressure;
2. faster warm-up time (2–3 minutes);
3. faster hot re-strike time (3–4 minutes);
4. higher lumen output, and better performance (35% better than universal burn metal halide);
5. longer life than the standard universal burn metal halide lamp.

In operation, the arc in the metal halide lamp is started by either a combination of heated electrode/high starting voltages from the ballast (standard) or by a high voltage pulse from the igniter (pulsed start) to ionize the mercury in the arc tube. Once the arc starts, the iodides gradually enter the arc stream, and the output of the lamp shifts from the blue–green of mercury vapor to white. Since the arc tends to curve upward when the lamp is horizontal, some lamps are designed with a special curved arc tube to burn horizontally. Others will burn in either a vertical or horizontal (universal) position. The pulse start lamps are designed to burn in the vertical position. In addition, the pulse start lamp requires a pulse rated socket. As with the mercury vapor lamp, the ballast required for metal halide operation must be closely matched to the lamp.

The first letter in a metal halide lamp designator is 'M'. The second letter is a manufacturer's letter to denote special features of the lamp. The number following that designator is the wattage of the lamp. A forward slash usually follows the wattage, and following the slash is abbreviated data concerning the required burning position, or the base type of the lamp. A M175/U lamp then, is a 175 W metal halide lamp which will burn in universal (either horizontal or vertical) position.

Metal halide lamps offer many benefits, including:

1. highest efficacy of any 'white' light producing lamp;
2. long life (10 000–20 000 hours);
3. available in a variety of bases and shapes;
4. available in sizes from 50 to 2000 W;
5. available in a broad range of correlated color temperatures (2900–6000 K);
6. high CRI (60–90);
7. insensitive to ambient temperature.

There are also a few drawbacks associated with metal halide lamps, including:

1. requires warm-up time of 2–3 minutes;
2. must be used in a shielded fixture. Outer envelope breakage can allow emission of high levels of UV light;
3. long restrike time after power outage (5–7 minutes – standard, 2–3 minutes – pulse start);
4. relatively expensive lamp.

Metal halide is a first choice for a wide variety of applications requiring various mounting heights, including:

1. industrial facilities with high ceilings;
2. sports facilities;
3. warehouses needing high CRI;
4. retail facilities;
5. downlighting, uplighting, and accent lighting in commercial facilities;
6. ambient lighting in facilities with high ceilings;
7. outdoor building lighting.

There is a metal halide option available for almost every type of indoor luminaire. In sensitive areas, luminaires can be equipped with a second, incandescent quartz lamp, to provide illumination during the cool-down period in the event of a power dip. This is called 'quartz restrike'.

### High pressure sodium lamps

High pressure sodium has the highest efficacy of any member of the true HID family, and equals or exceeds the efficacy of fluorescent in all wattages. The drawback of the high pressure sodium lamp is a low CRI, since almost 100% of the output light is in the yellow–orange region of the spectrum. Construction of the high pressure sodium lamp is similar to that of the other HID lamps, with some notable exceptions. The arc tube is made of ceramic material, rather than quartz, to withstand the corrosive effects of sodium and the extremely high temperatures required for operation. The arc tube is filled with sodium, and a small amount of mercury and xenon gas for arc starting. No starting electrodes are used in the standard sodium vapor lamp, so the ballast includes an electronic igniter circuit that works in conjunction with the transformer to provide the high voltage to start the lamp.

A hybrid sodium vapor lamp that includes the starting circuitry inside the lamp has been developed to replace mercury vapor lamps in street lighting luminaires. This lamp will start on a standard mercury vapor lamp ballast.

In operation, the lamp goes through a warm-up period of 3–4 minutes before reaching full brightness. During warm-up, the lamp undergoes several color changes as the various elements ionize. As the pressure in the arc tube increases, the lamp comes to full brightness. A power outage or voltage dip will require only approximately 1 minute for restrike.

In operation in an open luminaire, the sodium vapor lamp produces a 'narrow spotlight' light distribution pattern which focuses most of the light in a small area. A diffusion coating on the outer envelope is available for most sodium vapor lamps to spread the distribution pattern. This is the 'coated' version of the lamp. Unfortunately, this coating does not improve the CRI, which is only about 21 for most lamps, unless phosphor coatings are used.

Phosphor coatings are available for the outer envelope of lamps up to 400 W to improve the CRI to about 65, and for lamps up to 100 W to produce a CRI as high as 85.

Manufacturers have chosen to call their high pressure sodium vapor lamps by another name, ending in 'lux'. Philips, for example, calls theirs Ceramalux, while Osram–Sylvania calls theirs Lumalux. The manufacturer descriptor for Osram–Sylvania, then, would start with 'LU', for Lumalux. Following that is a number indicating wattage, and after that, a forward slash. If a 'D' follows the slash, this indicates a diffusion coating. If not, the following letters indicate base type. For example, LU50/D/MED denotes a 50 W high pressure sodium, coated, medium base lamp.

Benefits of high pressure sodium lighting are:

1. high efficacy;
2. available in sizes up to 1000 W;
3. no dangerous mercury arc in the event of outer envelope breakage;
4. normal restrike in 1 minute or less – instant restrike available;
5. insensitive to ambient temperature;
6. long life (24 000+ hours).

The main drawbacks of high pressure sodium lighting are:

1. low CRI (21 without phosphor coating);
2. relatively expensive lamp and ballast;
3. extremely high operating temperatures – must be used in fixtures which are ANSI designated for high pressure sodium use.

High pressure sodium lamps are widely used where CRI is not of great importance, but where high lumen output is. These areas include:

1. parking lots;
2. warehouses;
3. loading docks;
4. streets and highways;
5. building floodlighting.

*Low pressure sodium lamps*

Although not a true high intensity source, the low pressure sodium lamp is a discharge source, and is included here for completeness. Low pressure sodium is a monochromatic source, having almost 100% of its light output in two narrow bands at 589 and 589.6 nanometers in wavelength, which is in the yellow portion of the spectrum near the peak of the eye sensitivity curve. This gives low pressure sodium the lowest CRI, and the highest efficacy in the industry.

The low pressure sodium lamp is constructed much the same as the high pressure sodium lamp, except that the ceramic arc tube is bent into a 'U' shape, and dimpled to assure proper flow of the vaporized sodium.

In operation, the lamp is started on neon–argon gas, and after a warm-up period of 7–9 minutes the sodium vapor conducts, and the lamp delivers full output. If a power outage occurs, the lamp will re-strike in less than 1 minute.

The strong points of low pressure sodium are:

1. very high efficacy (as high as 200 lumens per watt);
2. low re-strike time

The drawbacks of low pressure sodium are:

1. monochromatic light output (very low CRI);
2. long warmup time;
3. relatively expensive.

Due to its low CRI, low pressure sodium has no interior applications. Its uses are primarily in large, outdoor applications where CRI is of no consequence, such as railroad yards and parking lots. Table 2.2 is a performance comparison of the four types of HID lamps.

**Table 2.2. High intensity discharge (HID) performance comparison chart.**

| Light source | Warm-up time | Restrike time | Efficacy (lumens/watt) | Color rendering |
|---|---|---|---|---|
| Mercury vapor | 5–7 min | 3–6 min | 30–55 | Poor |
| Metal halide | 2–5 min | 10–20 min | 60–95 | Good, CRI 60–90 |
| Pulse start metal halide | 2–3 min | 3–4 min | 60–95 | Good, CRI 60–90 |
| High pressure sodium | 3–4 min | 1–2 min | 60–125 | Moderate |
| Low pressure sodium | 7–10 min | 3–12 s | 80–180 | Bad |

## Ballasts

Although not specifically a light source, ballasts are an integral part of the fluorescent and HID lighting systems. The ballasts used with these lamps serve two functions: (1) *they start the arc in the tube* and (2) *they regulate the current flow in the arc tube.*

Lamps and ballasts must be properly matched in order to achieve optimum (or any) operation. Let's take a look at some of the different ballasts.

### Fluorescent ballasts

In general, there are three types of ballasts used with fluorescent lamps. These are: *magnetic, hybrid*, and *electronic.*

*Magnetic fluorescent ballasts*
The magnetic ballast is the oldest of the three types, and has been in service as long as fluorescent lamps have. The magnetic ballast is basically a transformer consisting of a laminated steel core wrapped with insulated copper or aluminum windings, and housed in a steel case which is filled with a heat dissipating 'potting' compound. In operation, the magnetic ballast supplies either line voltage to the electrode coil to start a rapid start lamp, or high voltage across the electrodes to start an instant start lamp. The windings of the transformer serve as a choke coil to limit current flow in the arc while the lamp is operating. Magnetic ballasts lower the power factor of an electrical system, and capacitors are often built into magnetic ballasts to counteract this. Magnetic ballasts containing capacitors are called *high power factor* ballasts. A magnetic ballast can control a maximum of two lamps. There is a heating, or 'ballast' loss associated with the use of magnetic ballasts. Ballast losses from magnetic ballasts can account for up to 25% of total luminaire energy consumption.

*Hybrid fluorescent ballasts*
Hybrid ballasts are used only for rapid start lamps. A magnetic ballast applies power to the starting coil during lamp start, and that power continues to be applied while the lamp operates. The hybrid ballast is identical in construction to the magnetic ballast, except that it contains electronic circuitry to remove the coil power during lamp operation. This reduces energy consumption by about three watts per lamp.

*Electronic fluorescent ballasts*
Electronic ballasts start and control current flow through fluorescent lamps through the use of electronic circuitry, rather than transformer

windings. Electronic ballasts are similar to the power supplies found in computers and other electronic devices: the incoming AC power is rectified, regulated, and re-introduced into the system with different characteristics. The electronic ballast operates the lamp at a much higher frequency (20 000 Hz or greater) than the 60 Hz magnetic ballasts, and thus eliminates lamp flicker and a host of other lamp inefficiencies. Since there is no heavy steel core or windings involved in the electronic ballast, it can be made smaller and lighter than the magnetic ballast, and in operation, there is no ballast loss. One electronic ballast can control up to four lamps.

The electronic circuitry used in the ballasts tends to distort the waveform of the incoming power by injecting harmonics into the system, notably the third harmonic. To counteract this, larger power conditioning components are used in what are called *low THD* (total harmonic distortion) ballasts. A word of caution here – if the THD is *too* low (THD < 10%), the ballast inrush current goes up dramatically, and there could be switching problems with the luminaires. A ballast with a maximum THD of 15% is appropriate for most systems.

The high frequency of the electronic ballast also generates electromagnetic interference (EMI) which affects some sensitive electronic equipment such as library book detectors, power line carrier control systems, central time clock systems, and medical equipment. Hybrid ballasts should be considered for use in areas where such equipment is present. Otherwise, electronic ballasts should be the first choice for all fluorescent applications.

## High intensity discharge (HID) ballasts

High intensity discharge (HID) ballasts perform the same functions that fluorescent ballasts do: they start the arc, and they regulate the arc current once the arc is started. High intensity discharge ballasts are magnetic, and each is designed to operate a particular lamp. In general, there are four types of HID ballast: *reactor, high reactance autotransformer, constant wattage*, and *constant wattage autotransformer.* Let's look at some of the features of each type.

### Reactor ballast
The reactor ballast is essentially just a choke coil in series with the lamp which limits current through the lamp. The reactor ballast is used with lamps which require only line voltage to start the arc, such as mercury vapor lamps. The pros are that the reactor ballast has low inrush current, and low cost; the cons are that the ballast is very sensitive to line voltage, and lamp output can vary up to

12% with a 5% variance in voltage. With about a 10% dip in line voltage, the lamp arc will extinguish. Reactor ballasts have a low power factor, and are often used in conjunction with capacitors to remedy this. The manufacturer's symbol for a reactor ballast is 'R'.

### High reactance autotransformer ballast

This ballast is an autotransformer, which means that it has two windings connected electrically, and coupled magnetically. The high reactance autotransformer ballast can boost line voltage to any starting voltage required by the lamp. The windings also serve to limit current through the lamp, as in the reactor ballast. Like the reactor ballast, the high reactance autotransformer ballast is susceptible to low line voltage, and has poor power factor. Since there is an additional winding, this ballast is more expensive that the reactor ballast. This ballast has a manufacturer's symbol of 'HX'.

### Constant wattage ballast

The constant wattage ballast is a true two-winding transformer, having only magnetic coupling between the coils. This configuration is inherently insensitive to drops in the input voltage, and the ballast can sustain up to a 50% dip in line voltage before the lamp arc extinguishes. The constant wattage ballast contains a capacitor for high power factor, and the starting current does not exceed the lamp operating current. Unfortunately, these ballasts are expensive, often costing three times as much as a reactor ballast. The ballast has a manufacturer's symbol of 'CW'.

### Constant wattage autotransformer ballast

This ballast is an autotransformer, with the two windings electrically connected and magnetically coupled to provide excellent voltage regulation. This ballast can withstand a 30% dip in line voltage before the lamp arc is extinguished. With its built-in capacitor for high power factor, and its low inrush current, this ballast offers the best compromise between cost and performance. Its symbol is 'CWA'.

For the more visual oriented among you, Figure 2.8 is a schematic comparison of the four types.

You should always specify constant wattage (CW or CWA) ballasts for HID luminaires which are fed from panels that also contain heavy motor loads. Industrial motors have an inrush current of 5–10 times their running current, and the inrush can last for several seconds. If the motor is large enough, this can cause a momentary voltage dip of 15–20% at the panel. Reactor ballasts connected to the panel will drop their lamps out, and the place will go dark – and remain so for the cool-down period of the lamp. It

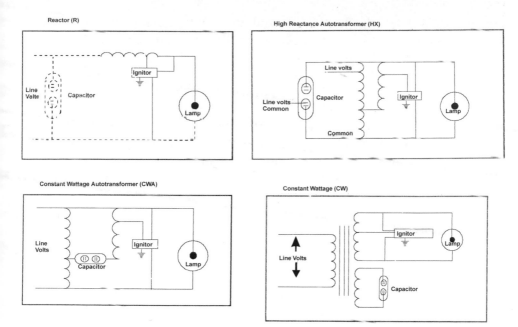

Figure 2.8. Ballast schematics (source: Advance Transformer).

is especially embarrassing to the lighting designer when the lights go out every time the compressor comes on.

## Luminaires

A luminaire is a complete system for providing the kind of light we want, where we want it. The luminaire consists of the lamp, the lamp socket, the ballast, if required, the reflector, diffuser, shielding, and the housing for all those things. A luminaire needs only a mounting spot and a source of electricity to produce light.

To be an efficient producer of light, a luminaire must provide balanced, glare-free illumination. Luminaires accomplish this through the use of reflectors, which shape the light output, and shielding and diffusion devices that reduce glare and distribute the light evenly. Let's take a look at these components.

### Reflectors

Reflectors are bent painted metal panels in an inexpensive fluorescent luminaire, or precisely cut or vacuum-formed mirror finish optical systems in an expensive architectural downlight, or anything in between. Reflectors serve a twofold purpose: they *shape the light pattern* of the luminaire, and they *improve the efficiency* of the luminaire (defined as the amount of light delivered by the luminaire, divided by the light produced by the lamps) by directing the usable light out of the luminaire onto the work surface. Glossy or mirrored reflectors produce direct light, while matte finish reflectors produce scattered, or diffuse light.

Factors affecting the performance of a reflector are *the reflectivity of the reflector material* and *the optical geometry of the reflector*. Reflectivities range from about 60% for standard white paint to 95% for polished aluminum. The geometry of the reflector largely determines the shape of the light pattern emitted by the luminaire.

Over the years, luminaire manufacturers have experimented with various shapes of reflectors to improve performance. One shape in particular which has proven effective in ceiling-mounted fluorescent fixtures is the parabola: a reflective trough in the shape of a parabola is installed lengthwise in the luminaire, and the fluorescent tube is mounted at the apogee of the parabola. Figure 2.9 is a ray trace of the parabolic reflector.

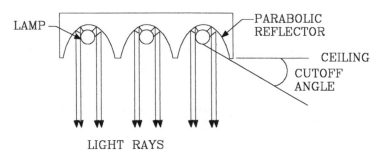

Figure 2.9. Parabolic reflector ray trace.

Another interesting geometric shape used for lighting purposes is the ellipse. When used in a recessed downlight, an elliptical reflector can be designed such that its focal point is exactly at ceiling level. The focal point in an elliptical reflector is that point at which all reflected light rays converge. Past that point, they diverge again. This allows all the light produced by the lamp to be passed through a very small opening in the ceiling. This is called a 'pin-hole' luminaire, and its ray trace is shown in Figure 2.10.

## Shielding and diffusion devices

Shielding devices are used to reduce the glare produced by a luminaire; diffusion devices are used to diffuse the produced light, and to conceal the bare lamps of the luminaire.

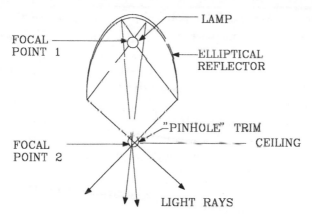

Figure 2.10. Elliptical reflector ray trace.

## Baffles

An example of a shielding device is the baffles used in conjunction with the parabolic reflectors in a recessed parabolic luminaire. Figure 2.11 illustrates this luminaire.

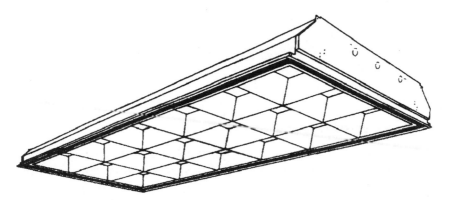

Figure 2.11. Baffles in a parabolic luminaire (source: Lithonia Lighting).

It is seen that the baffles running along and perpendicular to the sides of the parabolic trough form multiple 'cells' along the length of the luminaire. These cells provide sharp cut-off glare shielding when the luminaire is viewed at an angle. The smaller the cells are, the better the control. However, since the baffles are made of opaque material, small cells tend to mask much of the light produced by the luminaire, and thereby reduce its efficiency.

## Diffusers

Diffusers used with fluorescent fixtures also influence the distribution pattern of the luminaire. The most common diffuser employed in fluorescent fixtures used for general area lighting is a flat plastic sheet embossed with some sort of diffusion pattern. The two most common plastics used for this purpose are *polycarbonate* and *acrylic*. Polycarbonate is exceptionally strong, but unless specially treated to resist the UV emissions from the lamp, will turn yellow after a period of time. Acrylic, though not as strong, is impervious to UV. Acrylic diffusers are made as thin as 0.10 in. (0.25 cm) thick for a 2 ft × 4 ft (0.6 m × 1.2 m) unit, but since this material is not especially strong, sheets this thin will tend to sag in the luminaire. It is always best to specify UV-treated polycarbonate, or acrylic diffusers no less than 0.125 in. (0.318 cm) thick. There is a variety of diffuser styles available in either material which can produce a desired distribution pattern.

## Luminaire housings

Housings for luminaires can be metallic, such as sheet steel or aluminum for general purpose, or plastic for corrosive or high abuse areas. Luminaires that fit into standard 2 ft × 4 ft grid (0.6 m × 1.2 m) ceilings are available with air-handling housings, which can serve as return or supply registers for the building HVAC (heating, ventilating, and air conditioning system). A benefit of this type of housing is that luminaire heat is removed before entering the space.

## Luminaire classifications

Luminaires used for area lighting have been classified by the Illuminating Engineering Society (IES) according to their distribution pattern and by the downward and upward component of the light generated. In general, there are five categories of luminaire: *direct, semi-direct, general diffuse, semi indirect,* and *indirect.* The distinguishing patterns of four of these types is shown in Figure 2.12a–d.

The general diffuse lighting pattern is one that contains equal components in all directions.

### Direct illumination
Direct illumination, as shown in Figure 2.12a, is the most efficient use of the luminaire to provide lumens to the work area. It also produces the greatest amount of glare, unless shielded. Direct illumination is best used where you desire a high ambient lighting level, such as kitchens, day care centers, and general work spaces. Direct

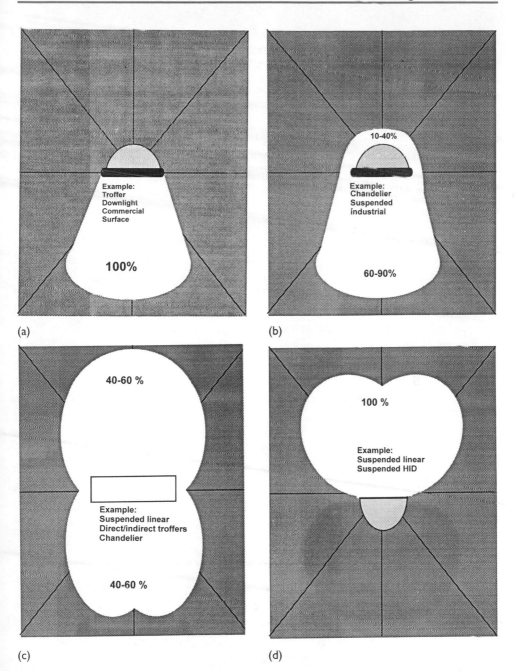

Figure 2.12. (a) Direct illumination, (b) semi-direct illumination, (c) semi-indirect illumination, (d) indirect illumination (source: F.H. Jones).

illumination is also used for task lighting, where a high lighting level in a limited area is desired. When used with parabolic reflectors and baffles, or cutoff shielding, direct illumination is acceptable for use in areas where glare is undesirable, such as offices containing video data terminals. Direct luminaires may be recessed into the ceiling, surface-mounted on it, or suspended below it.

### Semi-direct illumination

Semi-direct luminaires have an upward component, as shown in Figure 2.12b, so they must be either surface-mounted or suspended. They are used for the same purpose as direct luminaires, except that when suspended, the upward component lights the ceiling, and eliminates the 'cave effect' produced by the suspended direct luminaire.

### Semi-indirect and indirect illumination

The semi-indirect and indirect luminaires must be suspended to utilize the upward reflected component of light. These luminaires differ only in the percentages of uplight and downlight that they produce. Indirect luminaires produce almost 100% uplight, which provides a diffuse, glare-free illumination particularly well suited for use in a room with heavy computer terminal usage. Most manufacturers offer an indirect luminaire which produces about 10% direct illumination to eliminate the dark area created by the luminaire housing.

   Wall-mounted luminaires are classified in the same manner as ceiling-mounted ones, and have basically the same illumination patterns. One type of wall-mounted luminaire that receives considerable use in modern buildings is the cove light, which is usually mounted near the ceiling, and is used in an indirect pattern to illuminate the ceiling. Figure 2.13 shows a typical cove lighting installation.

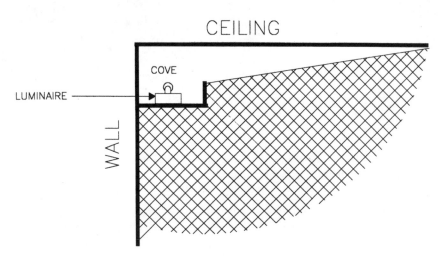

Figure 2.13. Cove lighting (source: author).

## Luminaire photometric data

Manufacturer's published information for their luminaires includes a graphic representation, in two axes, of the distribution pattern produced by each luminaire. This representation is called a *candlepower distribution curve* for the luminaire. It is also called a 'batwing' curve in the industry, because of the distinctive shape of some curves. Figure 2.14a and b shows candlepower distribution curves for a recessed HID direct pattern downlight luminaire, and a recessed fluorescent direct pattern downlight luminaire, respectively.

These curves were generated in test laboratories by taking candlepower measurements around a test luminaire and plotting them on polar coordinates. In applications where all the light from the luminaire is projected in one direction, such as the direct distribution of the luminaires in Figure 2.13, only half of the polar graph is shown. We can use this published data to give us some idea of the light pattern which a particular luminaire will provide. Distribution curves are a valuable tool for quickly evaluating luminaire suitability, especially for task lighting applications.

**AHZ 100MHC 6AR**, (1) MPC100/C/U/MED lamp, 8700 rated lumens, 1.3 s/mh, test no. LTL8066

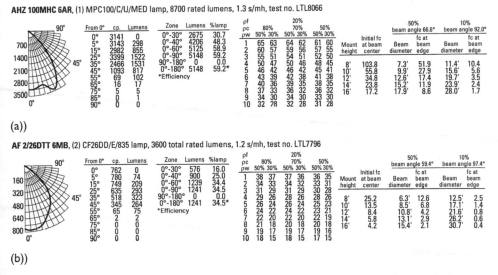

(a))

**AF 2/26DTT 6MB**, (2) CF26DD/E/835 lamp, 3600 total rated lumens, 1.2 s/mh, test no. LTL7796

(b))

Figure 2.14. Candlepower Data for (a) HID and (b) fluorescent downlights (source: Lithonia Lighting).

## Lighting calculations

The distance from the center point of the candlepower distribution curve to any point on the curve gives the lumens of the luminaire in that direction. For asymmetric luminaires, such as a 2 ft × 4 ft (0.6 m × 1.2 m) fluorescent luminaire, data is given for each of the two axes of the fixture: along the length of the luminaire, or parallel to it; and across the luminaire, or perpendicular to it.

Now, if you remember the inverse square law and the cosine law of incidence from Chapter 1, and you know the distance and direction of your chosen luminaire from a surface that you want to examine, you can calculate the footcandles on that surface. For example, suppose you were interested in finding the footcandles produced on a desk by the fluorescent luminaire in Figure 2.14b. The center of the desk is 10 ft from the luminaire, and at an angle of 30°. You can look at the distribution curve and read the lumens at 30°. Using the cosine law of incidence, $I = L \cos X/D^2$, the illumination produced at the center of the desk by this one luminaire is: Lumens × 0.866/100, for footcandles on the square foot at the center of the desk.

Some manufacturers produce the lumen values at the various angles in tabular form, and the total lumens in an angular zone are shown. The angles listed are measured from the vertical, or normal to the luminaire. Figure 2.15 is one such set of distribution data.

As seen in Figure 2.15, a lot of other luminaire information is included in the manufacturer's data, or 'cut' sheets. The coefficient of utilization (CU) of the luminaire, as shown in the table, is a ratio of the lumens which reach an assigned working plane to the total lumens generated by the lamps in the luminaire. Many factors contribute to the coefficient of utilization, including luminaire efficiency and distribution pattern, room geometry and reflectances, and luminaire mounting height. The coefficient of utilization is useful for general area calculations, such as the lumen, or zonal cavity calculation method, which yields an approximation of the average lumens per square foot of the space under examination.

## Zonal cavity calculations

Zonal cavity calculations are a quick manual method of estimating average footcandles, or the number of luminaires required to produce a desired average footcandle level within a defined space. To get some idea of how a zonal cavity calculation works, let's take a hypothetical office space which is 20 ft × 20 ft, with a 9 ft ceiling. The interior designer has selected a white ceiling tile with a reflectance of about 0.8, a light gray semi-gloss wall paint with a reflectance of about 0.7, and a darker gray carpet with a reflectance of 0.2. The work plane is the desktop, at 2.5 ft above the floor.

## FEATURES

- Door is fully gasketed flush steel with mitered appearance. Corners screwed together for rigidity, easy lens maintenance.
- Optional aluminum door frames available.
- Fixture sides feature rolled edge for safer handling.
- T-hinges die-formed for maximum strength. Hinges or latches from either side.
- Opposing, rotary-action cam latches standard for secure door closing. Latches painted after fabrication for smooth finish.
- Integral T-bar safety clips hold T-bar securely, no fasteners required to install.
- Snap-in socket tracks allow socket replacement without tools.
- Full-depth end plates embossed for rigidity.
- Urethane foam gasket eliminates light leaks between door frame and housing.
- Guaranteed for one year against mechanical defects in manufacture.

## SPECIFICATIONS

BALLAST — Thermally protected, resetting, Class P, HPF, non-PCB, UL listed, CSA-certified ballast is standard. Energy saving and electronic ballasts are sound rated A. Standard combinations are CBM approved and conform to UL 935.

WIRING & ELECTRICAL — Fixture conforms to UL 1570 and is suitable for damp locations. AWM, TFN or THHN wire used throughout, rated for required temperatures.

MATERIALS — Housing formed from cold-rolled steel. Acrylic shielding material 100% UV stabilized. No asbestos is used in this product.

FINISH — Five-stage iron-phosphate pretreatment ensures superior paint adhesion and rust resistance. Painted parts finished with high-gloss, baked white enamel.

LISTING — UL listed and labeled. Listed and labeled to comply with Canadian and Mexican Standards (see options).

Specifications subject to change without notice.

| Catalog Number | Type |
|---|---|
|  |  |

**Static Grid Troffer**

# GT 2'x4'

2, 3 or 4 lamps

## PHOTOMETRICS

Calculated using the zonal cavity method in accordance with IESNA LM41 procedure. Floor reflectances are 20%. Lamp configurations shown are typical. Full photometric data on these and other configurations available upon request. LER (Luminaire Efficiency Rating) calculated in accordance with NEMA standard LE-5 See sheet number LER for details.

### 2GT 3 32 A12
### Report LTL 5546
S/MH (along) 1.2   (across) 1.3

**Coefficient of Utilization**

| Ceiling | 80% | | | 70% | | | 50% | | |
|---|---|---|---|---|---|---|---|---|---|
| Wall | 70% | 50% | 30% | 70% | 50% | 30% | 50% | 30% | 10% |
| 1 | 82 | 79 | 76 | 80 | 77 | 75 | 74 | 72 | 70 |
| 2 | 76 | 70 | 66 | 74 | 69 | 65 | 66 | 63 | 60 |
| 3 | 70 | 63 | 58 | 68 | 62 | 57 | 60 | 55 | 52 |
| 4 | 65 | 56 | 51 | 63 | 56 | 50 | 54 | 49 | 45 |
| 5 | 59 | 50 | 44 | 58 | 50 | 44 | 48 | 43 | 39 |
| 10 | 40 | 30 | 24 | 39 | 30 | 24 | 29 | 24 | 20 |

**Zonal Lumens Summary**

| Zone | Lumens | %Lamp | %Fixture |
|---|---|---|---|
| 0-30 | 2019 | 23.2 | 31.1 |
| 0-40 | 3320 | 38.3 | 51.1 |
| 0-60 | 5495 | 63.2 | 84.7 |
| 0-90 | 6491 | 74.6 | 100.0 |
| 90-180 | 0 | 0.0 | 0.0 |
| 0-180 | 6491 | 74.6 | 100.0 |

LER = 62

### 2GT 4 32 A12
### Report LTL 4866
S/MH (along) 1.2   (across) 1.3

**Coefficient of Utilization**

| Ceiling | 80% | | | 70% | | | 50% | | |
|---|---|---|---|---|---|---|---|---|---|
| Wall | 70% | 50% | 30% | 70% | 50% | 30% | 50% | 30% | 10% |
| 1 | 84 | 81 | 78 | 82 | 79 | 76 | 76 | 74 | 72 |
| 2 | 77 | 72 | 67 | 75 | 70 | 66 | 68 | 64 | 61 |
| 3 | 71 | 64 | 59 | 70 | 63 | 58 | 61 | 57 | 53 |
| 4 | 66 | 58 | 52 | 64 | 57 | 51 | 55 | 50 | 46 |
| 5 | 60 | 51 | 45 | 59 | 51 | 45 | 49 | 44 | 40 |
| 10 | 41 | 31 | 25 | 40 | 30 | 25 | 30 | 24 | 21 |

**Zonal Lumens Summary**

| Zone | Lumens | %Lamp | %Fixture |
|---|---|---|---|
| 0-30 | 2757 | 23.8 | 31.3 |
| 0-40 | 4532 | 39.1 | 51.4 |
| 0-60 | 7476 | 64.5 | 84.9 |
| 0-90 | 8809 | 75.9 | 100.0 |
| 90-180 | 0 | 0.0 | 0.0 |
| 0-180 | 8809 | 75.9 | 100.0 |

LER = 62

### 2GT 4 40 A12
### Report LTL 4432
S/MH (along) 1.2   (across) 1.3

**Coefficient of Utilization**

| Ceiling | 80% | | | 70% | | | 50% | | |
|---|---|---|---|---|---|---|---|---|---|
| Wall | 70% | 50% | 30% | 70% | 50% | 30% | 50% | 30% | 10% |
| 1 | 76 | 73 | 70 | 74 | 71 | 69 | 69 | 67 | 65 |
| 2 | 70 | 65 | 61 | 68 | 64 | 60 | 61 | 58 | 56 |
| 3 | 65 | 58 | 53 | 63 | 57 | 53 | 55 | 51 | 48 |
| 4 | 60 | 52 | 47 | 58 | 51 | 46 | 50 | 45 | 42 |
| 5 | 55 | 47 | 41 | 54 | 46 | 41 | 45 | 40 | 36 |
| 10 | 37 | 28 | 23 | 36 | 28 | 22 | 27 | 22 | 19 |

**Zonal Lumens Summary**

| Zone | Lumens | %Lamp | %Fixture |
|---|---|---|---|
| 0-30 | 2739 | 21.7 | 31.6 |
| 0-40 | 4494 | 35.7 | 51.8 |
| 0-60 | 7372 | 58.5 | 85.0 |
| 0-90 | 8675 | 68.8 | 100.0 |
| 90-180 | 0 | 0.0 | 0.0 |
| 0-180 | 8675 | 68.8 | 100.0 |

LER = 62

**LITHONIA LIGHTING**
COMMERCIAL & INDUSTRIAL FLUORESCENT LIGHTING

Figure 2.15. Candlepower data, Lithonia 2GT432 (source: Lithonia Lighting).

The luminaire that we will use for the office is the 2GT432A12 fluorescent luminaire of Figure 2.15. The luminaire will be mounted in the ceiling grid.

The zonal cavity method entails dividing the space into three cavities: the ceiling cavity, which is that space above the luminaire; the room cavity, which is the space between the luminaire and the work plane; and the floor cavity, which is the space below the work plane. We have no ceiling cavity in this example, since the luminaires are mounted directly in the ceiling. The room cavity extends from the desk top to the ceiling, a distance of 9 ft - 2.5 ft = 6.5 ft. The floor cavity is 2.5 ft deep. We are primarily interested in illuminating the room cavity, since that's where all the activity will take place. Let's hypothesize a little further and say that we have defined the lighting criteria for this office space to be 70 footcandles (70 fc) at the desktop. (You'll learn more about defining lighting criteria in the next chapter.) What we need to find out by our zonal cavity calculation is how many luminaires will be required in this space to provide that 70 fc.

The zonal cavity method allows us to calculate a cavity ratio for the cavity under examination, which in our case is the room cavity. By calculating the room cavity ratio (RCR) for our space, we will be able to select the proper coefficient of utilization using the manufacturer's tables. The formula for calculating a cavity ratio is:

$$CR = \frac{5H \times (\text{room length} + \text{room width})}{(\text{room length} \times \text{room width})}$$

where $H$ is cavity height.

In our case, the RCR of the office is:

$$\frac{5(6.5 \text{ ft}) \times (20 \text{ ft} + 20 \text{ ft})}{(20 \text{ ft} \times 20 \text{ ft})} = 3.5$$

Now, if we go back to the coefficient of utilization table in Figure 2.15, we see that, with a RCR of 3.5, the CU can be interpolated as 0.645, using the reflectances mentioned earlier.

The zonal cavity method calculates average illumination by determining the total number of usable lumens produced by the lighting system, and dividing that number by the number of square feet in the space. This yields an average lumen/sq. ft (sq. m), or footcandle value at the work surface. Conversely, if the desired footcandles is given, the number of luminaires required to produce those footcandles may be calculated.

The usable lumens produced by a luminaire in a space depends upon a number of things: first is the maximum number of lumens that the luminaire is capable of producing. This depends on how many lamps the luminaire has, how many lumens each lamp will produce, and the ballast factor (BF) of the ballast. Manufacturer's data for the luminaire indicates the number of lamps in the luminaire, and often lists lamp lumens, as well. If not, lumens can be found in lamp catalogs.

Easy for me to say, right? OK – you won't find a lamp catalog in the public library, maybe not even in a college library, but this is the twenty-first century. We've got the Internet. Most lamp manufacturers have a Web site, and we'll find out how to access them a little later.

Ballast factor is a decimal number, usually less than 1, which compares the performance of the ballast in the luminaire with that of the standard laboratory ballast used to determine the lamp lumen output listed in the lamp catalogs. Ballast factor can be specified, or obtained from the luminaire manufacturer. A BF of less than 1 means that the lamps will produce less than the listed lumen output, since actual output is obtained by multiplying listed output times the BF. Ballasts are available with BF of greater than 1, which means that the lamps will produce *more* than listed output. In general, the lower the BF, the less the lumen output, and the less energy the luminaire will use. If the BF cannot be obtained from the manufacturer, use 0.80.

Using all of the above factors, we can design for an approximate *initial* lighting level in the space. If we want to design for a *maintained* lighting level, we must consider some things that will happen after the system is installed. The first of these is lamp lumen depreciation (LLD). As a lamp ages, its light output becomes less. The major lamp manufacturers have done sufficient testing to be able to predict fairly accurately how their lamps will perform over time. The normal design point is at 70% of the lamp's lifetime, and the manufacturers usually list a *mean*, or *design* lumen value for their lamps at that point. A typical 32 W T8 fluorescent lamp which operates 12 hours per start has an LLD of 0.86. Another factor which affects fixture performance over time, but which is harder to quantify, is lumen dirt depreciation (LDD) of the fixtures and the surfaces in the space. No matter how many charts and graphs you may see on this subject, LDD is dependent on the use of the space, and the quality of the maintenance of that space. In our office space, if the maintenance staff changes the HVAC filters regularly, and if they clean the luminaire each time they change a lamp, we could expect an LDD of about 0.85.

Now that we've thought through how our system will perform in the office space, let's use the zonal cavity method to calculate the number of luminaires of the type shown in Figure 2.15 that will be required to give us an average of 70 fc throughout the space.

We have said that the average FC level =

$$\frac{\text{Total usable lamp output (lumens)}}{\text{Area of the space (sq. ft)}}$$

And the usable lamp output is:

(No. of Luminaires) × (Lamps/Luminaire) × (Lumens/Lamp)
× BF × CU × LLD × LDD

So, by doing a little mathematical gymnastics, we can say that

$$\text{(No. of Luminaires)} = \frac{\text{Desired Average FC Level} \times \text{Area}}{\substack{\text{(Lamps/Luminaire)} \times \text{(Lumens/Lamp)} \\ \times \text{ BF} \times \text{CU} \times \text{LLD} \times \text{LDD}}}$$

All we have to do now is plug in the values, and turn the crank:

Desired footcandles = 70
Area = 20 ft × 20 ft = 400 sq ft
Lamps/Luminaire = 4
Lumens/Lamp = 2850 lumens – from lamp catalog data (we will see later how to get this).
BF = 0.80 (assumed BF for generic electronic ballast)
CU = 0.68
LDD = 0.80
LLD = 0.86

Plugging all this in, we get:

$$\text{No. of luminaires} = \frac{70 \times 400}{\substack{4 \times 2850 \times 0.8 \times 0.68 \\ \times \ 0.80 \times 0.86}} = 6.56$$

Since we cannot buy 0.56 of a luminaire, we'll use six luminaires and get a little less than 70 fc.

Now that we've done all this high level calculating, with decimal points and everything, we can be sure that, if we put six luminaires in the ceiling of that space, we'll have 70 fc at 30 in. above the floor everywhere in the room, right? *Wrong*! If you took a light meter into the space after the six luminaires were installed, you would be hard pressed to find a 70 fc reading anywhere in the room. That would only happen if the output from those six luminaires were spread evenly throughout the ceiling of the room, because what we have calculated is *average* illumination. As it is, we have 48 square feet of light source (six individual 2 ft × 4 ft luminaires) to place in 400 square feet of ceiling. If we put them all in one corner of the room, we would have considerably more than 70 fc there, but much less everywhere else. We would still have an *average* of 70 fc in the room.

What we try to do in general area illumination situations like this is to distribute the six luminaires such that their spacing from each other, and from the walls is approximately equal, to get the best distribution of light possible. Since the luminaires are asymmetric, we measure our spacing from the center point of the luminaire.

The manufacturer's information for our luminaire, shown in Figure 2.15, indicates the spacing criterion, S/MH, or maximum distance between luminaires, is 1.2 × mounting height in the long direction. Note that this is the mounting height above the working plane, not the floor. In our case, this is 1.2 × 6.5 ft, or approximately 8 ft. It is always best to maintain the published spacing criterion in your designs, in order to achieve relatively even illumination throughout the space.

Now let's lay out those six luminaires in the space, and calculate the illumination at 2.5 ft from the floor on a 1 ft × 1 ft grid everywhere in the space, using the cosine law of incidence. Bear in mind, that there are six luminaires, so we have to calculate the contribution of each luminaire at each point on the grid, and then add them up to get the total illumination at that point. This is a nice little exercise to do on a two-week holiday, *or* we can call in the computer.

## Computerized calculations

There are a number of commercially available point-by-point lighting calculation programs, and many of them are based around the ray tracing calculation engine of the LUMEN MICRO program, the original calculation program developed by Lighting Technologies, Inc. Others are based around a luminous rate of transfer calculation engine, such as the one that we will use in our calculations. For our computer calculations throughout this book, we will use VISUAL BASIC, a program developed by Lithonia Lighting, a large lighting company located in Conyers, GA, USA. VISUAL BASIC is available as a free download on Lithonia's Web site, www.lithonia.com. We should now get on the Internet, and go to this site.

From the opening welcome screen, we select the VISUAL download section, and in this section, we see the software options. Several offerings appear here, the 'Visual 2.0 SP1' option, the 'Photometrics' option, and the 'Basic edition users guide' option. The Visual 2.0 SP1 package contains the calculation package, and we should download the package that is compatible with our computer's operating system. Once this download is complete, we need to download the users guide for reference. This will give us a calculation program, and instructions for using it. The next thing we need to do is to download some photometrics for luminaires, which we will use in our exercises. We go back to the 'Photometrics option', and go into the Lithonia library of IES files. These files provide the performance data for the luminaires, which we will be using in our examples. This data is in digital IES format, and it is the same data that is published in the manufacturer's performance charts, such as that shown in Figure 2.15.

In the Lithonia library, we see that data is available for many different types of lighting. You can download the whole library if you have a lot of free hard drive space, and you will be able to run calculations using any luminaire made by the Lithonia group. For our immediate purposes, though, we only need to download the fluorescent lighting group, which contains architectural downlights, lensed troffers, parabolics, etc.

Once you complete these downloads, you need to install the VISUAL BASIC program by clicking on the self-extracting '.exe' file which was downloaded. Now we are ready to run a computer calculation. (Note: Web sites are changed from time to time, so the procedure above may not exactly fit the Web site as it exists when you visit it. All of the information will be there, though, and a little navigating around the site should allow you to download the latest versions of the files.)

The VISUAL program was developed by the Lithonia Lighting group as a design tool for lighting designers and specifiers. It is intended for use on personal computers, utilizing catalog and photometric data for all the luminaires manufactured by the Lithonia group, both interior and exterior. It will also use photometric data from other manufacturers, provided that the data is in IES format. The design program contains calculation engines and layout modeling capabilities.

There is an advanced version of the VISUAL program, called VISUAL PRO, which contains many more features than the VISUAL BASIC program, including the ability to import digital drawing files to allow lighting layout and calculations to be performed using actual floor plans, or site plans. VISUAL PRO is available for purchase from Lithonia Lighting.

For our example office space, we will use the internal layout editor in our downloaded VISUAL BASIC program. Just fill in the blanks in the layout screens to input the room geometry and reflectances. For a ceiling type, we'll use a 2 ft × 2 ft grid.

The program allows the designer to select photometric data from any of the luminaires in the Lithonia group to use for the calculations by entering an electronic 'catalog' and selecting the desired luminaire. You do this by selecting the file icon next to the 'photometric file' input field. Then select 'fluorescent', 'lensed troffers', and you'll see all the lensed troffers that Lithonia makes. Select a 2GT432 troffer as seen in Figure 2.15, and the program will load the same photometric data that is represented in tabular form in Figure 2.15, except that it is in IES standard format electronic form. We select the standard LLF for a lensed troffer, and put in the desired 70 fc. Our layout now shows six luminaires, which is about what we would expect, since our zonal cavity calculation showed that we need 6.3 luminaires. We can now run the calculator and find out the point-by-point footcandle values for each point on our 1 ft × 1 ft grid in a flash. How clever we are. Wait a minute... the average footcandle reading for this arrangement is higher than the zonal cavity calculation! How could this be? Well, this is because the particular luminaire that we selected has a little better performance than the generic one we conjured up for the zonal cavity calculations. This is another benefit of computer calculations. You can use the real IES performance data for the real luminaire instead of interpolating from a generic photometric chart. The results of our computer run are shown in Figure 2.16.

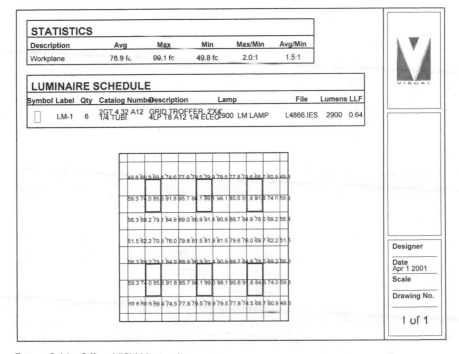

Figure 2.16. Office VISUAL results.

Clearly, the computer is the way to go if you are looking for accurate illumination values for a space, especially if you're using more than one type of luminaire, or if there are obstacles in the space. Still, the zonal cavity calculation method can provide a quick average illumination estimate if you don't have your computer handy.

## Controls

Controls are what you use to make your lighting system behave as you want it to. Controls are discussed in depth in Chapter 4, but it will be useful to know the basic concept behind the devices, so you can do the exercises in the next chapter. The most commonly used forms of lighting control are:

- Switches – on–off manual devices that control the power to the system.
- Dimmers – devices that reduce the light output of the luminaires in the system; incandescent lamps are dimmed directly, fluorescent lamps require a special dimming ballast.
- Photocells – devices that turn the luminaires on when no light is present, and off when light is present.
- Timers – devices that turn the lights on and off at preset times.
- Occupancy sensors – devices that turn the lights on when the space is occupied, and off when it isn't.

And that about covers it. A quick bagful of nifty devices that you can use to make your lighting system function just as you want it to. You're only limited by budget and imagination.

Now let's look at another kind of limits – those imposed by the regulatory agencies that impact our design and the way that it is built.

## Standards, codes, and design guidelines

(Note: The following section lists only US Codes and Standards, and is included to illustrate the application of standards in lighting design. The reader should become familiar with the comparable codes and standards which govern in his or her locale, and apply them in a similar fashion)

Although not design tools *per se*, lighting standards, codes, and design guidelines influence the selection of lamps and luminaires, and can set the tone for the design. It is useful to the designer to be familiar with the standards and codes with which the design must comply. It is also useful to be aware of recognized design guidelines, and to take advantage of the many hours spent by others in developing them. Let's look now at some of these standards and guidelines that apply directly to interior lighting design.

### NFPA-70: The National Electric Code

The National Electric Code (NEC) is the governing code for all the electrical installations within the US. Virtually every county in every state has adopted the NEC in whole, or in part, into their own electrical code. You can be sure, as a designer, that the installation of your design will come under the scrutiny of the building inspector – who is looking for violations of the NEC.

The NEC is a product of NFPA (The National Fire Protection Association), and therefore the focus of the document is the prevention of fires. Luminaires are heat-producing devices, so this is a genuine concern. Article 410 of the NEC addresses both the installation and the manufacture of luminaires. It contains a lot of useful information with which the designer should become familiar.

### NFPA 101 – The Life Safety Code

The Life Safety Code is another NFPA product, so the focus is on personnel safety in the event of a building fire. The code covers several of the most common building types, and specifies the type and amount of exit and emergency lighting required for each. Every lighting design must meet these requirements, so every designer needs to be familiar with them.

### ASHRAE/IES Standard 90.1

Standard 90.1 is a US National energy standard developed jointly by the American Society of Heating, Refrigeration, and Air conditioning Engineers (ASHRAE), and the Illuminating Engineering Society of North America (IESNA). It mandates the maximum energy consumption allowable for HVAC and lighting systems in new construction. Meeting the Standard 90.1 requirements is mandatory for all new US Government construction, and most states have adopted it as well. It is only a matter of time before all building departments adopt the Standard, so lighting designers should begin complying with the standard now.

Standard 90.1 addresses lighting by establishing lighting power budgets for various types of buildings and activities. The interior

**Table 2.3. Lighting power densities using the building area method**

| Building type | Lighting power density (W/ft²) |
|---|---|
| Automotive facility | 1.5 |
| Convention center | 1.4 |
| Court house | 1.4 |
| Dining: bar lounge/leisure | 1.5 |
| Dining: cafeteria/fast food | 1.8 |
| Dining: family | 1.9 |
| Dormitory | 1.5 |
| Exercise center | 1.4 |
| Gymnasium | 1.7 |
| Hospital/health care | 1.6 |
| Hotel | 1.7 |
| Library | 1.5 |
| Manufacturing facility | 2.2 |
| Motel | 2.0 |
| Motion picture theater | 1.6 |
| Multi-family | 1.0 |
| Museum | 1.6 |
| Office | 1.3 |
| Parking garage | 0.3 |
| Penitentiary | 1.2 |
| Performing arts theater | 1.5 |
| Police/fire station | 1.3 |
| Post office | 1.6 |
| Religious building | 2.2 |
| Retail | 1.9 |
| School/university | 1.5 |
| Sports arena | 1.5 |
| Town hall | 1.4 |
| Transportation | 1.2 |
| Warehouse | 1.2 |
| Workshop | 1.7 |

(Reprinted by permission from ASHRAE Standard 90.1-1999. Copyright 1999, American Society of Heating, Refrigerating and Air-Conditioning Engineers, Inc. Contact ASHRAE at 404-636-8400 or www.ashrae.org to purchase the standard in its entirety)

lighting power allowance may be determined for the building as a whole, or on a space-to-space basis. Figure 2.17 lists the allowable lighting power density for various types of buildings.

In calculating power usage either within a building or a space, the total allowable luminaire wattage may be found by multiplying the square footage of the space times the allowable power density. When using the space-by-space calculation method, the total building installed lighting power is the sum of the power calculated for all the spaces.

Certain types of lighting are exempt from the allowable power calculations. Specifically, these are:

1. display lighting in museums, galleries, and monuments.

Figure 2.17. Lighting power densities using the space-by-space method. (Reprinted by permission from ASHRAE Standard 90.1-1999. Copyright 1999, American Society of Heating, Refrigerating and Air-Conditioning Engineers, Inc. Contact ASHRAE at 404-636-8400 or www.ashrae.org to purchase the standard in its entirety)

Space-by-Space Method LPDs

| Building Type | Office—Enclosed | Office—Open Plan | Conference Meeting/Multipurpose | Classroom/Lecture/Training | Audience/Seating Area | Lobby | Atrium—first three floors | Atrium—each additional floor | Lounge/Recreation | Dining Area | Food Preparation | Restrooms | Corridor/Transition | Stairs—Active | Active Storage | Inactive Storage | Electrical/Mechanical | Building Specific Space Types and LPDs (W/ft²) | Additional Power Allowance (see 9.3.1.2) |
|---|---|---|---|---|---|---|---|---|---|---|---|---|---|---|---|---|---|---|---|
| **Office Buildings** — Office | 1.5 | 1.3 | 1.5 | 1.6 | | 1.8 | 1.3 | 0.2 | 1.4 | 1.4 | 2.2 | 1.0 | 0.7 | 0.9 | 1.1 | 0.3 | 1.3 | Banking Activity Area 2.4; Laboratory 1.8 | ✓ ✓ |
| **Penitentiary Buildings** — Penitentiary | 1.5 | 1.3 | | 1.4 | 1.9 | 1.8 | 1.3 | 0.2 | 1.4 | 1.4 | 2.2 | 1.0 | 0.7 | 0.9 | 1.1 | 0.3 | 1.3 | Confinement Cells 1.1 | |
| **Religious Buildings** — Religious Buildings | 1.5 | 1.3 | 1.5 | 1.6 | 3.2 | 1.8 | 1.3 | 0.2 | 1.4 | 1.4 | 2.2 | 1.0 | 0.7 | 0.9 | 1.1 | 0.3 | 1.3 | Worship-Pulpit, Choir 5.2; Fellowship Hall 2.3 | ✓ ✓ |
| **Retail Buildings** — Retail | 1.5 | 1.3 | 1.5 | | | 1.8 | 1.3 | 0.2 | 1.4 | 1.4 | 2.2 | 1.0 | 0.7 | 0.9 | 1.1 | 0.3 | 1.3 | General Sale Area 2.?; For accent lighting, see 9.3.1.2.1(e); Mall Concourse 1.? | ✓ ✓ |
| **Sports Arena Building** — Sports Arena | 1.5 | 1.3 | 1.5 | | 0.5 | 1.8 | 1.3 | 0.2 | 1.4 | 1.4 | 2.2 | 1.0 | 0.7 | 0.9 | 1.1 | 0.3 | 1.3 | Ring Sports Arena 3.?; Court Sports Arena 4.?; Indoor Playing Field Area 1.9 | |
| **Storage Buildings** — Warehouse | 1.5 | | 1.5 | | | 1.8 | 1.3 | 0.2 | | | | | 0.7 | 0.9 | 1.1 | 0.3 | 1.3 | Fine Material Storage 1.5; Medium/Bulky Material Storage 1.1 | |
| Parking Garage | 1.5 | | | | | 1.8 | | | | | | 1.0 | 0.7 | 0.9 | 1.1 | 0.3 | 1.3 | Parking Area - Pedestrian 0.2; Parking Area - Attendant only 0.1 | |
| **Theater Buildings** — Performing Arts | 1.5 | | | | 1.8 | 1.2 | 1.3 | 0.2 | 1.4 | 1.4 | | 1.0 | 0.7 | 0.9 | 1.1 | 0.3 | 1.3 | | ✓ ✓ |
| Motion Picture | | | | | 1.3 | 0.8 | | | 1.4 | | | 1.0 | 0.7 | 0.9 | 1.1 | 0.3 | 1.3 | | |
| **Transportation Buildings** — Transportation | 1.5 | 1.3 | 1.5 | | 1.0 | 1.8 | 1.3 | 0.2 | 1.4 | 1.4 | 2.2 | 1.0 | 0.7 | 0.9 | 1.1 | 0.3 | 1.3 | Airport - Concourse 0.7; Air/Train/Bus - Baggage Area 1.3; Terminal - Ticket counter 1.8 | ✓ ✓ |

2. lighting integral to equipment, such as medical and food service equipment;
3. lighting for plant growth;
4. lighting in spaces specifically designed for use by the visually impaired;
5. lighting in retail display windows, provided that the display area is enclosed by ceiling height partitions;
6. lighting in interior spaces that have been specifically designated as a registered interior historic landmark;
7. lighting that is an integral part of advertising, directional, or exit signage;
8. lighting that is for sale or lighting educational demonstration systems;
9. lighting for theatrical purposes, including performance, stage, film and video production;
10. lighting for athletic playing areas with permanent facilities for television broadcasting;
11. casino gaming areas.

Normally, you would only use the whole building method for projects where the whole building *is* the space, such as a warehouse, or retail facility. The space-by-space method allows tradeoffs between the spaces, as long as the total building installed power usage does not exceed the interior lighting power allowance for the building type.

In some cases, Standard 90.1 allows an increase in the interior power allowance when using the space-by-space method for calculating installed interior lighting power. These increases are:

1. 1 watt per square foot when decorative lighting, such as chandeliers and sconces, are installed for highlighting purposes in addition to the general lighting;
2. 0.35 watts per square foot when required to meet task requirements in video display terminal areas;
3. 1.6 watts per square foot for highlighting general merchandise, and 3.9 watts per square foot for highlighting jewelry in retail facilities.

Standard 90.1 also includes a clause which makes it mandatory that every building larger than 5000 square feet contains an automatic control device to shut off building lighting in all spaces. This device can be controlled by either a timer, an occupancy sensor, or a switch. In addition, each space must contain a control device for the lighting within the space. The Standard also specifies additional control requirements for specialized applications.

Control of the space lighting within the space is standard good practice, and is currently in use throughout the industry. A central control point for all building lighting will require some forethought on the part of the designer.

A 100% adoption of Standard 90.1 by building codes agencies will create some changes in current lighting design practices, and the fledgling designer will do well to become familiar with this Standard.

## EPACT 92

EPACT 92, or the national Energy Policy Act of 1992, is a US Government sponsored program to reduce the amount of lighting energy consumed by the country. Where Standard 90.1 restricts the energy consumption of the space, EPACT 92 restricts the energy consumption of the lamps themselves. After November 1, 1995, the manufacture or import of many of the then most popular lamp types was forbidden by the Federal government. Failure to comply with the Act could result in fines of up to $100.00 per lamp. At that rate, getting caught with a truckload of illegal lamps would put most people out of business. Needless to say, the lamp manufacturers have taken this Act to heart, and only those lamps with acceptably high efficacy are available now.

This posed a lamp replacement dilemma for those who already had thousands of fluorescent luminaires installed with ballasts designed for the outlawed low efficacy fluorescent lamps. Lamp manufacturers responded by producing lower wattage lamps that would fit into the same luminaire, and operate on the same ballast. But alas, there is no free lunch, even in lighting, and less watts meant less light output. In installations that were already borderline, this resulted in some dimly lit office space.

There is a great movement toward retrofitting existing luminaires with high efficacy lamps and electronic ballasts in order to get acceptable lighting levels out of existing luminaire installations. We'll discuss this in greater detail in Chapter 6 'The Second Time Around'.

## IES recommended footcandles

The IES has, since 1947, published a lighting handbook which recommends ambient footcandle levels for most kinds of interior spaces. Later versions of the handbook take into account light quality as well as quantity, and different categories of illumination have been developed. Such factors as age of occupants, contrast of

## Table 2.4. Determination of illuminance categories.* (lx = lux, fc = footcandle.)

*Orientation and simple visual tasks.* Visual performance is largely unimportant. These tasks are found in public spaces where reading and visual inspection are only occasionally performed. Higher levels are recommended for tasks where visual performance is occasionally important.

| A | Public spaces | 30 lx (3 fc) |
|---|---|---|
| B | Simple orientation for short visits | 50 lx (5 fc) |
| C | Working spaces where simple visual tasks are performed | 100 lx (10 fc) |

*Common visual tasks.* Visual performance is important. These tasks are found in commercial, industrial and residential applications. Recommended illuminance levels differ because of the characteristics of the visual task being illuminated. Higher levels are recommended for visual tasks with critical elements of low contrast or small size.

| D | Performance of visual tasks of high contrast and large size | 300 lx (30 fc) |
|---|---|---|
| E | Performance of visual tasks of high contrast and small size, or visual tasks of low contrast and large size | 500 lx (50 fc) |
| F | Performance of visual tasks of low contrast and small size | 1000 lx (100 fc) |

*Special visual tasks.* Visual performance is of critical importance. These tasks are very specialized, including those with very small or very low contrast critical elements. Recommended illuminance levels should be achieved with supplementary task lighting. Higher recommended levels are often achieved by moving the light source closer to the task.

| G | Performance of visual tasks near threshold | 3000–10 000 lx (300–1000 fc) |
|---|---|---|

*Expected accuracy in illuminance calculations are given in Chapter 9 of the IESNA Lighting Handbook, Lighting Calculations. To account for both uncertainty in photometric measurements and uncertainty in space reflections, measured illuminances should be within ±10% of the recommended value. It should be noted, however, that the final illuminance may deviate from these recommended values due to other lighting design criteria. (Source: IESNA Lighting Handbook, 9th edn.)

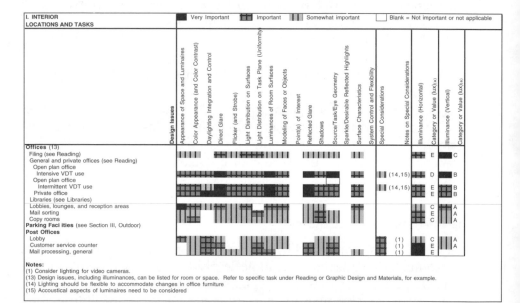

Figure 2.18. IESNA Lighting Design Guide excerpt (source: *IESNA Lighting Handbook*, 9th edn).

tasks, use of video display terminals (VDTs) and degree of detail are weighted into the recommendations for various activities within those spaces. These recommendations are updated periodically to take into account such factors as new technology and energy trends.

These recommendations are the result of a great number of hours of work and research by IES members, and they are recognized industry wide as a design standard. A good lighting designer is well advised to take advantage of the work that has been done, and to use these recommendations as a basis of design. Table 2.4 is an excerpt from the *IES lighting Handbook Reference and Application, Ninth Edition,* which explains the illumination categories of the IES Guidelines. Figure 2.18 shows the IES design guide listings for offices.

Well, OK, now you know enough to select a lamp to provide the proper type and color of light for your design space, and you can select a type of luminaire to put the light into the work area. You can control the luminaires, and you know where to find the proper illumination level for your space and activity. Then you can calculate the number of luminaires required to give you that level. So, armed with this knowledge, you are ready to venture forth into the world of design techniques to try your wings. But first, as always, let's try a few of these fun-filled exercises to refresh our memory.

## Exercises

1. What are the two most important methods by which we can produce light by using electricity?
2. How does incandescence produce light?
3. How does photoluminescence produce light?
4. What do spectral energy distribution curves tell us about a light source?
5. What does the CRI number tell us about a light source?
6. Why is an incandescent lamp envelope filled with inert gas?
7. What is the diameter of a PAR30 lamp?
   a. 4 3/4 in.
   b. 3 3/4 in.
   c. 3 in.
   d. 4 in.
8. Approximately what percentage of commercial space in the US is illuminated by fluorescent lamps?
9. A fluorescent lamp with one pin in each end is:
   a. Hot cathode rapid start
   b. Instant start

10. What is the diameter of an F40T12 lamp?
    a. 1.2 in.
    b. 1–1/2 in.
    c. 1 in.
    d. 12 in.
11. What type of lamp is a M250/U
12. What are the two functions of a ballast?
13. You are designing HID lighting for BigBucks Inc.'s new manufacturing facility. A 50 horsepower centrifuge is connected to your panel. What type of ballast should you specify?
    a. R
    b. CW
    c. HX
14. What are two purposes of reflectors?
15. Which type of diffuser would you use in a high abuse area?
    a. Acrylic
    b. Polycarbonate
16. How many lumens will a lamp rated 3000 lumens produce in a luminaire having a ballast factor of 0.85?
17. What is the RCR of a 12 ft × 12 ft room which has a ceiling height of 10 ft, and a working plane of 2 1/2 ft?
18. How many luminaires of the type shown in Figure 2.15 are required to produce 65 fc in the space of exercise 17?
19. You want to turn off the lights in the offices of BigBucks Inc. when nobody is in them. What type of control should you use?
    a. Photoelectric
    b. Occupancy sensors
    c. Dimmers
20. Why do we need to be familiar with building Codes and Standards?

**Big bonus exercise**

Download the VISUAL Basic Edition Tutorial from the www.lithonia.com/visual Web site. Print it out, and go through the tutorial exercise. This will take you through all the features available in the program. When you have finished the tutorial, print out your results.

# 3 The design process

A space design is an accumulation of ideas coupled with a knowledge of physical entities directed toward providing a particular environment. The same can be said for lighting design, which plays a major part in accomplishing the goals of the space designer. The design usually begins with the architect's concept of the ambience of the space. This is conveyed to the lighting designer, who makes suggestions for lighting the space based on the architect's ideas, the physical geometry of the space, the intended use of the space, the intended occupants of the space, the budget for the space, and a knowledge of the tools at his disposal.

This is likely to become an iterative process when those grandiose first thoughts bump into the realities of budget constraints. Even so, the design guidelines established at the first meeting with the architect should be followed as closely as possible.

At that first meeting, the architect will present his or her ideas to the designer, along with the floor plans for the space, and (hopefully) elevations and sections showing wall and ceiling construction. If not shown, the designer must ask the architect to define the ceiling height and type for each area within the space. This is usually done on a reflected ceiling plan, which shows the actual ceiling plan and grid layout that the architect intends to use. Wall and floor finishes should also be discussed, since these will affect lighting calculations later on. If not obvious, the intended use for each area should be discussed, as well as furniture layout, the type of tasks to be performed there, and the average age of the expected occupants. Finally, special effects lighting, owner's preferences for types of specialty luminaires, and an order-of-magnitude budget should be discussed. With all of this information firmly in hand, the lighting designer is prepared to take a first cut at designing a lighting system for the space.

Most lighting system designs include four basic types of lighting. These are: *ambient lighting, task lighting, accent lighting,* and *lighting for life safety*. Let's now take a look at each of these four types of lighting.

## Ambient lighting

Ambient lighting illuminates the entire space to a level which allows intended tasks to be performed comfortably. This generally means an even distribution of some quantity of light on an imaginary *work plane* selected for the tasks to be performed. The work plane in an office, for example, is 30 in. (76 cm) above the floor, the height of the average desktop.

The ambient light in a space also needs to have the proper quality to support the tasks being performed. In a space with a high use of computers, for instance, minimizing glare is very important. If the tasks being performed are color sensitive, such as matching fabrics or paint, a high color rendering index (CRI) is desired.

Color is also important in creating mood: an efficient office space needs the 'cool', efficient environment created by a 4100 K lamp; a cozy restaurant needs the warm, intimate feel produced by a 2700 K lamp; and public spaces are usually best suited for the neutral color of a 3000 K lamp. Let's now take a specific example and try to develop an approach to ambient lighting design.

### 1. Luminaire selection

Let's say that we have just been commissioned by an architect to provide the lighting design for a new office/conference room addition to Bullmoose Industries' office building. We sit down with the architect to go over the project requirements. He presents us with the floor plan shown in Figure 3.1.

He explains to us that the office area will have no permanent partitions in it. The workspaces will be separated with movable partitions, and the arrangement will be changed from time to time, as employees move in and out. The architect says that he will select the partitions, so we request that he select a light color, with a reflectance value as high as possible. The architect further informs us that the ceiling will be a suspended grid ceiling, using white acoustical tile, 9 ft above the floor, with a 2 ft by 2 ft grid spacing throughout the addition. With a little prodding, we discover that the average age of the office workers will be about 45 years, and that their work will be with normal contrast black and white text, with a high usage of computer video data terminals (VDT). The architect goes on to tell us that we have a moderate budget for this project, and the client wants to put most of his money into the conference room, which will be seen by prospective clients. The conference room will be multipurpose, and will be used for corporate training and board meetings, as well as for high tech presentations to those

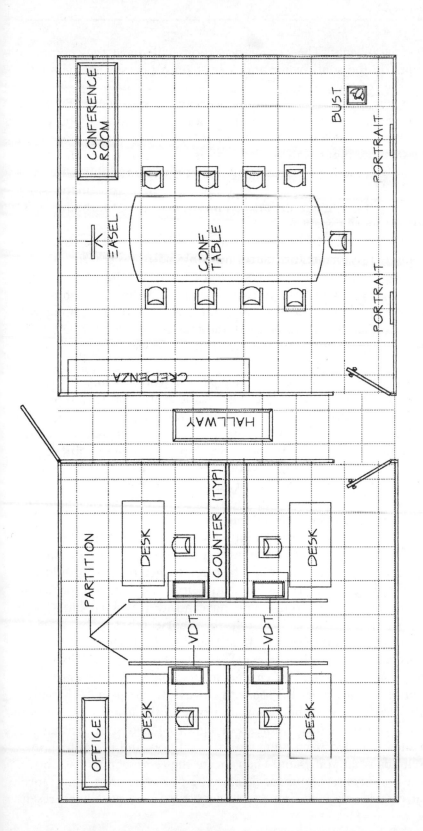

Figure 3. . Bullmoose office addition floor plan.

prospective clients. The conference room will have two company portraits on the south wall, and a bust on a pedestal in the southeast corner of the room.

With this information in hand, we can start to work. First, we will select our light source for the ambient lighting of the office space. We will use a fluorescent source, since it provides more light for less energy input, and we can choose the color temperature and CRI rating to suit our needs. For this application, a color temperature of 4100 K and a CRI of about 75 will be adequate, since color rendering isn't of primary importance, and the higher CRI lamps are more expensive.

Now we need to select the luminaire which we will be using. An ideal luminaire for the low glare, movable partition environment of the office space would be a suspended indirect luminaire, since indirect lighting would be uniform throughout the space and there would be no glare. Unfortunately, our budget and low ceiling will not allow the use of suspended luminaires, so we are left with troffers which fit into the ceiling grid. Now we need to find out what is available, and what the costs are.

Here's where we will call in the person who is going to be one of our closest allies during the design: the manufacturer's representative. In general, manufacturers' representatives are highly knowledgeable sales professionals who have a greater in-depth understanding of their products than most designers can get by skimming through catalogs. Many representatives also have well-developed design skills. A lot of time can be saved, and a lot of good ideas can be brought to the table by involving the manufacturer's representatives early in the design.

How you find these reps is relatively easy: if you know the brand name of a luminaire that you want to use, you can go to www.lightsearch.com and search by manufacturer to get their Web address. Then, simply go to their Web site, do a little clicking, and locate the representative nearest you. If you have no idea of what brand will suit your needs, you can search the lightsearch site by product description (i.e. 'fluorescent troffers'), and find out who makes what you need, and then go to their Web site. For a general listing of lighting manufacturers, you can go to www.lighting-inc.com/search-man.html. If all else fails, you can always to go to a Web search engine such as www.yahoo.com, and type in the generic description of what you want (i.e. 'fluorescent lighting') and start visiting the Web sites returned from your search until you find a manufacturer who builds what you want. Then locate the representative as above.

When you contact your local rep, it helps to remember that representatives are salesmen first and foremost, so they usually represent a full line of luminaires and associated equipment. They probably have an offering for every luminaire type that you need, and they are happy to provide you with manufacturers' literature and product application guidance. In return, they expect the opportunity to bid their equipment on your project.

Let's say that our local rep has been found, has stopped by with a few softback quick selector catalogs, and we are ready to select luminaires. We sit with the rep and look together at the criteria for ambient lighting of the office space:

1. the system should have low glare, because of the high use of VDTs;
2. the system should provide even light distribution throughout the space, because of the owner's desire for flexibility in furniture placement;
3. the system should provide 'cool' light, in order to get an efficient 'feel' within the space; color rendering is of secondary importance;
4. the system should provide adequate light for workers over the age of 45 to perform tasks comfortably;
5. as always, the system should comply with the energy requirements of ASHRAE 90.1;
6. the system should fit within a moderate budget – not 'cheap', but not 'gold plated', either.

When the lighting rep reviews these criteria, he will likely suggest a fluorescent troffer with either a parabolic diffuser or a direct/indirect diffuser which approximates the suspended indirect luminaire. Fluorescent is a good choice, for reasons mentioned previously. Troffers are suitable because of ceiling type, and the flexibility in arrangement to provide good light distribution. Either parabolic or direct/indirect diffusers will reduce glare. Now, if we take a little time to reflect on the salient features of the parabolic diffuser mentioned in Chapter 2, we will realize that parabolics are not ideal for our application because of the sharp light cutoff that they provide. Cutoff is beneficial in an open plan office, but when partitions are introduced into the space, light below the cutoff angle is blocked by the partitions, as shown in Figure 3.2. This prevents an even contribution of light from all luminaires to the working plane.

We will go with the direct/indirect luminaire to get the best distribution possible.

We now need to select a direct/indirect luminaire that will fit into the 2 ft by 2 ft ceiling grid of the office space. Let's say that we happen to have the Lithonia lighting catalog. Lithonia's softback

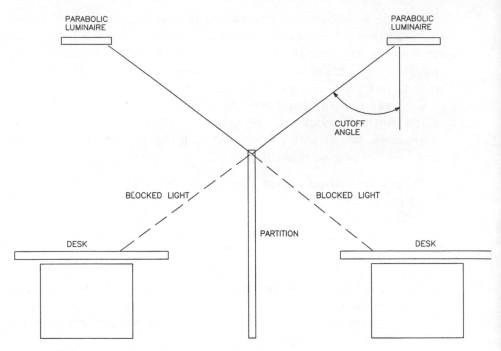

Figure 3.2. Parabolic luminaire distribution blocked by partition.

catalog is called the Product Selection Guide, and in it we find their recessed direct/indirect luminaire, which is called the 'Avante' luminaire, type AV. We see from the catalog that we have a choice of either 2 ft × 4 ft, or 2 ft × 2 ft type AV luminaire that will fit easily into the grid. Figure 3.3 is the catalog sheet for the type AV. To determine which size will best meet our performance criteria, we need to see the photometric distribution curves of each luminaire for comparison.

The Product Selection Guide is like most manufacturer's softback catalogs, and doesn't contain all the photometric distribution curves. We can either request that the rep send us the full data sheets for the type AV, or, we can go online and download them ourselves. Since we're in the twenty-first century, and have computers and all, let's download the data.

If you remember, in Chapter 2, we downloaded Lithonia's VISUAL BASIC program and photometric data to perform some calculations. Let's go back to Lithonia's central Web address, www.lithonia.com, and see what we can find. When we get to the opening screen, we see that there is a Product Info selection, which will take us to the Online Catalog. We can click through the Indoor and Fluorescent sections, until we find Recessed AV, which is what we're looking for. This will work for us, but it will take quite a bit

## FEATURES

### OPTICAL SYSTEM

- Twin matte white polyester powder paint finished reflectors provide uniform light distribution. Optional diffuse aluminum stepped reflectors available.
- All diffusers control direct light distribution and glare by shielding lamps from direct view.
- All shieldings snap into place by pivoting on light trap for easy lamp access.
- Injection molded light traps prevent light leaks between shielding and endplates.

### SHIELDING OPTIONS

- Metal Diffuser staggered Round holes (MDR) 52% open perforated metal with .075" diameter holes backed with white acrylic diffuser.
- Straight Blade Louver (SBL) sides of perforated metal with staggered round holes and solid blade louvered center. Sides and louver backed with white acrylic diffuser.
- Metal Diffuser aligned Mini slots (MDM) 46% open perforated metal backed with white acrylic diffuser.
- Acrylic Diffuser Prismatic lens (ADP) extruded acrylic lens backed with white acrylic diffuser.
- Metal Diffuser staggered Linear slots (MDL) 45% open perforated metal backed with white acrylic diffuser.

### ELECTRICAL SYSTEM

- Class P, Thermally protected, resetting, HPF, Non-PCB, UL Listed, CSA-certified electromagnetic ballast is standard. Energy saving and electronic ballast are sound rated A. Standard combinations are CBM approved and conform to UL 935.

### HOUSING

- Housing is powder painted cold rolled steel. All edges hemmed or rounded.
- Trims available for standard 1" tee bar, mini-tee bar, screw slot grids.
- Drywall ceiling adapters available.
- Fixtures can be row mounted end to end.

### LISTING

- UL listed and labeled. Listed and labeled to comply with Canadian and Mexican Standards (see options).

Specifications subject to change without notice.

| Catalog Number | | Type |
|---|---|---|
| | | |

**Direct/Indirect General Lighting System**

# AV 2'x4'

T8, T5 or T5HO
1 or 2 lamp
Compact Fluorescent
1 lamp in cross section

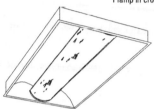

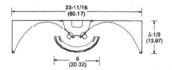

23-11/16
(60.17)

5-1/2
(13.97)

8
(20.32)

*Avante*
Recessed Direct/Indirect Lighting

## ORDERING INFORMATION

Example: **2AV G 2 32 MDR 120 GEB**

**2AV**

| Series | Lamps in Cross Section | Lamp Type | Voltage | Options |
|---|---|---|---|---|
| **2AV** 2' wide symmetric distribution | **1, 2, 3** | **32** 32W T8 (48") | **120, 277, 347** Others available | **GEB** Electronic ballast, <20% THD. |
| | | **28T5** 28W T5 (46") | | **GEB10IS** Electronic ballast, ≤10% THD, Instant Start. |
| | | **54HOT5** 54W T5 HO (46") | | **GEB10RS** Electronic ballast, ≤10% THD, Rapid Start. |
| **T2AV** 2' wide CF lamps in tandem (T). | | **CF40** 40W TT5 (24")[1] | | **ADEZ** Advance Mark X two-wire dimming ballast. (T8 only) |
| | | **CF50** 50W TT5 (24")[1] | | **EL** Emergency battery pack (nominal 300 lumens, see Life Safety section). |
| | | **CF55** 55W TT5 (24")[1] | **Diffuser** | **GLR** Internal fast-blow fuse. |

| Trim Type | Diffuser | Options |
|---|---|---|
| **G** Grid trim | **MDR** Metal diffuser, round holes. | **GMF** Internal slow-blow fuse. |
| **ST** Screw slot | **SBL** Straight blade louver, round holes. | **LP** Lamped. Specify lamp type and color. |
| | **MDM** Metal diffuser, mini slots. | **PWS1836** 6' prewire, 3/8" dia., 18-gauge, 3 wires. |
| *Accessories* | **ADP** Acrylic diffuser, linear prismatic lens. | **RIF** Radio interference filter. |
| Order as separate catalog number. | | **HTC** T-bar safety clips (snap-on). |
| **DGA24**[2] Flanged grid to drywall adapter, unit installation. | **MDL** Metal diffuser, staggered linear slots. | **LATC** T-bar safety clips (screw-on). |

Notes:
1 1 lamp in cross section, 2 lamps end to end in fixture.
2 Use G trim plus DGA accessory for fixture trim flange and fixture support in plaster or plasterboard ceilings.

**CSA** Listed and labeled to comply with Canadian Standards.

**NOM** Listed and labeled to comply with Mexican Standards.

**Reflector Option**

**ASR** Aluminum stepped reflector.

**LITHONIA LIGHTING**

AV 2x4

Figure 3.3. Lithonia type AV luminaire (source: Lithonia Lighting).

of time to evaluate all the photometric options for the AV. Since we're looking for comparative photometrics, there's an easier way to go. Back at the opening screen, we see a Lighting Software section. Go there. You'll see, among the options, a Photometric Viewer Download. We want to download this viewer. It's a few megabytes, so the download may take a little while. Once the download is finished, you'll need to unzip the file, and install the program according to the instructions. Now you're ready to go.

When you bring up the viewer, you'll see a folder called Lithonia Photometrics, Open it, then go to the Fluorescent folder. Open it, and go to the Architectural folder. Here you'll see the AV folder. Open it, and it brings up all the photometric files available for the AV luminaire. We're looking for a comparison between a 2 ft × 2 ft and a 2 ft × 4 ft luminaire, so we choose a luminaire with two CF40 lamps for the 2 ft × 2 ft, and one with three 32 W lamps for the 2 ft × 4 ft luminaire. Use the Set Compare button, select the two lamps, and hit the View button. When the photometric summary comes up, select the CP Curve option. You'll see the comparative photometrics for the two luminaires. This should look like Figure 3.4.

We see two curves, one looking down the axis of the lamps, and one looking perpendicular to the lamp axis. You can see from Figure 3.4 that the distribution pattern for both luminaires is symmetrical, regardless of viewing angle. This means that if either of the luminaires were installed, and all the desks were rotated 90°, the lighting levels at the desktop would remain the same.

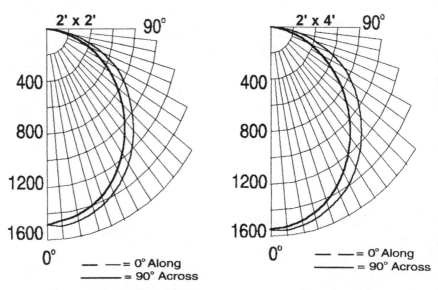

Figure 3.4. Comparison of 2 ft × 2 ft and 2 ft × 4 ft luminaire photometric curves (source: author/Lithonia Lighting).

Either is a good choice, then, since we need flexibility of furniture placement and orientation. Now what lamp should we use?

To make our final selection, we have to investigate further. Certainly energy consumption will be one of our main concerns, since we will have to meet ASHRAE 90.1 energy constraints with our design. We also don't want to encumber the client with high energy bills, or waste the country's valuable energy resources. In this vein, let's compare the lamps available to us, and see where our best value is.

Looking at the catalog information, we see that the 2 ft × 2 ft luminaire will accept the 40 W compact T5, the 50 W compact T5 and the 55 W compact T5. The 2 ft × 4 ft luminaire will accept the F32T8 lamp, the F28T5 lamp, and the F54T5 HO lamp. We have a choice of the number of lamps.

If our lighting rep has brought us a lamp catalog, we can look up each lamp to see which has the best efficacy, or lumens per watt in our color range. If we don't have a lamp catalog, we can go back to the Internet, and pull up a lamp manufacturer's Web site. Several lamp manufacturers have their product catalogs online, and if you have a reasonably recent browser, you're all set. Let's go to the Philips site, www.philips.com.

When we get to the opening screen, we see that we have a multitude of information at our fingertips, some useful, and some not applicable to our project. We click on the Professional Product heading at the top of the screen, and select Lighting, because that's what we're interested in. Under lighting, we see catalogs, and we have a choice of their European or North American offerings. We'll select North American, and we see that there is a Full Lighting Product Catalog available. This is Philips' on-line catalog. Click on that. We can download the entire catalog, and have it on hand for future projects, or we can just retrieve the information we need, and get on with the job. For now, we'll just get what we need. Under the Advanced Search section, in the 'Lamps' window, select compact fluorescent, and click on search. Philips' entire line of compact fluorescent lamps will be returned. Scroll through offerings down to 50W, and you'll see several 50W PL-L long fluorescent lamps. That's what we want. The 'long' indicates that the lamp is 24 inches long. Each offering is for a different color temperature. If you select the first lamp, you'll see that it is 4100K, which is what we want, with an initial lumen output of 4300 lumens, and a CRI of 82. The listing also includes 'design lumens', which is the initial lumens multiplied by the Lamp Lumen Depreciation (LLD). The 50W lamp appears to fit our needs. Now, in order to find the 40W lamp, we'll refine our search efforts somewhat.

Back-up to the search menu and again select compact fluorescent. This time, pick the 'watts' scroll arrow, and scroll down to 40W. Philips doesn't offer a 40W lamp. The closest to it is a 38W lamp. Select that, and only the 38W offerings are returned. Select the first one, and you'll see that it has an output of 3300 lumens, a color temperature of 4100K, and a CRI of 82. Also a good choice. Now go back to the search menu, select Fluorescent, 32W, and T8 for tube type. This zeros right in on the lamp we are looking for. We select the 4100K offering, and find that it has an initial lumen output of 2950 lumens, and a CRI of 86. Another winner. We could go through the same process and obtain the data for our T5 lamps, the 28WT5 and the high output, F54T5 HO, but the T5 lamps are expensive, and not readily obtainable in some places, so we'll deem them not suitable for our purposes. T5 lamps are very efficient, however, and should be considered for projects which will be maintained professionally. One thing to remember is that the high output F54T5 HO lamps produce so many lumens (5000), that they can only be used in indirect luminaires, where the source is not visible.

Now we have all the lamp data we need to see which lamp best fits our needs, so let's look at the efficacies (lumens per input watt ratio) for each of our selected lamps using initial lumens.

For the 50W compact fluorescent, efficacy is 4300 lumens/50W=86; and for the compact 40WT5, we have 3300 lumens/40W=82.5, and for the F32T8, efficacy is 2950 lumens/32W=92. Clearly, the lamp with the best efficacy in this lot is the F32T8 lamp, so we would like to use F32T8's if the design will allow. Any of these lamps will operate efficiently on an electronic ballast, which we will use to eliminate flicker.

Now we have two luminaires complete with lamps and ballast that we can use to provide ambient lighting for the office space. The question is, how many of each of these luminaires will we need, and where should they be placed?

To determine the number of luminaires, we first need to determine the number of footcandles we need on the desktops in the space. This requires going to the IES-recommended footcandle tables and looking up our space. The proper classification for our space is open plan office with intermittent VDT use. If we look back at Figure 2.18, we see that our office falls into category E, which corresponds to a value of 50 footcandles (50 fc). For a first cut at determining quantities of each type of luminaire, we can run some zonal cavity calculations. If you don't remember the zonal cavity calculation procedure from Chapter 2, now would be a good time to take a couple of minutes for review. Or. . . we can run our zonal cavity calculations using that nifty photometric viewer program that we just downloaded. For those of you who enjoy grinding through the math by hand, have fun. You have all the data needed to evaluate a 2 ft × 4 ft luminaire with two, and three lamps, and a 2 ft × 2 ft luminaire with two 40 W lamps. Our office space is 20 ft × 20 ft, with a 9 ft ceiling, and a work plane 2.5 ft from the floor.

O.K. – just to humor the math whizzes among you, here's how the zonal cavity calculation would go for the 2 ft × 2 ft luminaire with two CF40W lamps:

$$\text{Room cavity ratio, or RCR} = \frac{5(6.5) \times (20 \text{ ft} + 20 \text{ ft})}{(20 \text{ ft} \times 20 \text{ ft})} = 3.5$$

We can pull up the photometric viewer, and find our coefficient of utilization (CU) is 0.5. We use a generic electronic ballast factor of 0.80, and since an office is a relatively clean environment, a luminaire dirt depreciation (LDD) of 0.90. That gives us a total light loss factor (LLF) of 0.9 × 0.8, or 0.72. Use the design lumens of 2970 to account for LLD, and for the desired 50 fc, we have:

$$\text{Luminaires} = \frac{50 \times (20 \text{ ft} \times 20 \text{ ft})}{2(\text{lamps}) \times 2970 \times 0.72 \times 0.5} = 9.35$$

For the rest of us, let's look back at the photometric viewer and click on the AV luminaire. Then let's select the 2AVG2CF40ADP luminaire. That's the one with two CF40 lamps. Next, hit the View button to bring up the photometric summary. This time, we will select the Room Estimator option. The estimator asks us to fill in the room dimensions, and the luminaire mounting height. The viewer provides default working plane height, lamp lumens, and light loss factor. The RCR, and CU are also provided by the program, having been calculated from room data input and stored luminaire data. We can type in the desired footcandles at 50, and the program runs a zonal cavity calculation to show that we need... What? 6.2 luminaires? Why is this so different from the 9.35 luminaires that we calculated by hand? Let's go back and look at the estimator screen. The default LLF is 1.0, which doesn't take into account the ballast factor, or the LDD, which brings our LLF down to 0.72. We type that in, and see that we still only need 8.7 luminaires. The program also defaults to 3100 lumens/lamp, which is lamp initial lumens. If we go back and plug in the design lumens of 2970, we see that the required luminaires goes to 9.2, which is close enough.

This is a good point to remember. When using quick calculator computer programs, always check to make sure that the default values make sense. We know that the LLF is *always* less than 1.0, and that we use design lumens to assure better system performance over the long run.

Now let's run the program for the 2 ft × 4 ft luminaire, with both two and three F32T8 lamps. The program will retain our inputs, so all we need to change is the lamp photometrics, and the LLF to a more reasonable value of 0.65. After a bit of mousework, we see that we need 9.4 of the two-lamp luminaires, and 6.7 of the three lamp.

If cost was our major concern, the quick estimator would suffice, and at this point we would choose to go with six of the three lamp luminaires. But we have light distribution as a major concern, so we will now have to run point-by-point calculations to see how our three choices perform. The easiest way to do this, of course, is by using the computer, so now we will pull up the VISUAL BASIC program that we downloaded in Chapter 2.

We will enter the geometry of the room, the ceiling grid type, use standard reflectances, and select the photometric file for the luminaires and lamps that we have chosen. Below the photometric file selector is a light loss factor selector which offers either standard or IES calculated light loss factors. None of the standard values fit our needs, so click on the icon beside the 'LLF Value' window. For the 40 W compact fluorescent lamp, we will use 'fluorescent' and 'T5 Long CF'. For the

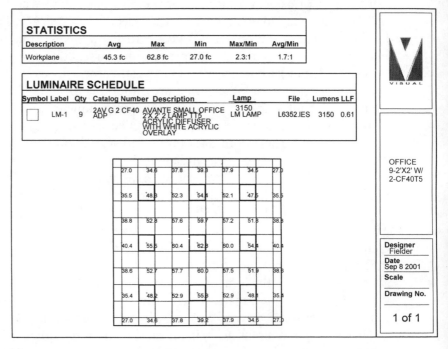

(a)

Figure 3.5. Bullmoose office VISUAL results using (a) 9.2CF40 luminaires, (b) 9.2F32T8 luminaires, and (c) 6.3F32T8 luminaires.

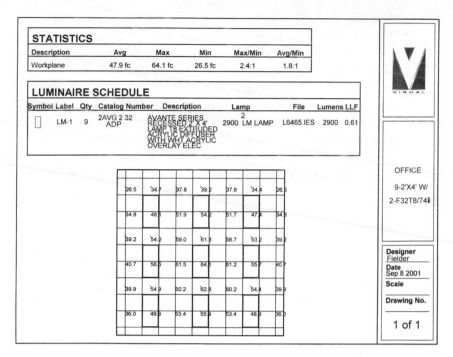

(b))

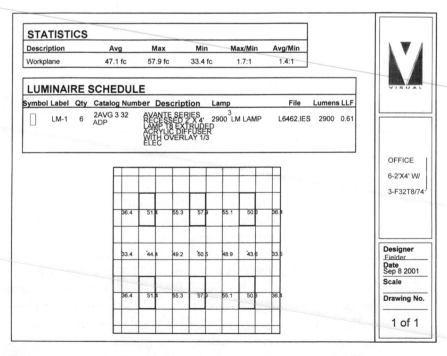

(c))

F32T8s we will use 'Fluorescent' and 'T8 Rapid Start'. Both lamps have a calculated LLF of 0.9. Continue on through the windows: For LDD we will use 'Indirect Luminaires'; and for the ballast we will use 'Electronic'. None of the 'other' factors apply to our situation. We are now left with a combined LFF of 0.61. Now run the calculator using a desired illumination of 50 fc for each luminaire.

As the results come up for each luminaire type, make a note of the calculated power density for later comparison with the ASHRAE 90.1 allowances. Make a calculation run for each of the three types, using the 'Tools' menu, and the 'Lumen Method' button to return to the calculation program to change lamps. The program will retain the other settings. When you have finished, the results should look like Figure 3.5a–c.

The first thing that we notice about the results is that the average footcandles computed for the six 3F32T8 luminaires is more than that for the nine 2CF42 luminaires, and about equal to that of the nine 2F32T8 luminaires. The max/min ratio is seen to be better for the six three-lamp luminaires, which indicates a better light distribution. The higher the ratio, the less uniform the light, and vice versa. Further examination shows that the six three lamp luminaires provide approximately 50 fc in the central three-quarters of the room, where the work will most likely take place. This is a good distribution of light, and the lamps with the highest efficacy are being utilized. We will use the six 2 ft × 4 ft luminaires with three F32T8 lamps each. It is interesting to note again, that at no point on the grid do we get 47.1 fc, which was our average illumination, and what we would expect from a zonal cavity calculation. As you become more accustomed to computerized lighting calculations, you will probably abandon zonal cavity calculations altogether, and instead utilize the more advanced programs, such as VISUAL PRO, which will allow you to model spaces much more accurately.

### Energy code compliance

OK, now that we have settled in on luminaires for the office, let's run a quick ASHRAE 90.1 check to make sure that the power density is within the guidelines.

From Figure 2.17 in Chapter 2, we can see that our power density allowance for this space is 1.3 W per square foot, or 520 W for the whole 400 sq. ft space. We can look back at the notes we took during the calculations and see that the power density for the six 3F32T8 luminaires is 1.3, so we are just within the guidelines. Our notes show that the other two options exceed 1.3 W/sq ft, and are not within the guidelines. We have made a good choice.

## FEATURES

- Full family of specification grade static luminaires offers choice of sizes, ceiling trims, door frames and other options to provide general illumination for every recessed application.
- Standard door is fully gasketed flush steel with mitered appearance — completely frames shielding. Corners screwed together for rigidity, easy lens replacement.
- Urethane foam gasket eliminates light leaks between door frame and housing.
- Overlapping flange and modular ceiling trims factory installed with standard swing-gate hangers or field convertible with optional trim and hanger kits.
- Aluminum door frames available — flush or regressed.
- Integral T-bar safety clips hold T-bar securely, no fasteners required to install.
- Optional spring-loaded latches
- Guaranteed for one year against mechanical defects in manufacture.

## SPECIFICATIONS

BALLAST — Thermally protected, resetting, Class P, HPF, non-PCB, UL listed, CSA certified ballast is standard. Energy-saving and electronic ballasts are sound rated A. Standard combinations are CBM approved and conform to UL 935.

WIRING & ELECTRICAL — Fixture conforms to UL 1570 and is suitable for damp locations. AWM, TFN or THHN wire used throughout, rated for required temperatures.

MATERIALS — Housing formed from cold-rolled steel. Acrylic shielding material 100% UV stabilized. No asbestos is used in this product.

FINISH — Five-stage iron-phosphate pretreatment ensures superior paint adhesion and rust resistance. Painted parts finished with high-gloss, baked white enamel.

LISTING — UL listed and labeled. CSA certified (see options). NOM labeled (see options).

Specifications subject to change without notice.

Type ...................   Catalog number ...................

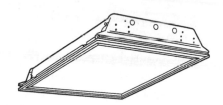

**Specification Premium Static Troffer**

# SP 2'x2'

## Compact Fluorescent

2, 3 or 4 lamps

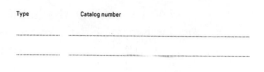

---

## PHOTOMETRICS

Calculated using the zonal cavity method in accordance with IESNA LM41 procedure. Floor reflectances are 20%. Lamp configurations shown are typical. Full photometric data on these and other configurations available upon request.

### 2SP 2 CF40 A12
**Report LTL 3606**
**S/MH (along) 1.2  (across) 1.3**
#### Coefficient of Utilization

| Ceiling | 80% | | | 70% | | | 50% | | |
|---|---|---|---|---|---|---|---|---|---|
| Wall | 70% | 50% | 30% | 70% | 50% | 30% | 50% | 30% | 10% |
| 1 | 75 | 72 | 70 | 73 | 71 | 68 | 68 | 66 | 64 |
| 2 | 69 | 64 | 60 | 67 | 63 | 59 | 61 | 58 | 55 |
| 3 | 64 | 58 | 53 | 62 | 57 | 52 | 55 | 51 | 48 |
| 4 | 59 | 52 | 47 | 58 | 51 | 46 | 49 | 45 | 42 |
| 5 | 54 | 46 | 41 | 53 | 46 | 41 | 44 | 40 | 36 |
| 10 | 37 | 28 | 23 | 36 | 28 | 23 | 27 | 22 | 19 |

#### Zonal Lumens Summary

| Zone | Lumens | %Lamp | %Fixture |
|---|---|---|---|
| 0-30 | 1395 | 22.5 | 33.2 |
| 0-40 | 2255 | 36.4 | 53.7 |
| 0-60 | 3587 | 57.9 | 85.4 |
| 0-90 | 4202 | 67.8 | 100.0 |
| 90-180 | 0 | 0 | 0 |
| 0-180 | 4202 | 67.8 | 100.0 |

### 2SP 3 CF40 A12
**Report LTL 3334**
**S/MH (along) 1.2  (across) 1.3**
#### Coefficient of Utilization

| Ceiling | 80% | | | 70% | | | 50% | | |
|---|---|---|---|---|---|---|---|---|---|
| Wall | 70% | 50% | 30% | 70% | 50% | 30% | 50% | 30% | 10% |
| 1 | 69 | 66 | 64 | 67 | 65 | 63 | 62 | 60 | 59 |
| 2 | 63 | 59 | 55 | 62 | 58 | 54 | 56 | 53 | 50 |
| 3 | 59 | 53 | 48 | 57 | 52 | 48 | 50 | 47 | 44 |
| 4 | 54 | 47 | 43 | 53 | 47 | 42 | 45 | 41 | 38 |
| 5 | 50 | 42 | 37 | 49 | 42 | 37 | 41 | 36 | 33 |
| 10 | 34 | 26 | 21 | 33 | 25 | 21 | 25 | 20 | 17 |

#### Zonal Lumens Summary

| Zone | Lumens | %Lamp | %Fixture |
|---|---|---|---|
| 0-30 | 1901 | 20.4 | 32.9 |
| 0-40 | 3076 | 33.1 | 53.3 |
| 0-60 | 4926 | 53.0 | 85.3 |
| 0-90 | 5772 | 62.1 | 100.0 |
| 90-180 | 0 | 0 | 0 |
| 0-180 | 5772 | 62.1 | 100.0 |

### 2SP 3 U31 A12125
**Report LTL 3365**
**S/MH (along) 1.2  (across) 1.4**
#### Coefficient of Utilization

| Ceiling | 80% | | | 70% | | | 50% | | |
|---|---|---|---|---|---|---|---|---|---|
| Wall | 70% | 50% | 30% | 70% | 50% | 30% | 50% | 30% | 10% |
| 1 | 69 | 67 | 65 | 68 | 65 | 63 | 63 | 61 | 60 |
| 2 | 64 | 60 | 56 | 63 | 59 | 55 | 57 | 54 | 52 |
| 3 | 59 | 54 | 50 | 58 | 53 | 49 | 51 | 48 | 45 |
| 4 | 55 | 49 | 44 | 54 | 48 | 43 | 46 | 42 | 39 |
| 5 | 51 | 44 | 38 | 49 | 43 | 38 | 42 | 37 | 34 |
| 10 | 34 | 26 | 21 | 33 | 26 | 21 | 25 | 21 | 18 |

#### Zonal Lumens Summary

| Zone | Lumens | %Lamp | %Fixture |
|---|---|---|---|
| 0-30 | 1745 | 20.8 | 33.2 |
| 0-40 | 2871 | 34.2 | 54.6 |
| 0-60 | 4631 | 55.1 | 88.1 |
| 0-90 | 5257 | 62.6 | 100.0 |
| 90-180 | 0 | 0 | 0 |
| 0-180 | 5257 | 62.6 | 100.0 |

---

 **LITHONIA LIGHTING**
COMMERCIAL & INDUSTRIAL FLUORESCENT LIGHTING        **SP 2x2 CF**

---

Figure 3.6. Lithonia type SP luminaire (source: Lithonia Lighting).

Now let's look at the hallway, which also requires ambient lighting. Hallways, or paths of egress, are required by code to have illumination of 10 fc at the floor. No visual tasks are being performed there, so glare and color rendition are of minimal concern. Ten footcandles doesn't represent many watts/square foot, so if we have reasonable efficacy, we're not likely to get into trouble with the energy code. This means that a 'plain Jane' luminaire will suffice. Looking back at our catalog, we see the 2 ft × 2 ft troffer shown in Figure 3.6.

This luminaire has an acrylic diffuser, and will fit nicely into our ceiling grid. We want to use a 3000 K lamp here, to produce a 'neutral' feel to the space and we'll use the 40 W compact fluorescent lamp because of the efficacy. To calculate the number of luminaires required to produce the 10 fc, we can return to the computer. This time, the working plane is at 0 in. (the floor). Running the program, we find that we can get 10 fc with only one luminaire. This would make our hallway look like a cave, so we'll try two luminaires and see what happens. The VISUAL results are shown in Figure 3.7. The light distribution is fairly even, and we have more than the required 10 fc, so we'll consider this a good fit, and move on.

Now let's take a look at task lighting for the conference room.

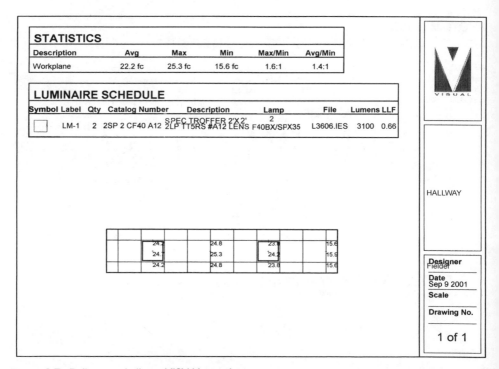

Figure 3.7. Bullmoose hallway VISUAL results.

## Task lighting

Task lighting means putting the light where you need it to do the job. This infers that task lighting is almost always direct lighting. If we look at the conference room in Figure 3.1, we can see a couple of areas where task activity will take place: the conference table, of course, where everybody will be sitting; and the side table area, where people will refill their coffee cups from time to time. Now, how should we light these areas?

If we look back at our criteria for the conference room, we see that we need enough light on the table to be able to perform detailed work in training sessions (60–70 fc), but we need only enough light to take notes (10–20 fc) when a presentation video is in use. The same is true for the side table: when everybody is up getting coffee and pastries, they need to see what they're doing; when the video is on, you only need minimum light in that part of the room.

This situation will be accommodated by putting dimming controls on the luminaires that we use for the table and the side table. Now what luminaires should we use?

### Luminaire selection

Again, for color temperature, CRI, and amount of light produced, we should consider fluorescent first for the lighting of the conference table. We will mount the luminaires directly over the conference table, since that is where the tasks will be performed. The spacing of the luminaires should offer even light distribution on the entire surface of the table, and glare should be minimized. There are no partitions involved in this application, so parabolic diffusers should work well – perhaps better than the direct/indirect luminaires, because light trespass into areas other than the task area will be limited by the cutoff characteristics of the parabolic diffusers.

This time, instead of calculating a number of luminaires required to provide a specific light level, we will locate our luminaires directly above the task area, and space them evenly to provide good light distribution. If we look at the geometry of the conference table, we see that three 2 ft × 4 ft luminaires will fit nicely into the ceiling grid above the table. The long axis of the luminaires will be parallel to the line of vision of those sitting at the sides of the table, which will further reduce glare for the majority of those at the table. If necessary, we can adjust the maximum light output of the luminaires by choosing the number of lamps in the luminaires. We now go back to the catalog to select the luminaire. Figure 3.8 shows the catalog data for an appropriate parabolic luminaire selection.

## FEATURES
- Choice of low iridescent diffuse or specular louver finishes. Ideal for use with triphosphor lamps.
- Black reveal provides floating louver appearance, conceals optional air-supply slots.
- Overlapping flange and modular ceiling trims factory installed with standard swing-gate hangers or field convertible with optional trim and hanger kits.
- Optional heat-removal dampers and air-pattern control blades allow airflow control.
- T-hinges die-formed for maximum strength. Latches spring-loaded, concealed in reveal.
- Guaranteed for one year against mechanical defects in manufacture.

## SPECIFICATIONS
BALLAST — Thermally protected, resetting, Class P, HPF, non-PCB, UL listed, CSA certified ballast is standard. Energy-saving and electronic ballasts are sound rated A. Standard combinations are CBM approved and conform to UL 935.

WIRING & ELECTRICAL — Fixture conforms to UL 1570 and is suitable for damp locations. AWM, TFN or THHN wire used throughout, rated for required temperatures.

MATERIALS — Housing formed from cold-rolled steel. Louvers formed from anodized aluminum. No asbestos is used in this product.

FINISH — Five-stage iron-phosphate pretreatment ensures superior paint adhesion and rust resistance. Painted parts finished with high-gloss, baked white enamel.

LISTING — UL listed and labeled. Listed and labeled to comply with Canadian and Mexican Standards (see Options).

Specifications subject to change without notice.

| Catalog Number | Type |
|---|---|
|  |  |

**PARAMAX® Parabolic Troffer**

# PM4 2'x4'

### 4" Deep Louver
3 lamps

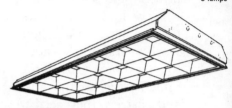

## ENERGY
Luminaire Efficacy Rating (LER) and Annual Energy Cost/1,000 Lumens

3-lamp, LD louver - LER.FP = 57. Annual energy cost = $4.21.
Based on 32W T8 lamp, 2850 lumens and electronic ballast.
Ballast factor = .88 and input watts = 86.

Calculated in accordance with NEMA standard LE-5.

## PHOTOMETRICS
Calculated using the zonal cavity method in accordance with IESNA LM41 procedures. Floor reflectances are 20%. Lamp configurations shown are typical. Full photometric data on these and other configurations available upon request.

**2PM4 G B 3 32 18LD**

**Report LTL 6757 - Lumens per lamp = 2850**

**S/MH (along) 1.2  (across) 1.5**

**Coefficient of Utilization**

| Ceiling | 80% | | | 70% | | | 50% | | | 0% |
|---|---|---|---|---|---|---|---|---|---|---|
| Wall | 70% | 50% | 30% | 70% | 50% | 30% | 50% | 30% | 10% | 0% |
| 0 | 78 | 78 | 78 | 76 | 76 | 76 | 73 | 73 | 73 | 66 |
| 1 | 73 | 71 | 68 | 71 | 69 | 67 | 66 | 65 | 63 | 58 |
| 2 | 68 | 63 | 60 | 66 | 62 | 59 | 60 | 57 | 55 | 51 |
| 3 | 62 | 57 | 52 | 61 | 56 | 51 | 54 | 50 | 47 | 45 |
| 4 | 58 | 51 | 46 | 56 | 50 | 45 | 48 | 44 | 41 | 39 |
| 5 | 53 | 46 | 41 | 52 | 45 | 40 | 44 | 40 | 36 | 34 |
| 6 | 49 | 42 | 36 | 48 | 41 | 36 | 40 | 35 | 32 | 30 |
| 7 | 46 | 38 | 33 | 45 | 37 | 32 | 36 | 32 | 29 | 27 |
| 8 | 43 | 35 | 29 | 42 | 34 | 29 | 33 | 29 | 26 | 24 |
| 9 | 40 | 32 | 27 | 39 | 31 | 27 | 31 | 26 | 23 | 22 |
| 10 | 37 | 29 | 24 | 37 | 29 | 24 | 28 | 24 | 21 | 20 |

**Zonal Lumens Summary**

| Zone | Lumens | %Lamp | %Fixture |
|---|---|---|---|
| 0-30 | 1911 | 22.3 | 34.1 |
| 0-40 | 3247 | 38.0 | 57.9 |
| 0-60 | 5313 | 62.1 | 94.8 |
| 0-90 | 5604 | 65.5 | 100.0 |
| 90-180 | 0 | 0 | 0 |
| 0-180 | 5604 | 65.5 | 100.0 |

**LITHONIA LIGHTING**
COMMERCIAL FLUORESCENT LIGHTING

PM4 2x4 3LP

Figure 3.8. Lithonia Type PM4 luminaire (source: Lithonia Lighting).

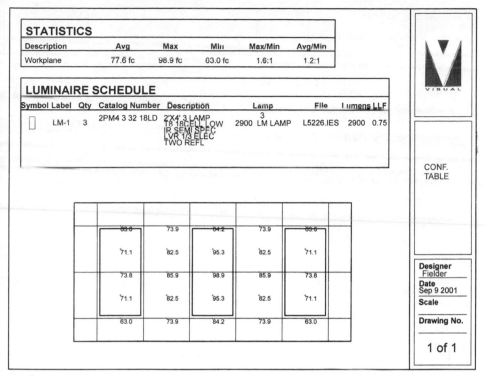

**STATISTICS**

| Description | Avg | Max | Min | Max/Min | Avg/Min |
|---|---|---|---|---|---|
| Workplane | 77.6 fc | 98.9 fc | 63.0 fc | 1.6:1 | 1.2:1 |

**LUMINAIRE SCHEDULE**

| Symbol | Label | Qty | Catalog Number | Description | Lamp | File | Lumens | LLF |
|---|---|---|---|---|---|---|---|---|
| | LM-1 | 3 | 2PM4 3 32 18LD | 2'X4' 3 LAMP T8 18CELL LOW IR SEMI SPEC LVR 1/3 ELEC TWO REFL | 3 2900 LM LAMP | L5226.IES | 2900 | 0.75 |

CONF. TABLE

| 63.0 | 73.9 | 84.2 | 73.9 | 63.0 |
| 71.1 | 82.5 | 95.3 | 82.5 | 71.1 |
| 73.8 | 85.9 | 98.9 | 85.9 | 73.8 |
| 71.1 | 82.5 | 95.3 | 82.5 | 71.1 |
| 63.0 | 73.9 | 84.2 | 73.9 | 63.0 |

**Designer**
Fielder
**Date**
Sep 9 2001
**Scale**
**Drawing No.**

1 of 1

Figure 3.9. Bullmoose conference table VISUAL Results.

A zonal cavity calculation for the area above the table wouldn't have much meaning, but we can run the computer analysis for that area. We assume that our space stops at the boundaries of the table, and that the wall reflectance is zero. This time we use the feature of the program which allows us to place the luminaires where we want them. First, let's try three F32T8/841 lamps per luminaire. Figure 3.9 shows the results of the computer run.

We can see that three lamps per luminaire will give us about 99 fc in the center of the table, and an average of 78 fc. This is about right for maximum illumination for close work at the table. We're going to use dimming ballasts on these luminaires, so we can reduce the level for less intensive tasks. Now, what about the side table?

Let's take a look at how we want the side table lighting to function: it must provide enough light to operate a coffee pot; it must be dimmable to reduce brightness during presentations; and we also don't want to be able to see the source of light while seated at the conference table. Because of the room layout, we can see that troffers aren't suitable for this job. Looking back into our toolbox from Chapter 2, we can see that a recessed downlight will give us direct lighting, and also hide the light source. We must go back to our

## FEATURES

### OPTICAL SYSTEM
- Reflector - Self-flanged, specular clear, highly specular or semi-diffuse reflector.
- Baffle/cone - Specular clear upper reflector. Microgroove baffle with white painted flange or specular black cone with flange that matches cone finish.
- Brightness control and high efficiency are optimally balanced.
- Controlled anodized coating suppresses iridescence.

### HOUSING
- 16-gauge galvanized steel mounting/plaster frame with friction support springs to retain optical system. Maximum 7/8" ceiling thickness.
- Expandable, self-locking mounting bars provide horizontal and vertical adjustment.
- Galvanized steel junction box with bottom-hinged access covers and spring latches. Two combination 1/2"–3/4" and four 1/2" knockouts for straight-through conduit runs. Capacity: 8 (4 in, 4 out) No. 12 AWG conductors, rated for 75°C.

### ELECTRICAL SYSTEM
- Die-cast aluminum socket housing. Ventilated top for convective cooling.
- Horizontally-mounted, positive-latch, thermoplastic socket(s) in single (centered) or double-lamp configuration.
- Class P (thermally protected) high power factor ballast(s) mounted to the junction box.

### LISTING
- Fixtures are UL listed for thru-branch wiring, recessed mounting and damp locations. Listed and labeled to comply with Canadian Standards (see Options).

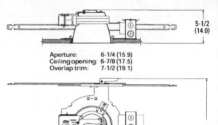

**Compact Fluorescent Downlights**

# 6" AF

## Open Reflector

Horizontal Lamp
Twin-Tube, Double Twin-Tube or Tri-Tube

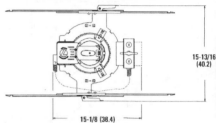

5-1/2 (14.0)

Aperture:        6-1/4 (15.9)
Ceiling opening: 6-7/8 (17.5)
Overlap trim:    7-1/2 (19.1)

15-13/16 (40.2)

15-1/8 (38.4)

All dimensions are inches (centimeters).
Maximum dimension for two ballast configuration. Actual configuration may vary.

## ENERGY

| LER.DOL | Annual Energy Cost | Lamps | Lamp Lumens | Ballast Factor | Input Watts |
|---|---|---|---|---|---|
| 29 | $8.23 | 2/26DTT | 3600 | .98 | 52 |

Calculated in accordance with NEMA standard LE-5.

## ORDERING INFORMATION

Example:   **AF 2/18DTT 6AR 120 GEB10 WLP**

Choose the boldface catalog nomenclature that best suits your needs and write it on the appropriate line. Order accessories as separate catalog numbers (shipped separately).

**AF**

| Series | Wattage/Lamp | | Reflector Type | | Finish | | Ballast[6] | | Options |
|---|---|---|---|---|---|---|---|---|---|
| **AF** | **1/9TT**[1] One 9W twin-tube | | **6AR** Clear | **(blank)** | Specular low iridescent | **EMB** | Electromagnetic ballast. Requires two-pin lamp. | **WLP** | With 35K lamp (shipped separately). |
| | **2/9TT**[1] Two 9W twin-tube | | **6PR** Pewter | **LD** | Semi-diffuse low iridescent | | | **TRW** | White painted flange. (Standard with 6MB.) |
| | **1/13TT**[1] One 13W twin-tube | | **6UBR** Umber | | | **GEB10** | Electronic ballast. Requires four-pin lamp. | **LRC**[7] | Provides compatibility with Lithonia Reloc System. Lithonia Reloc System can be installed **less this option** with connectors provided by others. Access above ceiling required. |
| | **2/13TT**[1] Two 13W twin-tube | | **6WTR** Wheat | **LS** | Highly specular | | | | |
| | **1/13DTT**[2] One 13W double twin-tube | | **6CR**[4] Champagne Gold | | | | | | |
| | **2/13DTT**[2] Two 13W double twin-tube | | **6GR**[4] Gold | | **Voltage** | **DMHL** | Lutron Hi-lume® electronic dimming ballast. (120V or 277V; 18DTT, 26DTT, 32TRT and 42TRT only.) Requires four-pin lamp. Minimum dimming level 5%. | | |
| | **1/18DTT**[2] One 18W double twin-tube | | **6BC**[5] Black Cone | | **120** | | | **GMF** | Single slow-blow fuse |
| | **2/18DTT**[2] Two 18W double twin-tube | | **6MB**[5] Black Micro-groove Baffle | | **277** | | | **RIF** | Radio interference filter. |
| | **1/26DTT**[2] One 26W double twin-tube | | | | **347** | | | **ELR**[9] | Emergency battery pack. Remote test switch provided. |
| | **2/26DTT**[2] Two 26W double twin-tube | | | | | | | | |
| | **1/18TRT**[3] One 18W tri-tube | | | | **Accessories** | **ADEZ** | Advance Mark X electronic dimming ballast. (120V or 277V; 26TRT, 32TRT and 42TRT only.) Requires four-pin lamp. Minimum dimming level 5%. | **EL**[9,9] | Emergency battery pack. Integral test switch provided. |
| | **1/26TRT**[3] One 26W tri-tube | | | | Order as separate catalog numbers. | | | | |
| | **1/32TRT**[3] One 32W tri-tube | | | **SC6** | Sloped ceiling adaptor. Degree of slope must be specified (10D, 15D, 20D, 25D, 30D). Ex: SC6 **10D**. | | | **QDS**[10] | Quick disconnect for easy ballast replacement. |
| | **1/42TRT**[3] One 42W tri-tube | | | | | | | **DS** | Dual switching. |
| | | | | **CTA6** | Ceiling thickness adaptor. Not available with TRT fixtures. (Extends mounting frame to accommodate ceiling thickness up to 2".) | | | **GSKT** | 1/8" x 3/8" foam gasketing. |
| | | | | | | | | **CSA** | Listed and labeled to comply with Canadian Standards. |

**NOTES**
1  Available with electromagnetic ballast only.
2  Available with electromagnetic or electronic ballast.
3  Available with electronic ballast only.
4  Not recommended for use with compact fluorescent lamps; consult factory.
5  Not available with finishes.
6  Refer to options and accessories tab for additional ballast types.
7  For compatible Reloc systems, refer to options and accessories tab.
8  For dimensional changes, refer to options and accessories tab.
9  Available in two-lamp units only.
10 Not available with ELR or EL option.

**LITHONIA LIGHTING**
DOWNLIGHTING & TRACK LIGHTING

100-AF6

Figure 3.10. Lithonia type AF luminaire (source: Lithonia Lighting).

catalog and find a downlight which suits our purpose. The catalog specification sheet for one such luminaire is shown in Figure 3.10.

A word of caution here: when selecting recessed downlights, always find out whether insulation is to be installed above the ceiling in which the luminaire is mounted. If it is, you must use an 'IC' (insulation contact) rated luminaire, or you will have excessive heat buildup.

Fortunately, the ceiling in our conference room has no insulation above it, so we can use a non-IC rated luminaire. We will use a fluorescent lamp, so that we may match the color of the lamps over the conference table, and we will use a dimming ballast to dim the luminaire when desired. Horizontal lamp mounting will be selected, so that the lamp will not be visible when the luminaire is viewed from an angle. To further camouflage the light source, we will use a black grooved baffle trim on the inside of the luminaire.

Our selected luminaire offers a 42 W triple tube compact fluorescent lamp as an option, so we will use this to get as much undimmed light as possible out of the luminaire. No particular footcandle level is required here, so three luminaires, mounted directly above the side table, should suffice.

Even though there is no furniture shown on the east wall, there is a big open space there, so there's a pretty good chance a telephone table, or something else, will eventually be put there. We will put two downlights over that area just for good measure.

We now have the task areas lit, so what else do we need? If we look around, we see a presentation easel in the front of the room, the company portraits on the back wall, and the bust of Commodore Bullmoose himself in the back corner. It would be nice to accent these features with light, so let's see what we can do.

## Accent lighting

Accent lighting, like task lighting, is direct lighting projected on a specific area for a specific purpose: *Task lighting illuminates an area so that a task may be performed there; accent lighting illuminates an object so that it may be seen clearly and dramatically.*

The majority of accent lighting that we encounter in the real world is done from above, with ceiling-mounted luminaires. Most of these luminaires, particularly in retail spaces, are track mounted. Track lights are versatile, easily adjusted, and easily re-lamped to perform a variety of accent lighting jobs. In our conference room, however, the architect feels that a surface-mounted track would 'clutter' his ceiling, and take away from the effect that he is trying

## FEATURES

### OPTICAL SYSTEM

- Internal housing components painted matte black. Lamp snoot minimizes stray light in housing.
- Self-flanged, specular clear, semi-diffuse or highly specular anodized cone maximizes output while minimizing room-side flash.
- Seamless cast faceplate is retained by self-aligning, constant force, torsion support springs.
- Accommodates up to two lens and/or louver filters.
- 0°-45° vertical adjustment and 360° horizontal adjustment.
- Tool-less vertical and horizontal lamp adjustments are made with optical system lowered below ceiling for simple focusing. Adjustment mechanism is lockable to maintain focus during relamping.
- Locking mechanism is visible from below ceiling with trim assembly lowered.
- Position indicators allow consistent aiming from fixture to fixture.
- Softening lens standard.

### MECHANICAL SYSTEM

- Re-lamp capability from above or below ceiling.
- Tool-less removal of step-down transformer.
- Tool-less removal of thermally-activated insulation detector.
- Universal housing with matte black finish and plaster flange will accommodate a wide variety of DLV series trims. Maximum 2" ceiling thickness.
- Expandable, self-locking adjustable mounting bars standard.
- Secondary housing adjustment system for precise, final ceiling to flange alignment.
- Painted steel junction box with bottom-hinged access covers and spring latches.

### ELECTRICAL SYSTEM

- Replaceable socket assembly. MR16 socket assembly standard (20W - 75W).
- Four combination 1/2"-3/4" knockouts for straight-through conduit runs. Capacity: 8(4 in, 4 out) No. 12 AWG conductors, rated for 90°C.

### LISTING

- UL listed for thru-branch wiring, recessed mounting and damp locations.
- Listed and labeled to comply with Canadian Standards.

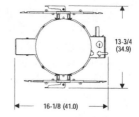

**Low Voltage Downlights**

# 3" DLV
### Adjustable Downlight
Seamless Cast Faceplate with Open Cone

GOTHAM®

6-3/4 (17.1)

Aperture:              3 (7.6)
Ceiling opening:    4-7/8 (12.4)
Overlap trim:       5-7/16 (13.8)

All dimensions
are inches
(centimeters).

13-3/4
(34.9)

16-1/8 (41.0)

## ORDERING INFORMATION

Example:  **DLV ADJ MR16 3BCT30 120 DWHG**

Choose the boldface catalog nomenclature that best suits your needs and write it on the appropriate line. Order accessories as separate catalog numbers (shipped separately).

**DLV   ADJ**

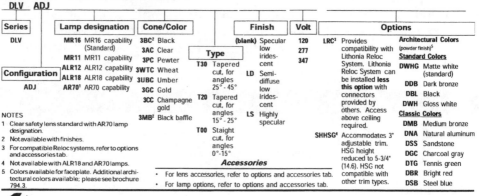

| Series | Lamp designation | Cone/Color | | Finish | Volt | Options | |
|---|---|---|---|---|---|---|---|
| **DLV** | **MR16** MR16 capability (Standard) | **3BC²** Black | | **(blank)** Specular low irides- | **120** | **LRC³** Provides compatibility with Lithonia Reloc System. Lithonia Reloc System can be installed **less this option** with connectors provided by others. Access above ceiling required. | **Architectural Colors** (powder finish)⁵ |
| | **MR11** MR11 capability | **3AC** Clear | | cent | **277** | | **Standard Colors** |
| | **ALR12** ALR12 capability | **3PC** Pewter | **Type** | **LD** Semi-diffuse low irides- | **347** | | **DWHG** Matte white (standard) |
| **Configuration** | **ALR18** ALR18 capability | **3WTC** Wheat | **T30** Tapered cut, for angles 25°- 45° | | | | **DDB** Dark bronze |
| **ADJ** | **AR70¹** AR70 capability | **3UBC** Umber | | cent | | | **DBL** Black |
| | | **3GC** Gold | **T20** Tapered cut, for angles 15°- 25° | **LS** Highly specular | | | **DWH** Gloss white |
| | | **3CC** Champagne gold | | | | | **Classic Colors** |
| **NOTES** | | **3MB²** Black baffle | **T00** Staight cut, for angles 0°-15° | | | **SHHSG⁴** Accommodates 3" adjustable trim. HSG height reduced to 5-3/4" (14.6). HSG not compatible with other trim types. | **DMB** Medium bronze |
| 1 Clear safety lens standard with AR70 lamp designation. | | | | | | | **DNA** Natural aluminum |
| 2 Not available with finishes. | | | | | | | **DSS** Sandstone |
| 3 For compatible Reloc systems, refer to options and accessories tab. | | | **Accessories** | | | | **DGC** Charcoal gray |
| 4 Not available with ALR18 and AR70 lamps. | | | • For lens accessories, refer to options and accessories tab. | | | | **DTG** Tennis green |
| 5 Colors available for faceplate. Additional architectural colors available; please see brochure 794.3. | | | • For lamp options, refer to options and accessories tab. | | | | **DBR** Bright red |
| | | | | | | | **DSB** Steel blue |

**⚫ LITHONIA LIGHTING**

Figure 3.11. Lithonia Type DLV luminaire (source: Lithonia Lighting).

to achieve. We must go back to the catalogs to find a recessed, adjustable downlight that will accommodate the type of lamps that we require for accent lighting. The catalog data for one such luminaire is shown in Figure 3.11.

The luminaire shown in Figure 3.11 is available in adjustable accent and wallwash accent configurations. It is designed to utilize the MR-16 low voltage lamp, which, as you should remember from Chapter 2, produces a brilliant, focused white light in several beam spreads. It is also very hot, but as said before, we have no insulation concerns in our space.

The rule of thumb for wallwashing in general is to space the luminaires equally from the wall and each other to obtain uniform illumination. In our case, we are using the wallwashers for accent lighting, so we will mount a wallwash luminaire in front of each of the portraits, so that the light strikes the portrait at an angle of 30° or less from the vertical. This is to prevent glare from the glass covering of the portraits. The mechanics of this is illustrated in Figure 3.12.

This will give us an oval of light on the portrait for an accent effect. The lamp position and beam spread are both selectable, and may be adjusted to frame the entire portrait. For our presentation easel, we will mount one of the accent downlights so that we can frame the entire presentation surface and maintain our 30° angle. For our pièce de résistance, we will locate two downlights over the bust of the Commodore, to accent his chiseled features and give

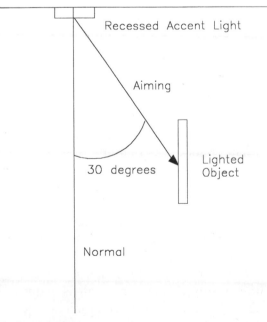

Figure 3.12. Accent luminaire mounting.

him a three-dimensional appearance (remember 'light and shadows' from Chapter 1). Again, these downlights should be placed so that the light hits the sculpture at an angle of 30° or less from the vertical to eliminate glare from the Commodore's prominent proboscis. All of the accent luminaires will be adjusted after installation to maximize their effect.

Finally, we will put dimmers on the accent lights so that they may be dimmed while presentations are taking place, and that about wraps it up for the new spaces – or does it?

What happens if the electricity goes off just when Bullmoose, Inc. has a conference room full of foreign investors? Those guys will be crashing around in the dark, trying to find their way out of unfamiliar surroundings, and not feeling very benevolent toward Bullmoose, Inc., unless. . . unless we have done a good job of lighting for life safety.

## Lighting for life safety

Lighting for life safety means providing enough light for occupants to safely exit a building in case of a power outage, which often accompanies a fire in the building. It also entails guiding those occupants out of the building by the use of exit signs.

The Life Safety Code (NFPA 101) requires that we provide an average of 1 fc at the floor on the paths of egress from the building in emergency conditions. Power for the emergency lighting can either be provided by a backup battery system, or an emergency generator with an automatic switch to transfer loads from the normal to the generator source. Battery systems can be purchased in two configurations: a centralized battery, charger, and transfer switch that powers a number of emergency luminaires; or individual batteries, charger, and transfer switch located within the luminaire itself. Most fluorescent luminaire manufacturers offer the emergency backup option, up to 1100 lumens per luminaire, as a standard package.

The Bullmoose building has no emergency generator, so that option is out for us. A centralized battery system is best suited for a large area in which a number of emergency luminaires are required, but would not be cost-effective for our small area, which will use only a few emergency luminaires. We are left with providing battery backup in a few selected luminaires in our space. Now, which luminaires should we choose?

Emergency lighting design is about 90% 'feel', and 10% calculation, if any calculation at all is required. Since 1 fc is very easy to obtain, we can place luminaires equipped with emergency packs almost anywhere within our small spaces and get there. In our case,

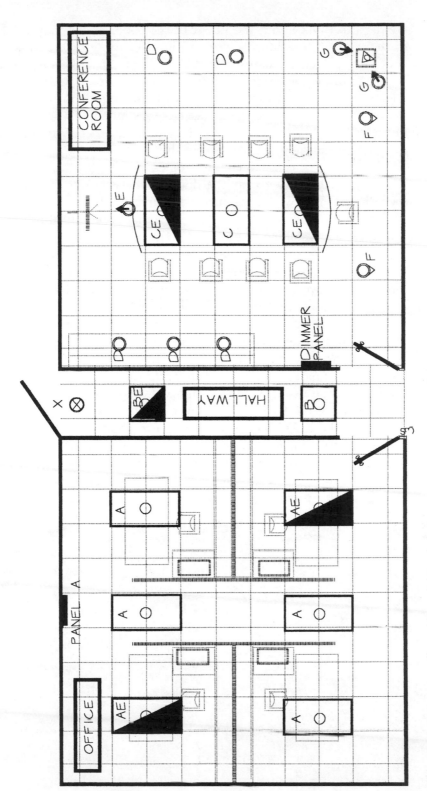

Figure 3.13. Bullmoose office addition lighting layout.

we need to think about the intent of the code, rather than the letter of the code, in order to do a good job of lighting for life safety. The code stresses that the paths of egress be illuminated. This means that, were you inside the space when the power went out, you would need enough light to safely exit the space without tripping over something.

Let's look at the office space. We have six luminaires spaced evenly in the ceiling. It would make sense to provide emergency lighting in the back of the room, farthest from the door, since the workers there have more obstacles in their path of egress. It also makes sense to put emergency lighting near the door, so the way out can be clearly seen. We can use two emergency luminaires and have the office space well covered. Now for the conference room, where all those foreign investors are watching a corporate presentation. We'll use a little overkill here, since it is important that the guests feel comfortable, even in the event of a power outage. We have three 2 ft × 4 ft luminaires above the conference table, so we will put battery backup in the two outboard ones. That will provide plenty of light. Out in the hallway, where everybody will be, once they leave their rooms, we need to provide both illumination and direction to the way out. We will provide battery backup in the luminaire closest to the door, and an illuminated exit sign with battery backup above the door leading to the outside. The exit sign will be illuminated by an LED (light emitting diode) source because of the low wattage and long lifetime of the LEDs. This should lead everybody out safely.

Now our layout for Bullmoose, Inc. is complete. Next, we need to identify the luminaires that we have used. To do this, we have assigned a luminaire type for each of the luminaires in our layout, and these are shown in the final layout of Figure 3.13.

**Table 3.1. Luminaire schedule.**

| Type | Manufacturer | Volts | Catalog no. | Lamps | Ballast | Comments |
|------|-------------|-------|-------------|-------|---------|----------|
| A | Lithonia | 120 | 2AVG332 ADP 120 | 3-F32T8/741 | 2-2F32 Elect | Office double sw |
| AE | Lithonia | 120 | 2AVG332 ADP EL 120 | 3-F32T8/741 | 2-2F32 Elect | W/emerg batt pack |
| B | Lithonia | 120 | 2SP2CF400A12 120 | 2-40WTT5 | Electronic | Hallway |
| BE | Lithonia | 120 | 2SP2CF40A12EL 120 | 2-40WTT5 | Electronic | W/emerg batt pack |
| C | Lithonia | 120 | 2PM433218L 120 | 3-F32T8/741 | Elec dimming | Conf table |
| CE | Lithonia | 120 | 2PM433218L EL 120 | 3-F32T8/741 | Elec dimming | W/emerg batt pack |
| D | Lithonia | 120 | LGF 42TRT 75B 120 ADEZ | 4-2WTRT | Elec dimming | Side table |
| E | Lithonia | 120 | DLVADJ MR'16 4ACT30 120 | 7-5WMR16 MF | | Accent-easel |
| F | Lithonia | 120 | DLV WSH MR16 4AC 120 | 7-5WMR16 FL | | Wallwash |
| G | Lithonia | 120 | DLVADJ MR'16 4ACT30 120 | 7-5WMR16 NS | | Accent-bust |
| X | Lithonia | 120 | LE S W 1 R 120ELNSD | Red led | | exit |

Adapted with permission from Bullmoose, Inc.

Now we need to prepare a luminaire schedule, which will tabulate the luminaires that we have selected. The schedule should list each type of luminaire, and contain all pertinent information on the luminaires that we have chosen. Table 3.1 is a good format for the luminaires in the Bullmoose project.

All we have to do now is send the architect an invoice for our services, and then go out and party, right? Not quite. We still have to figure out how to make the design work, and put it all on a set of plans, so that a contractor can build it.

The next chapters will help us to do this. First though, let's try a few of those fun exercises to see what you have remembered from this chapter.

## Exercises

1. What do you need to know about a space before you can begin lighting design?
2. What are the four basic types of lighting?
3. What does 'work plane' mean in ambient lighting?
4. What temperature lighting would normally be used for ambient lighting in a restaurant? Why?
5. If there were no constraints, which type of luminaire would you choose for low glare and even ambient light distribution?
6. What is a lighting manufacturer's representative, and why is he or she important to our design efforts?
7. What does it mean when the parallel and the perpendicular photometric curves of a luminaire are almost identical?
8. What is the efficacy of a CF40T5/741 lamp with an output of 3300 lumens?
9. Would a layout with a max/min ratio of 2.5:1 provide better light distribution than one with a max/min ratio of 1.5:1? Why?
10. How many footcandles at floor level are required for hallways?
11. What is the purpose of task lighting?
12. What does 'IC rated' mean for a recessed downlight?
13. What type of luminaire provides the majority of accent lighting in retail spaces?
14. How many footcandles do we need on the paths of egress during emergency conditions?
15. When would we use a centralized battery system for emergency lighting?
16. What are the two things we are trying to accomplish with emergency lighting?
17. Why do we specify an LED source in exit signs?

# 4 Powering and controlling the system

Now that we have selected the luminaires, and put together the layout for Bullmoose, Inc., we need to make the system function. To accomplish this, a number of decisions must be made, some of which could influence or negate our wonderful layout. The process of lighting design, as stated before, is an iterative process that offers the best compromise between the artist's vision and the real world.

Let's look now at some of the components involved in the power and control of lighting in general, and then we'll get back to the Bullmoose project.

## Power

### System voltage

One major concern is the voltage available to power the lighting system. Power companies in the US have standardized on three delivery voltages for buildings. These are 480Y/277 V three phase, 208Y/120 V three phase, and 240/120 V single or three phase. Larger buildings will typically have three-phase power available to drive the large heating and cooling loads, while smaller buildings often are served only with single-phase power.

Lighting loads are all single phase, so the lighting designer has the choice of 277 V or 120 V. Except in rare cases, incandescent lighting will only operate on 120 V, so if the building is served with 480Y/277 V, a step-down transformer must be used to power incandescent luminaires. We must then evaluate the economics of installing the transformer versus going to a ballasted light source, such as fluorescent or high intensity discharge (HID) lamps, which will operate on either 277 or 120 V.

Often, in a large project, 277 V is more economically attractive than 120 V, and to see how this works, let's consider how the system will be wired.

## Wire sizing

Standard practice is to limit the size of the wirir_
loads to No. 12 AWG, because anything larger than i_
cult to connect inside a luminaire, and labor costs increase
ically (as do muttered insults by contractors) when larger wiring
used. No. 12 wire is rated for 25 A current load by code, and this
limits the number of luminaires which can be connected to a circuit.
To complicate matters further, the circuit must be protected by a
20 A molded case circuit breaker, which can only be loaded to 16 A
by code. Each lighting circuit, therefore, is limited to 16 A load.
A quick review of the basic electrical relationship $P = I \times E$ will
reveal that for the allowable load of 16 A at 120 V, only nineteen
100 W luminaires can be connected on a circuit, while 16 A at
277 V will allow forty-four 100 W luminaires to be connected. If
you're not up on electrical relationships, $P$ is power in watts, $I$ is
current in amps, and $E$ is voltage. Simply put, watts equals amps
multiplied by volts.

It's easy to see that where the objective is to serve the maximum
number of luminaires with the least number of circuits, 277 V
would be the choice. This is applicable where it is desirable to serve
large areas, such as open office spaces, by a single panelboard,
which is limited to 42 breaker spaces by code.

## Panelboards

Panelboards which serve lighting loads are standard circuit breaker
panels, mounted either flush into, or on the surface of a wall. The
panelboard assembly consists of a metallic box, called a *can*, which
houses an interior section consisting of *vertical busses*, to distribute power, *circuit breakers* to serve the individual loads, and *termi-
nation blocks* for neutral and grounding connections. A three-phase
panel contains three vertical busses, and a single-phase panel
contains two busses.

The circuit breakers either snap or are bolted onto the busses to
provide the power, or 'hot' connection for the lighting circuit. The
circuit breakers are designed to automatically open the circuit in
the event of overcurrent, and also serve as an on–off switch for the
circuit. The final element of the panelboard is the *cover*, which
contains a lockable door to keep the public out of the circuits, and
which also trims out the opening made for the can.

Lighting loads may be directly controlled from the panelboard
itself if a special switching duty circuit breaker, called a SWD
breaker, is used. This is useful in large areas, such as in open plan
office spaces, where it is desirable to switch large areas as a block

load. In these cases, SWD breakers of either one, two, or three poles may be used to control the power from two or three panelboard busses simultaneously. For example, 132–100 W luminaires could be switched together using a three pole, 277 V, 20 A switching duty breaker, with 44 luminaires connected to each of the three panelboard busses.

It is always a good idea, when using a three-phase panelboard, to distribute the lighting loads as evenly as possible between the busses. This is called 'balancing the load'. Unbalanced loading can create large neutral currents and overheating within the panelboard. Figure 4.1 illustrates how the individual circuit breakers are connected to the busses in a three-phase panelboard.

Balancing the loads is an iterative operation that is most easily accomplished through the use of a panel schedule, such as the one shown in Figure 4.2.

Lighting loads are assigned to the busses sequentially, and then the total bus loads are calculated by adding the individual loads together. If the bus loads significantly differ from one another, the designer must reassign the loads until balance is obtained.

## PANEL CONNECTION DIAGRAM

| CKT NO | LOAD DESCRIPTION | LOAD (W) | BUSSES A B C | LOAD (W) | LOAD DESCRIPTION | CKT NO |
|--------|------------------|----------|--------------|----------|------------------|--------|
| 1 | | | | | | 2 |
| 3 | | | | | | 4 |
| 5 | | | | | | 6 |
| 7 | | | | | | 8 |
| 9 | | | | | | 10 |
| 11 | | | | | | 12 |
| 13 | | | | | | 14 |
| 15 | | | | | | 16 |
| 17 | | | | | | 18 |
| 19 | | | | | | 20 |
| 21 | | | | | | 22 |
| 23 | | | | | | 24 |
| 25 | | | | | | 26 |
| 27 | | | | | | 28 |
| 29 | | | | | | 30 |
| 31 | | | | | | 32 |
| 33 | | | | | | 34 |
| 35 | | | | | | 36 |
| 37 | | | | | | 38 |
| 39 | | | | | | 40 |
| 41 | | | | | | 42 |

Figure 4.1. Three-phase panelboard diagram.

| PANEL: _____ | | | | | | | | | TYPE: _____ | | | |
| VOLTAGE: _____ | | | | | | | | | MAINS: _____ | | | |

| PANEL NOTE NO. | SIZE CONDUIT | WIRE | | | GRD.WIRE SIZE | DEVICE | | | BRANCH CIRCUIT | | NO. | PHASE LOAD (VOLT-AMPS) | | | NO. | VOLT-AMPS | BRANCH CIRCUIT | DEVICE | | | GRD.WIRE SIZE | WIRE | | | SIZE CONDUIT | PANEL NOTE NO. |
| | | TYPE | SIZE | NO. | | KAIC | POLE | AMP.TRIP | DESIGNATION | VOLT-AMPS | | ØA | ØB | ØC | | | DESIGNATION | KAIC | POLE | AMP.TRIP | | TYPE | SIZE | NO. | | |
| | | | | | | | | | | | 2 | | | | 1 | | | | | | | | | | | |
| | | | | | | | | | | | 4 | | | | 3 | | | | | | | | | | | |
| | | | | | | | | | | | 6 | | | | 5 | | | | | | | | | | | |
| | | | | | | | | | | | 8 | | | | 7 | | | | | | | | | | | |
| | | | | | | | | | | | 10 | | | | 9 | | | | | | | | | | | |
| | | | | | | | | | | | 12 | | | | 11 | | | | | | | | | | | |
| | | | | | | | | | | | 14 | | | | 13 | | | | | | | | | | | |
| | | | | | | | | | | | 16 | | | | 15 | | | | | | | | | | | |
| | | | | | | | | | | | 18 | | | | 17 | | | | | | | | | | | |
| | | | | | | | | | | | 20 | | | | 19 | | | | | | | | | | | |
| | | | | | | | | | | | 22 | | | | 21 | | | | | | | | | | | |
| | | | | | | | | | | | 24 | | | | 23 | | | | | | | | | | | |
| | | | | | | | | | | | 26 | | | | 25 | | | | | | | | | | | |
| | | | | | | | | | | | 28 | | | | 27 | | | | | | | | | | | |
| | | | | | | | | | | | 30 | | | | 29 | | | | | | | | | | | |
| | | | | | | | | | | | 32 | | | | 31 | | | | | | | | | | | |
| | | | | | | | | | | | 34 | | | | 33 | | | | | | | | | | | |
| | | | | | | | | | | | 36 | | | | 35 | | | | | | | | | | | |
| | | | | | | | | | | | 38 | | | | 37 | | | | | | | | | | | |
| | | | | | | | | | | | 40 | | | | 39 | | | | | | | | | | | |
| | | | | | | | | | | | 42 | | | | 41 | | | | | | | | | | | |

Figure 4.2. Three-phase panelboard schedule.

## Load calculations

To calculate incandescent loads, we only need to add up the wattage of all the luminaires connected to the circuit. If amp loads are desired, total wattage divided by voltage (120 or 277) yields amps load. Ballasted loads are somewhat different in that a power loss associated with the ballast must be included in the total. For fluorescent fixtures with magnetic ballasts, this loss will be approximately 10% of the lamp wattage. For example, a fixture containing two 34 W F34T12 lamps would present a load of 110% of (2 × 34 W), or 75 W. Fluorescent fixtures operating on electronic ballasts, however, often present a load of *less* than the combined wattages of the included lamps. This is because the electronic ballast, which turns on and off at high frequency, is 'off' a lot of the time. Typically, a fixture containing two 32 W F32T8 lamps would present a load of only 56 W. High intensity discharge ballast losses vary with the type and wattage of lamp. It is always advisable to consult the manufacturer's information when calculating HID loads. In all cases, single-phase lighting amp loads can be found by dividing the total luminaire wattage by the voltage across the load ($I = P/E$).

## Wiring and raceways

Because of circuit breaker restrictions, the amp load on a single 20 A circuit breaker cannot exceed 16 A. This is safely within the 25 A allowed by code for a No. 12 AWG copper conductor wiring system. Copper conductors should always be used for lighting wiring systems, because of the corrosion and temperature problems associated with aluminum conductors.

Wiring insulation, or the jacket covering the copper conductor, is typically either rubber, denoted by an 'R' in the wire type designation, or thermoplastic, denoted by a 'T'. Thermoplastic insulation is thinner than the rubber, so the overall cable diameter is less in a 'T' type cable. The 'T' type cable allows more cables to be pulled in a given size conduit than the 'R', and generally results in lower job cost. For this reason, type 'T' insulation is recommended for lighting jobs requiring conduit.

Some luminaires generate a considerable amount of heat, so the wiring insulation has to have the ability to withstand heat. This capacity is denoted by an 'H' or 'HH' in the wire type designation, such as 'THHN'. The NEC table 310-16 lists three categories of temperature ratings for wiring insulation: 60, 75, and 90°C. For almost all lighting applications, the 75°C wire is adequate. In rare cases, such as enclosed high power fiber optic sources, 90°C wiring insulation is required.

**Table 310-16.** Allowable Ampacities of Insulated Conductors Rated 0 through 2000 Volts, 60°C through 90°C (140°F through 194°F) Not More than Three Current-Carrying Conductors in Raceway, Cable, or Earth (Directly Buried), Based on Ambient Temperature of 30°C (86°F)

| Size | Temperature Rating of Conductor (See Table 310-13) | | | | | | Size |
|---|---|---|---|---|---|---|---|
| | 60°C (140°F) | 75°C (167°F) | 90°C (194°F) | 60°C (140°F) | 75°C (167°F) | 90°C (194°F) | |
| AWG or kcmil | Types TW, UF | Types FEPW, RH, RHW, THHW, THW, THWN, XHHW, USE, ZW | Types TBS, SA, SIS, FEP, FEPB, MI, RHH, RHW-2, THHN, THHW, THW-2, THWN-2, USE-2, XHH, XHHW, XHHW-2, ZW-2 | Types TW, UF | Types RH, RHW, THHW, THW, THWN, XHHW, USE | Types TBS, SA, SIS, THHN, THHW, THW-2, THWN-2, RHH, RHW-2, USE-2, XHH, XHHW, XHHW-2, ZW-2 | AWG or kcmil |
| | COPPER | | | ALUMINUM OR COPPER-CLAD ALUMINUM | | | |
| 18 | — | — | 14 | — | — | — | — |
| 16 | — | — | 18 | — | — | — | — |
| 14* | 20 | 20 | 25 | — | — | — | — |
| 12* | 25 | 25 | 30 | 20 | 20 | 25 | 12* |
| 10* | 30 | 35 | 40 | 25 | 30 | 35 | 10* |
| 8 | 40 | 50 | 55 | 30 | 40 | 45 | 8 |
| 6 | 55 | 65 | 75 | 40 | 50 | 60 | 6 |
| 4 | 70 | 85 | 95 | 55 | 65 | 75 | 4 |
| 3 | 85 | 100 | 110 | 65 | 75 | 85 | 3 |
| 2 | 95 | 115 | 130 | 75 | 90 | 100 | 2 |
| 1 | 110 | 130 | 150 | 85 | 100 | 115 | 1 |
| 1/0 | 125 | 150 | 170 | 100 | 120 | 135 | 1/0 |
| 2/0 | 145 | 175 | 195 | 115 | 135 | 150 | 2/0 |
| 3/0 | 165 | 200 | 225 | 130 | 155 | 175 | 3/0 |
| 4/0 | 195 | 230 | 260 | 150 | 180 | 205 | 4/0 |
| 250 | 215 | 255 | 290 | 170 | 205 | 230 | 250 |
| 300 | 240 | 285 | 320 | 190 | 230 | 255 | 300 |
| 350 | 260 | 310 | 350 | 210 | 250 | 280 | 350 |
| 400 | 280 | 335 | 380 | 225 | 270 | 305 | 400 |
| 500 | 320 | 380 | 430 | 260 | 310 | 350 | 500 |
| 600 | 355 | 420 | 475 | 285 | 340 | 385 | 600 |
| 700 | 385 | 460 | 520 | 310 | 375 | 420 | 700 |
| 750 | 400 | 475 | 535 | 320 | 385 | 435 | 750 |
| 800 | 410 | 490 | 555 | 330 | 395 | 450 | 800 |
| 900 | 435 | 520 | 585 | 355 | 425 | 480 | 900 |
| 1000 | 455 | 545 | 615 | 375 | 445 | 500 | 1000 |
| 1250 | 495 | 590 | 665 | 405 | 485 | 545 | 1250 |
| 1500 | 520 | 625 | 705 | 435 | 520 | 585 | 1500 |
| 1750 | 545 | 650 | 735 | 455 | 545 | 615 | 1750 |
| 2000 | 560 | 665 | 750 | 470 | 560 | 630 | 2000 |

**CORRECTION FACTORS**

| Ambient Temp. (°C) | For ambient temperatures other than 30°C (86°F), multiply the allowable ampacities shown above by the appropriate factor shown below. | | | | | | Ambient Temp. (°F) |
|---|---|---|---|---|---|---|---|
| 21–25 | 1.08 | 1.05 | 1.04 | 1.08 | 1.05 | 1.04 | 70–77 |
| 26–30 | 1.00 | 1.00 | 1.00 | 1.00 | 1.00 | 1.00 | 78–86 |
| 31–35 | 0.91 | 0.94 | 0.96 | 0.91 | 0.94 | 0.96 | 87–95 |
| 36–40 | 0.82 | 0.88 | 0.91 | 0.82 | 0.88 | 0.91 | 96–104 |
| 41–45 | 0.71 | 0.82 | 0.87 | 0.71 | 0.82 | 0.87 | 105–113 |
| 46–50 | 0.58 | 0.75 | 0.82 | 0.58 | 0.75 | 0.82 | 114–122 |
| 51–55 | 0.41 | 0.67 | 0.76 | 0.41 | 0.67 | 0.76 | 123–131 |
| 56–60 | — | 0.58 | 0.71 | — | 0.58 | 0.71 | 132–140 |
| 61–70 | — | 0.33 | 0.58 | — | 0.33 | 0.58 | 141–158 |
| 71–80 | — | — | 0.41 | — | — | 0.41 | 159–176 |

*See Section 240-3.

Figure 4.3. NFPA-70 (NEC) ampacity table (source: NFPA).

The building codes for commercial installations usually require lighting circuits to be pulled through conduit. If the pulls are difficult, or extremely long, an overall nylon jacket is available to protect the insulation from damage. This is denoted by an 'N' in the wire type designation. Finally, to protect against condensate, or moisture in the conduit, the cable should be rated for wet applications. This is denoted by a 'W' in the type designation.

Fortunately for lighting designers, almost all cable manufacturers make a type THHN/THWN cable, which is ideally suited for lighting applications. Note: residential installations, which do not require conduit, are normally wired with type NM (Non-Metallic sheathed) cable. Type NM is only rated for 60°C, and could cause problems with high temperature luminaires. Since we are primarily concerned with commercial installations, we will only use 75°C wiring pulled in conduit. Figure 4.3 is the NEC ampacity table, which lists current carrying capacity of type THHN/THWN wire in conduit.

The conduits through which lighting circuits are pulled are nothing more than pipes, and they are manufactured from several materials, each having a special purpose. Underground or in-slab conduit runs utilize PVC (polyvinyl chloride) conduit because of its resistance to corrosion. Conduit runs inside the structure are made in electrical metallic tubing (EMT) for mechanical protection, and also to provide electromagnetic shielding for sensitive equipment. In extremely harsh environments, such as in heavy industrial plants, circuits may be run in PVC-covered metal conduits.

Lighting circuits are usually run in hard conduit to a junction box in the immediate vicinity of the luminaires served. The wiring from the junction box to the individual luminaires is run in flexible metallic conduit (greenfield) for ease of connection. The code allows six feet of flexible cable to be used for this purpose. Figure 4.4 is a NEC table that specifies the maximum number of wires that may be pulled in a conduit.

## Lighting controls

After the luminaires have all been selected and arranged within the space, it's time to consider how the system will be controlled to offer the occupants the most beneficial use of the space.

### Switches

The most common form of lighting control is by the single pole, manually operated toggle switch. Switches are typically located

**Table C1. Maximum Number of Conductors and Fixture Wires in Electrical Metallic Tubing**
(Based on Table 1, Chapter 9)

| | Conductor Size (AWG/kcmil) | | | | Trade Size (in.) | | | | | |
|---|---|---|---|---|---|---|---|---|---|---|---|
| Type | | ½ | ¾ | 1 | 1¼ | 1½ | 2 | 2½ | 3 | 3½ | 4 |
| **CONDUCTORS** | | | | | | | | | | | |
| RH | 14 | 6 | 10 | 16 | 28 | 39 | 64 | 112 | 169 | 221 | 282 |
| | 12 | 4 | 8 | 13 | 23 | 31 | 51 | 90 | 136 | 177 | 227 |
| RHH, RHW, RHW, | 14 | 4 | 7 | 11 | 20 | 27 | 46 | 80 | 120 | 157 | 201 |
| | 12 | 3 | 6 | 9 | 17 | 23 | 38 | 66 | 100 | 131 | 167 |
| RH, RHH, RHW, RHW-2 | 10 | 2 | 5 | 8 | 13 | 18 | 30 | 53 | 81 | 105 | 135 |
| | 8 | 1 | 2 | 4 | 7 | 9 | 16 | 28 | 42 | 55 | 70 |
| | 6 | 1 | 1 | 3 | 5 | 8 | 13 | 22 | 34 | 44 | 56 |
| | 4 | 1 | 1 | 2 | 4 | 6 | 10 | 17 | 26 | 34 | 44 |
| | 3 | 1 | 1 | 1 | 4 | 5 | 9 | 15 | 23 | 30 | 38 |
| | 2 | 1 | 1 | 1 | 3 | 4 | 7 | 13 | 20 | 26 | 33 |
| | 1 | 0 | 1 | 1 | 1 | 3 | 5 | 9 | 13 | 17 | 22 |
| | 1/0 | 0 | 1 | 1 | 1 | 2 | 4 | 7 | 11 | 15 | 19 |
| | 2/0 | 0 | 1 | 1 | 1 | 2 | 4 | 6 | 10 | 13 | 17 |
| | 3/0 | 0 | 0 | 1 | 1 | 1 | 3 | 5 | 8 | 11 | 14 |
| | 4/0 | 0 | 0 | 1 | 1 | 1 | 3 | 5 | 7 | 9 | 12 |
| | 250 | 0 | 0 | 0 | 1 | 1 | 1 | 3 | 5 | 7 | 9 |
| | 300 | 0 | 0 | 0 | 1 | 1 | 1 | 3 | 5 | 6 | 8 |
| | 350 | 0 | 0 | 0 | 1 | 1 | 1 | 3 | 4 | 6 | 7 |
| | 400 | 0 | 0 | 0 | 1 | 1 | 1 | 2 | 4 | 5 | 7 |
| | 500 | 0 | 0 | 0 | 0 | 1 | 1 | 2 | 3 | 4 | 6 |
| | 600 | 0 | 0 | 0 | 0 | 1 | 1 | 1 | 3 | 4 | 5 |
| | 700 | 0 | 0 | 0 | 0 | 0 | 1 | 1 | 2 | 3 | 4 |
| | 750 | 0 | 0 | 0 | 0 | 0 | 1 | 1 | 2 | 3 | 4 |
| | 800 | 0 | 0 | 0 | 0 | 0 | 1 | 1 | 2 | 3 | 4 |
| | 900 | 0 | 0 | 0 | 0 | 0 | 1 | 1 | 1 | 3 | 3 |
| | 1000 | 0 | 0 | 0 | 0 | 0 | 1 | 1 | 1 | 2 | 3 |
| | 1250 | 0 | 0 | 0 | 0 | 0 | 0 | 1 | 1 | 1 | 2 |
| | 1500 | 0 | 0 | 0 | 0 | 0 | 0 | 1 | 1 | 1 | 1 |
| | 1750 | 0 | 0 | 0 | 0 | 0 | 0 | 1 | 1 | 1 | 1 |
| | 2000 | 0 | 0 | 0 | 0 | 0 | 0 | 1 | 1 | 1 | 1 |
| TW | 14 | 8 | 15 | 25 | 43 | 58 | 96 | 168 | 254 | 332 | 424 |
| | 12 | 6 | 11 | 19 | 33 | 45 | 74 | 129 | 195 | 255 | 326 |
| | 10 | 5 | 8 | 14 | 24 | 33 | 55 | 96 | 145 | 190 | 243 |
| | 8 | 2 | 5 | 8 | 13 | 18 | 30 | 53 | 81 | 105 | 135 |
| RHH*, RHW*, RHW-2*, THHW, THW THW-2, | 14 | 6 | 10 | 16 | 28 | 39 | 64 | 112 | 169 | 221 | 282 |
| RHH*, RHW*, RHW-2*, THHW, THW | 12 | 4 | 8 | 13 | 23 | 31 | 51 | 90 | 136 | 177 | 227 |
| | 10 | 3 | 6 | 10 | 18 | 24 | 40 | 70 | 106 | 138 | 177 |
| RHH*, RHW*, RHW-2*, THHW, THW, THW-2 | 8 | 1 | 4 | 6 | 10 | 14 | 24 | 42 | 63 | 83 | 106 |

*Types RHH, RHW, and RHW-2 without outer covering.

**Table C1.  Continued**

| Type | Conductor Size (AWG/kcmil) | | | | Trade Size (in.) | | | | | |
|---|---|---|---|---|---|---|---|---|---|---|---|
| | | ½ | ¾ | 1 | 1¼ | 1½ | 2 | 2½ | 3 | 3½ | 4 |
| **CONDUCTORS** | | | | | | | | | | | |
| RHH*, RHW*, RHW-2*, TW, THW, THHW, THW-2 | 6 | 1 | 3 | 4 | 8 | 11 | 18 | 32 | 48 | 63 | 81 |
| | 4 | 1 | 1 | 3 | 6 | 8 | 13 | 24 | 36 | 47 | 60 |
| | 3 | 1 | 1 | 3 | 5 | 7 | 12 | 20 | 31 | 40 | 52 |
| | 2 | 1 | 1 | 2 | 4 | 6 | 10 | 17 | 26 | 34 | 44 |
| | 1 | 1 | 1 | 1 | 3 | 4 | 7 | 12 | 18 | 24 | 31 |
| | 1/0 | 0 | 1 | 1 | 2 | 3 | 6 | 10 | 16 | 20 | 26 |
| | 2/0 | 0 | 1 | 1 | 1 | 3 | 5 | 9 | 13 | 17 | 22 |
| | 3/0 | 0 | 1 | 1 | 1 | 2 | 4 | 7 | 11 | 15 | 19 |
| | 4/0 | 0 | 0 | 1 | 1 | 1 | 3 | 6 | 9 | 12 | 16 |
| | 250 | 0 | 0 | 1 | 1 | 1 | 3 | 5 | 7 | 10 | 13 |
| | 300 | 0 | 0 | 1 | 1 | 1 | 2 | 4 | 6 | 8 | 11 |
| | 350 | 0 | 0 | 0 | 1 | 1 | 1 | 4 | 6 | 7 | 10 |
| | 400 | 0 | 0 | 0 | 1 | 1 | 1 | 3 | 5 | 7 | 9 |
| | 500 | 0 | 0 | 0 | 1 | 1 | 1 | 3 | 4 | 6 | 7 |
| | 600 | 0 | 0 | 0 | 1 | 1 | 1 | 2 | 3 | 4 | 6 |
| | 700 | 0 | 0 | 0 | 0 | 1 | 1 | 1 | 3 | 4 | 5 |
| | 750 | 0 | 0 | 0 | 0 | 1 | 1 | 1 | 3 | 4 | 5 |
| | 800 | 0 | 0 | 0 | 0 | 1 | 1 | 1 | 3 | 3 | 5 |
| | 900 | 0 | 0 | 0 | 0 | 0 | 1 | 1 | 2 | 3 | 4 |
| | 1000 | 0 | 0 | 0 | 0 | 0 | 1 | 1 | 2 | 3 | 4 |
| | 1250 | 0 | 0 | 0 | 0 | 0 | 1 | 1 | 1 | 2 | 3 |
| | 1500 | 0 | 0 | 0 | 0 | 0 | 1 | 1 | 1 | 1 | 3 |
| | 1750 | 0 | 0 | 0 | 0 | 0 | 0 | 1 | 1 | 1 | 2 |
| | 2000 | 0 | 0 | 0 | 0 | 0 | 0 | 1 | 1 | 1 | 1 |
| THHN, THWN, THWN-2 | 14 | 12 | 22 | 35 | 61 | 84 | 138 | 241 | 364 | 476 | 608 |
| | 12 | 9 | 16 | 26 | 45 | 61 | 101 | 176 | 266 | 347 | 443 |
| | 10 | 5 | 10 | 16 | 28 | 38 | 63 | 111 | 167 | 219 | 279 |
| | 8 | 3 | 6 | 9 | 16 | 22 | 36 | 64 | 96 | 126 | 161 |
| | 6 | 2 | 4 | 7 | 12 | 16 | 26 | 46 | 69 | 91 | 116 |
| | 4 | 1 | 2 | 4 | 7 | 10 | 16 | 28 | 43 | 56 | 71 |
| | 3 | 1 | 1 | 3 | 6 | 8 | 13 | 24 | 36 | 47 | 60 |
| | 2 | 1 | 1 | 3 | 5 | 7 | 11 | 20 | 30 | 40 | 51 |
| | 1 | 1 | 1 | 1 | 4 | 5 | 8 | 15 | 22 | 29 | 37 |
| | 1/0 | 1 | 1 | 1 | 3 | 4 | 7 | 12 | 19 | 25 | 32 |
| | 2/0 | 0 | 1 | 1 | 2 | 3 | 6 | 10 | 16 | 20 | 26 |
| | 3/0 | 0 | 1 | 1 | 1 | 3 | 5 | 8 | 13 | 17 | 22 |
| | 4/0 | 0 | 1 | 1 | 1 | 2 | 4 | 7 | 11 | 14 | 18 |
| | 250 | 0 | 0 | 1 | 1 | 1 | 3 | 6 | 9 | 11 | 15 |
| | 300 | 0 | 0 | 1 | 1 | 1 | 3 | 5 | 7 | 10 | 13 |
| | 350 | 0 | 0 | 1 | 1 | 1 | 2 | 4 | 6 | 9 | 11 |
| | 400 | 0 | 0 | 0 | 1 | 1 | 1 | 4 | 6 | 8 | 10 |
| | 500 | 0 | 0 | 0 | 1 | 1 | 1 | 3 | 5 | 6 | 8 |
| | 600 | 0 | 0 | 0 | 1 | 1 | 1 | 2 | 4 | 5 | 7 |
| | 700 | 0 | 0 | 0 | 1 | 1 | 1 | 2 | 3 | 4 | 6 |
| | 750 | 0 | 0 | 0 | 0 | 1 | 1 | 1 | 3 | 4 | 5 |
| | 800 | 0 | 0 | 0 | 0 | 1 | 1 | 1 | 3 | 4 | 5 |
| | 900 | 0 | 0 | 0 | 0 | 1 | 1 | 1 | 3 | 3 | 4 |
| FEP, FEPB, PFA, PFAH, TFE | 14 | 12 | 21 | 34 | 60 | 81 | 134 | 234 | 354 | 462 | 590 |
| | 12 | 9 | 15 | 25 | 43 | 59 | 98 | 171 | 258 | 337 | 430 |
| | 10 | 6 | 11 | 18 | 31 | 42 | 70 | 122 | 185 | 241 | 309 |
| | 8 | 3 | 6 | 10 | 18 | 24 | 40 | 70 | 106 | 138 | 177 |
| | 6 | 2 | 4 | 7 | 12 | 17 | 28 | 50 | 75 | 98 | 126 |
| | 4 | 1 | 3 | 5 | 9 | 12 | 20 | 35 | 53 | 69 | 88 |
| | 3 | 1 | 2 | 4 | 7 | 10 | 16 | 29 | 44 | 57 | 73 |
| | 2 | 1 | 1 | 3 | 6 | 8 | 13 | 24 | 36 | 47 | 60 |
| PFA, PFAH, TFE | 1 | 1 | 1 | 2 | 4 | 6 | 9 | 16 | 25 | 33 | 42 |
| PFA, PFAH, TFE, Z | 1/0 | 1 | 1 | 1 | 3 | 5 | 8 | 14 | 21 | 27 | 35 |
| | 2/0 | 0 | 1 | 1 | 3 | 4 | 6 | 11 | 17 | 22 | 29 |
| | 3/0 | 0 | 1 | 1 | 2 | 3 | 5 | 9 | 14 | 18 | 24 |
| | 4/0 | 0 | 1 | 1 | 1 | 2 | 4 | 8 | 11 | 15 | 19 |
| Z | 14 | 14 | 25 | 41 | 72 | 98 | 161 | 282 | 426 | 556 | 711 |
| | 12 | 10 | 18 | 29 | 51 | 69 | 114 | 200 | 302 | 394 | 504 |
| | 10 | 6 | 11 | 18 | 31 | 42 | 70 | 122 | 185 | 241 | 309 |
| | 8 | 4 | 7 | 11 | 20 | 27 | 44 | 77 | 117 | 153 | 195 |
| | 6 | 3 | 5 | 8 | 14 | 19 | 31 | 54 | 82 | 107 | 137 |
| | 4 | 1 | 3 | 5 | 9 | 13 | 21 | 37 | 56 | 74 | 94 |
| | 3 | 1 | 2 | 4 | 7 | 9 | 15 | 27 | 41 | 54 | 69 |
| | 2 | 1 | 1 | 3 | 6 | 8 | 13 | 22 | 34 | 45 | 57 |
| | 1 | 1 | 1 | 2 | 4 | 6 | 10 | 18 | 28 | 36 | 46 |

*Types RHH, RHW, and RHW-2 without outer covering.

Figure 4.4. NFPA-79 (NEC) conduit allowable fill table (source: NFPA).

within the space to be lighted, usually on the lock side of the entrance door, mounted 48 in. (122 cm) above the floor. If there is more than one door leading into a space, three-way, and four-way toggle switches are available to allow switching from any door location. Figure 4.5 contains connection diagrams for toggle switching.

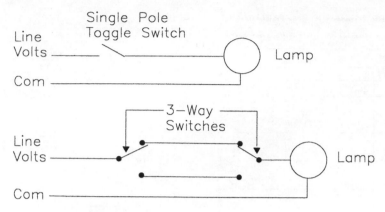

Figure 4.5. Single and three-way switching.

Switches are available for both 277 V and for 120 V, and are available in either 15 or 20 A ratings. The 20 A switches have slightly heavier contacts, and their use is recommended. The total luminaire amperage on a circuit cannot exceed the switch rating, so if you push the circuit to its full 16 A capacity, you *must* use 20 A rated switches.

Toggle switches are either on or off, and have no light dimming capability. When used in conjunction with multi-ballasted fluorescent fixtures, however, they can offer some lighting level control. Figure 4.6 is a connection diagram of a double-switched, two-ballast, three-lamp fluorescent luminaire that provides three light levels: one, two, or all three lamps can be illuminated by the use of the two toggle switches.

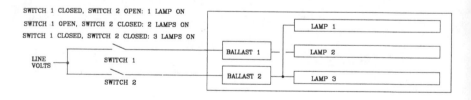

DOUBLE SWITCHING OF A 3–LAMP LUMINAIRE

Figure 4.6. Double-switched three-lamp luminaire.

## Dimmers

When continuous dimming of the light level is desired, such as in churches, restaurants, theaters, etc., specialized dimming equipment, tailored to the luminaires to be dimmed must be used. Since dimmers are much more expensive than on–off switching, economics will often dictate the type and amount of dimming used.

Dimmers do as their name implies: they dim, or reduce the light output of the luminaires attached to them. The type of dimmer most widely in use today is the simple rheostat, or variable resistor, which is used to dim incandescent loads. Rheostats directly vary the amount of current flowing through the incandescent filament, and by doing so, they vary the light output of the lamp. Rheostats are line voltage, and come in various sizes, which indicate the maximum amount of wattage that they are capable of controlling. Common rheostats are available up to 2000 W, which is the maximum load that can be put on a 20 A circuit at 120 V.

Dimming a fluorescent luminaire requires a special dimming ballast. Magnetic fluorescent dimming ballasts operate from line voltage, and are normally used in conjunction with a single circuit, wall-mounted dimming control. Electronic fluorescent dimming ballasts are controlled by a separate low voltage dimming circuit, usually 24 V. This arrangement is usable with locally mounted controls, and is also suitable for zone control of block loads. This is particularly applicable to computerized dimming control that offers multiple preset 'scenes' such as that used in churches and performing arts theaters. In operation, this system allows the user to set the level of several groups of luminaires through the use of slide or rotary switches, and then 'save' the settings as a scene. To recall a scene, the user needs only to push a single button, and the saved settings are automatically retrieved by the dimming system.

If groups of high wattage luminaires are to be dimmed, such as in large churches or theaters, specialized dimming systems capable of handling large power loads are used. These systems are designed, building block style, from rack-mounted components specifically selected for the purpose.

## Contactors

When it is desired to control a larger group of luminaires than a single-switched circuit can handle, and multiple switches are undesirable, as is common in large office spaces, lighting contactors are used. Contactors are essentially solenoids, or relays, in which a low power coil actuates, or 'pulls in' a set of heavy contacts that applies power to the lighting load. The coil itself can be activated by a single manual control device such as a wall-mounted toggle switch, or it may be activated by automatic controllers, such as photocells, occupancy sensors, timers, or building management computer systems. Contactors typically have from 2 to 12 sets of contacts, rated at 30 A each. Larger contactor ratings are available for specialized purposes. Each set of contacts can be used to power a separate lighting circuit.

Once activated, contactors are either mechanically or electrically held in the closed position. In the mechanically held contactor, the mechanism closes when the coil is activated, and it is held closed by a mechanical 'latch' until a second coil is activated, which 'unlatches' the mechanism. This requires only a momentary application of coil power through the switching device. The electrically held contactor requires that coil power be continuously applied while the contactor is closed. Removal of coil power allows the spring-loaded contactor to open. The electrically held contactor is less expensive than the mechanically held, but it can produce an undesirable buzzing sound while the coil is energized.

## Photocells

Photocells are basically devices used to turn the lights on when it is dark, and to turn them off when daylight is present. A photocell consists of a photovoltaic cell connected to a current-sensitive relay. When light strikes the cell, an electric current proportional to the intensity of the light is generated. When that current is sufficient to activate the relay coil, normally closed relay contacts open, and interrupt the power to a lighting circuit or contactor. When low light is present, the contacts close, and apply power to the lighting circuit. The contacts are normally rated for only 5 or 10 A, so the photocell can directly control only a small number of luminaires. When the lighting circuit load exceeds the rating of the photocell contacts, a contactor must be used in conjunction with the photocell.

Some luminaires designed for outside use offer an attached photocell as an option for controlling that luminaire only. Photocells are normally used to control exterior lighting by turning the luminaires on at dusk and off at dawn. Common practice is to mount the photocell on the north side of the building. For obvious reasons, photocells should not be used in extremely dirty environments.

## Timers

A lighting timer is a device that turns luminaires on or off at preset times. Timers can be simple electro-mechanical devices with manual stops that provide on–off switching, or sophisticated devices programmed for sunrise and sunset times throughout the year. They can be built into the program of a building management system so that the lights come on at the beginning of the workday, and go off at quitting time. There is a timer available to provide any desired schedule for any facility. The only limiting factor in providing timer control is economics.

## Occupancy sensors

Occupancy sensors are devices used to turn on lights when people are in the space, and to turn them off when the space is vacant. They are widely used in office spaces to conserve energy, and in sensitive areas for security.

Three types of occupancy sensors are available: *passive infrared*, or PIR, *ultrasonic*, and *dual technology*, which is a combination of the two. All of the sensors contain timers that delay the action of the switch for a preset time.

The PIR sensor detects the difference in infrared energy between a human body and the surrounding space. When sufficient infrared energy is detected, it closes a set of normally open contacts to turn on the lights. The PIR sensor controls an area within its line of sight, and within its coverage zone. False 'ON' can occur when someone passes a doorway within sight of the sensor, and a false 'OFF' can occur if the room occupant is obscured by a piece of furniture.

PIR sensors are best suited for:

1. enclosed open offices;
2. warehouses;
3. hallways;
4. areas with high air flow;
5. high ceiling areas.

They are poorly suited for:

1. restrooms;
2. storage areas with shelving;
3. offices with partitions.

The ultrasonic sensor uses the Doppler principle to sense motion within the space: it bounces ultrasonic sound waves off objects in the space, and measures the time that it takes the waves to return. Movement by a person in the space disrupts the return time, and the sensor reacts by closing a set of contacts. False 'ON' can occur if there is high air flow within the space. False 'OFF' can occur if the occupant of the space is inactive for longer than the delay period.

Ultrasonic sensors work best in:

1. open plan office spaces;
2. conference rooms;
3. restrooms;
4. large areas.

They work poorly in:

1. spaces with high air flow;
2. spaces with high ceilings;
3. unenclosed areas.

The dual technology sensor utilizes both PIR and ultrasonic sensing, and will not react unless *both* sensors detect occupancy, or the lack thereof. This eliminates false 'ON' and false 'OFF' in most instances.
Dual technology sensors perform best in:

1. classrooms;
2. office spaces with partitions;
3. areas with high ceilings;
4. large open areas.

The only drawback to dual technology sensors is their relatively high cost.
Occupancy sensors range in size from those which replace toggle switches for single office applications, to those which are designed to cover warehouses.

## Light-sensitive controls

Light-sensitive controls are sensors that measure the ambient daylight in a space, and then control the artificial lighting to maintain a preset lighting level. These sensors contain an internal photo conductive cell that measures light levels, and after a preset delay, they either cut off luminaires, or dim them to a preset level. These controls have been used with limited success, largely in areas of the country that have minimal cloud cover. Obviously, the timing of cloud passage cannot be predicted, so the setting of the delay can present problems in areas with a high incidence of cloud cover. These sensors contain internal switches for on–off control, and electronic circuitry for electronic ballast control.

## Example

Now that we're all experts in powering and controlling lighting systems, let's take another look an Bullmoose Inc.'s office and conference room addition, and see how we would make the system work. The maintenance supervisor for Bullmoose has told us that the voltage available in the building is 208Y120 V, three phase, and the nearest panelboard to our area is several hundred feet away.

Instead of pulling all of our lighting circuits for that distance, it makes more sense to install a small new 208Y120 V panelboard in our area, and only pull one feeder circuit from the existing panelboard. We know that there will be other circuits needed in the area for receptacles, equipment, etc., so the cost of the new panelboard is justified. We will flush mount the new panelboard in the office space, out of sight from visitors. Now let's wire up the lighting.

Let's look at the office first. We can see from Figure 3.5 that we have chosen six direct/indirect luminaires, each with three 32 W lamps, for a total load of 576 W, assuming no reduction for our electronic ballasts. At 120 V, our amp load is 576/120, or about 5 A. This is well within the capacity of one 20 A circuit, which you will remember, is good for 16 A. This is also within the capacity of a single 20 A switch, so we could control all the lights in the office with one switch. For just a little extra money, we can put in another switch, and allow control of the lighting level with the switching scheme shown in Figure 4.6. Let's do that. That will give the office workers some control over the brightness in their room.

Standard practice is to put the switch on the lock side of the door, so that a person entering the room may reach in and turn on the lights without searching for the switch. We will assign one circuit to the office lighting by itself, even though we have spare capacity, so if the circuit breaker serving the office is tripped off, only the office area will be affected.

Now what happens when everybody in the office goes to lunch for an hour? Do you think that one of the workers will remember to turn the lights off to save energy? Don't count on it. With just a little more expense, we can install an occupancy sensor to do the job every time. We will need a dual technology sensor, since the office has partitions in it, and the sensor will work best if it is mounted in the ceiling in the center of the room so that it can 'see' into all the cubicles. The sensor should be wired ahead of, and in series with the wall switches, so that it is only active when a switch is in the 'on' position.

That done, let's have a look at the conference room, which will be a bit trickier. In reviewing the requirements for that room we see that we will need full light over the conference table and credenza when meetings are taking place, but only minimal light there when a presentation is in progress. Our accent light for the easel should be on full when a presenter is showing charts, but off when a video presentation is in progress. Our accent lighting for the art features, such as the portraits and the bust of the Commodore should be on full for meetings, but subdued for presentations. Upon closer examination, several scenes begin to arise from these requirements.

Our architect has told us that Bullmoose Inc. doesn't mind spending a little money on the conference room lighting, so we will use a small computerized dimming system that has the capacity for, say, four scenes. We will start with each of our lighting functions separately circuited, each adjustable from a slider switch. The conference table and credenza fluorescent ballasts will require a low voltage signal, and the MR16 accent lights will be dimmed directly from line voltage. The load for the 3–3 lamp conference table luminaires will be 3 × 3 × 32 W, or 288 W. The credenza lighting and east wall load will be 5 × 42 W, or 210 W. The accent luminaires all use 75 W MR16 lamps, so the easel light load will be 75 W, and the artwork lighting will be 4 × 75 W, or 300 W. Obviously, none of these loads, or even all of these loads combined, will exceed the 2000 W limit of one circuit. So instead of using four power circuits for the conference room, we will use only two circuits, one for the fluorescent dimmers, and one for the MR16 dimmers.

Now, would there be any benefit to using occupancy sensors for the conference room? Probably not, because the same company ferret who is assigned to make sure that the coffee and pastries arrive on time, will also be responsible for turning the lights on and off.

That takes care of circuiting the office and conference room. What about the hallway – how will we handle the switching for that, since the hallway can be entered from either end? This is where we use three-way switching, as shown in Figure 4.5. We will use one circuit for the hallway, and we have two luminaires with two 40 W lamps each, or a total of 160 W.

Now what about power for the battery chargers for the emergency and exit lighting? We will want to put those on their own circuit, and label the circuit 'EMERGENCY LIGHTS – DO NOT TURN OFF' by placing a sticker inside the panelboard beside the circuit breaker serving this load. Each charger requires about 50 W, and we have a total of five chargers, so the load is 250 W.

Now let's take a look at the loads that we have assigned to circuit breakers:

- 720 W office lighting
- 288 W + 210 W = 498 W conference room fluorescent lighting

- 75 W + 300 W = 375 W conference room MR16 accent lighting
- 160 W hallway lighting
- 250 W emergency lighting battery chargers

Let's use the panel schedule shown in Figure 4.2 to assign these loads. If we put the 720 W office lighting load on circuit 1, which is on bus A, we find that if we use circuit 2, also bus A, we will have a heavy imbalance on bus A. So we will skip circuit 2, and assign the conference room fluorescent lighting to circuit 3, which is on bus B. We can assign the hallway lighting to circuit 4, also on bus B, and have a total of 754 W on bus B. We assign the 375 W conference room MR16 accent lighting load to circuit 5, on bus C, and the 250 W emergency battery charger load to circuit 6, which gives us a total load of 625 W load on bus C. This is not an ideal balance of loads, but it is as close as we can get with the loads that we have. This arrangement is shown in Figure 4.7. You will also see in Figure 4.7 that the panel schedule has space to define the circuit breaker and the circuit serving each load. As planned, we will use single pole circuit breakers and no. 12 AWG wire run in 0.5 in. conduit.

PANEL: A (SEE NOTE 3)  TYPE: _____
VOLTAGE: 120/208 3P 4W  MAINS: 100A MAIN LUGS

| PANEL NOTE NO. | SIZE CONDUIT | POWER CABLE NO.RUNS | NEUT | PHASE | GRD | DEVICE AMP.TRIP | POLE | KAIC | BRANCH CIRCUIT DESIGNATION | AMPS | NO. | #A | #B | #C | NO. | AMPS | BRANCH CIRCUIT DESIGNATION | KAIC | POLE | AMP.TRIP | GRD | PHASE | NEUT | NO.RUNS | SIZE CONDUIT | PANEL NOTE NO. |
|---|---|---|---|---|---|---|---|---|---|---|---|---|---|---|---|---|---|---|---|---|---|---|---|---|---|---|
| ½ | 1 | 12 | 12 | 12 | 20 | 1 | 10 | OFFICE LIGHTS | | 1 | 720 | | | 2 | | SPACE | | | | | | | | | |
| ½ | 1 | 12 | 12 | 12 | 20 | 1 | 10 | DIMMER PANEL | | 3 | | 658 | | 4 | | HALLWAY | 10 | 1 | 20 | 12 | 12 | 12 | 1 | ½ | |
| ½ | 1 | 12 | 12 | 12 | 20 | 1 | 10 | DIMMER PANEL | | 5 | | | 625 | 6 | | EXIT + EMERG | 10 | 1 | 20 | 12 | 12 | 12 | 1 | ½ | |
| | | | | | | | | | | 7 | | | | 8 | | | | | | | | | | | |
| | | | | | | | | | | 9 | | | | 10 | | | | | | | | | | | |
| | | | | | | | | | | 11 | | | | 12 | | | | | | | | | | | |
| | | | | | | | | | | 13 | | | | 14 | | | | | | | | | | | |
| | | | | | | | | | | 15 | | | | 16 | | | | | | | | | | | |
| | | | | | | | | | | 17 | | | | 18 | | | | | | | | | | | |
| | | | | | | | | | | 19 | | | | 20 | | | | | | | | | | | |
| | | | | | | | | | | 21 | | | | 22 | | | | | | | | | | | |
| | | | | | | | | | | 23 | | | | 24 | | | | | | | | | | | |
| | | | | | | | | | | 25 | | | | 26 | | | | | | | | | | | |
| | | | | | | | | | | 27 | | | | 28 | | | | | | | | | | | |
| | | | | | | | | | | 29 | | | | 30 | | | | | | | | | | | |
| | | | | | | | | | | 31 | | | | 32 | | | | | | | | | | | |
| | | | | | | | | | | 33 | | | | 34 | | | | | | | | | | | |
| | | | | | | | | | | 35 | | | | 36 | | | | | | | | | | | |
| | | | | | | | | | | 37 | | | | 38 | | | | | | | | | | | |
| | | | | | | | | | | 39 | | | | 40 | | | | | | | | | | | |
| | | | | | | | | | | 41 | | | | 42 | | | | | | | | | | | |

Figure 4.7. Bullmoose office addition panel schedule.

We've now put together a luminaire layout, a luminaire schedule, and a panel schedule for our space. What next? We'll find out, just as soon as we get these pesky exercises out of the way.

## Exercises

1. What are the two voltages in the US of concern in lighting design?
2. What size wire is normally used for lighting circuits?
3. What is the allowable amperage load on a 20 A molded case circuit breaker?
4. What is the maximum number of circuit breaker spaces that can be built into a panelboard?
5. How many amps of current would be drawn by 12–100 W lamps connected to a 120 V circuit? A 277 V circuit?
6. What are the five components of a panelboard?
7. What are the functions of a circuit breaker in a lighting circuit?
8. What is meant by 'balancing the load' in a three-phase panelboard?
9. What percent of fluorescent lamp wattage do most magnetic ballasts consume as power loss?
10. What does the 'T' in the wiring designation THWN mean? The 'W'?
11. What does 'EMT' mean when referring to conduit?
12. Can a single ballast fluorescent luminaire be double switched? Why?
13. When an incandescent lamp is dimmed, does its color temperature increase or decrease? Why?
14. What is a 'scene' as applied to computerized fluorescent dimming?
15. What are the two methods for holding lighting contactors closed?
16. What is the most common use for the photocell?
17. How many types of occupancy sensors are there? What are they?
18. What type of occupancy sensor would you use in a school classroom?
19. Would you use light-sensitive control of ambient lighting in an area with a high incidence of cloud cover? Why?

# 5 The contract documents

OK, we have created a design for the Bullmoose, Inc. office addition using catalogued lamps and luminaires, and we have selected controls and circuits to make it work. Where do we go from here?

Now we have to pull all of these elements together, and create a set of contract documents that an electrical contractor can use to bid on, and then use to build the system. The designer's part of the contract documents usually consists of a set of drawings, which describe the system graphically, and the technical specifications, which describe the components that the designer wants the contractor to use. Terms of contract, bidding procedures, etc., are usually prepared by the lead professional, which in our case is the architect.

The contractor will first use these documents to prepare a bid for constructing the project. His bid will include his cost for the components shown on the drawings, plus his cost for labor to construct the project, plus any overhead and materials markup that he chooses to include. If the contractor is the successful bidder, he will then use the documents to build the system.

If you as a designer have left a required component off of the plans, or out of the specifications, it's a sure bet that the contractor will find the oversight. The same holds true for components that are sized incorrectly, and for designs that fail to meet the applicable codes. During construction, the contractor will request a change order to correct the deficiencies. Change orders invariably mean an increase in contract price, and they tend to make owners and architects very unhappy – even to the point of requiring the delinquent designer to pay for the changes out of his own pocket. You can see, then, how imperative it is for contract documents to be complete and accurate.

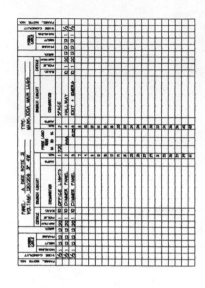

Figure 5.1. Bullmoose plan sheet start.

## The project plans

So far, we've been pretty concise with our Bullmoose, Inc. design: life safety codes have been met; luminaires have been selected to meet our design requirements; and circuiting has all been well within NEC requirements. So let's put our luminaire schedule and panel schedule on a single sheet to begin our plan set. This is shown in Figure 5.1.

In order to show graphically what we intend the contractor to do, we need to develop a legend of symbols that have specific meaning on this drawing set. Standardized drawing symbols are published in many architectural reference books, such as *Architectural Graphic Standards* published by John Wiley and Sons, and other texts, all available from the IES publications division. Many computerized drawing programs also contain libraries of standard symbols. For our project, we will use the legend shown in Figure 5.2

Now we need to show the wiring on the plan view, and identify all the luminaires. This is shown in Figure 5.3.

All the wiring is identified in the panel schedule, with the exception of the extra conductor needed in the run between the two

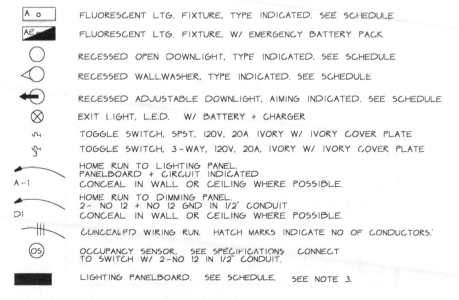

Figure 5.2. Legend for Bullmoose office addition plan.

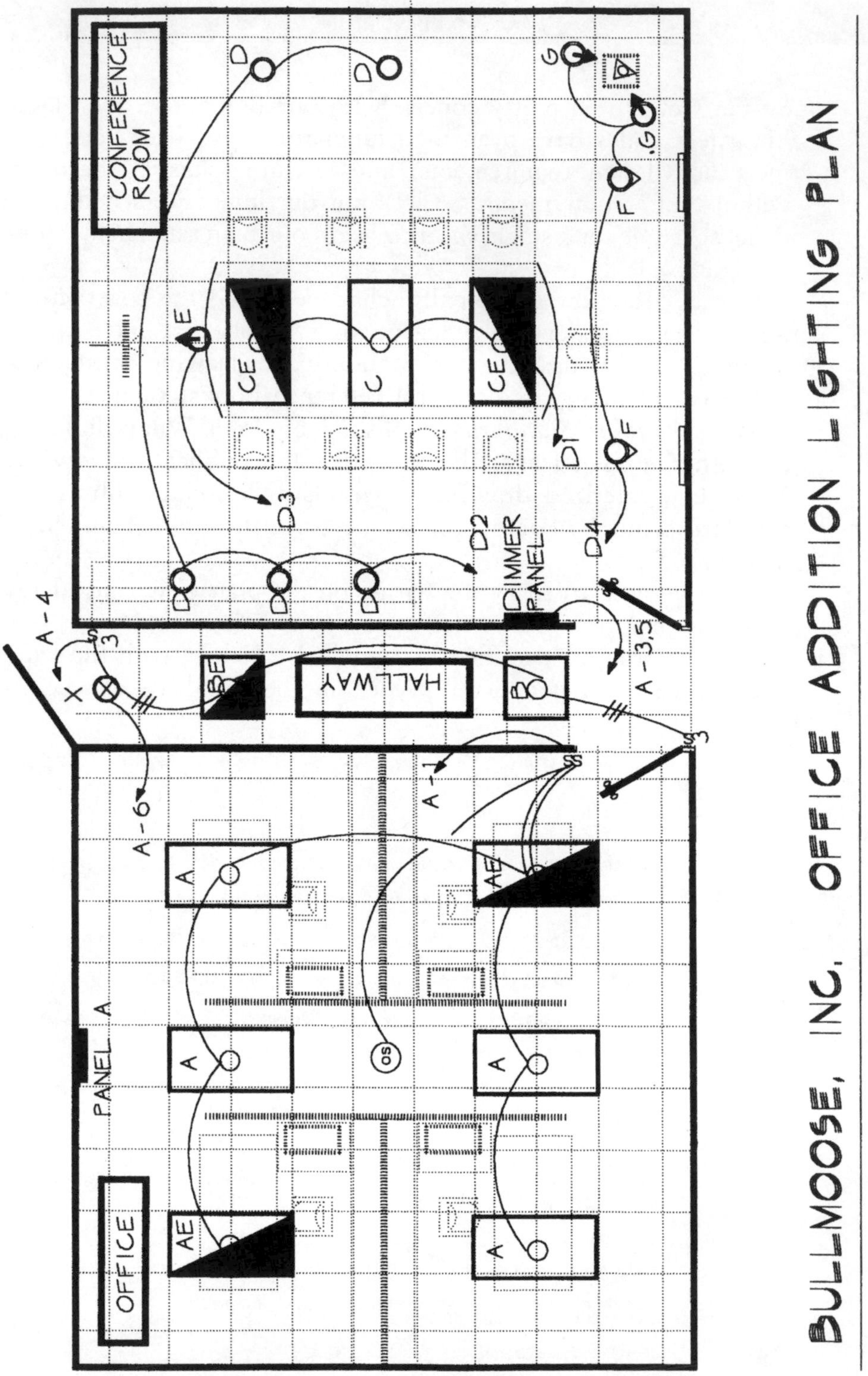

Figure 5.3. Wiring for Bullmoose office addition plan.

hallway three-way switches, which, if you recall from Figure 4.3, runs from switch to switch. We will indicate this extra conductor by hatching the switch leg of the hallway lighting, as shown in Figure 5.3.

The circuit breaker in the existing building panelboard, the feeder run from that panelboard to our new panelboard A, and new panelboard A should be provided by the electrical contractor, if he is different from the lighting contractor.

All the components and wiring are identified now, so our plans are almost complete. All we need to add are some clarifying notes and details to make sure that our intent is understood by the contractor. We need to make sure that the contractor installs the system in compliance with all applicable codes, so we'll make this note 1. Wiring is usually included in the electrical (rather than the lighting) specifications, but it's a good idea to specify the lighting wiring on the lighting plans as well. This will be note 2. We want to clarify responsibility for the circuit breaker in the existing building panel, the feeder to the new panel, and the new panel itself, so we will make this note 3. The note will indicate that the electrical contractor is to coordinate with the owner to identify the existing panel from which our new panel will be fed. Note 3 will also make reference to the electrical specification section which covers circuit breakers and panelboards. Note 4 will require that the contractor provide the exact luminaires that we show in the luminaire schedule. This is necessary, because this is where a lot of lighting designs go awry: most contractors have long-standing relationships with suppliers who only carry specific brands of equipment. If our contractor buys from a supplier who does not carry the luminaires that we have specified, he will want to substitute luminaires that his supplier does carry. The contractor is not aware of the design criteria, and he may ask to substitute a luminaire that is unsuitable for the job. Since the contractor's bid is final, it is incumbent upon the designer to make sure, before the bids, that he has the opportunity to accept or reject a contractor's request for substitutions. In our case, since we have worked closely with only one manufacturer's representative and products, it is safer to not allow any substitutions. A substitution clause will be included in note 4.

(A note on substitutions is appropriate here. If your project is funded by public money, such as schools or government buildings, it is mandated by law in most locations that you allow at least three manufacturer's products to be bid, in order to obtain the best competitive price. If the project is privately funded, as is the case with our Bullmoose, Inc. project, the designer may require that the project be built as shown on the plans and specifications, with no substitutions.)

# NOTES

1.  ALL WIRING TO BE PER N.E.C. AND THE LOCAL COUNTY AND STATE ELECTRICAL CODES
2.  ALL WIRING TO BE 75 DEG C, TYPE THHN/THWN INSULATION COPPER CONDUCTOR, UNLESS NOTED OTHERWISE (U.N.O.)
3   ELECTRICAL CONTRACTOR WILL PROVIDE NEW PANEL A, NEW FEEDER CIRCUIT AND NEW CIRCUIT BREAKER IN OWNER DESIGNATED PANEL UNDER SECTION 16402. LIGHTING CONTRACTOR TO PROVIDE LIGHTING BRANCH CIRCUITS.
4   LUMINAIRES ARE TO BE PROVIDED AS SHOWN IN THE LUMINAIRE SCHEDULE.  NO SUBSTITUTIONS WILL BE ACCEPTED.
5.  LUMINAIRES ARE TO BE PROVIDED COMPLETE WITH ALL REQUIRED TRIM, HANGERS, AND OTHER APPURTENANCES NECESSARY FOR THEIR INSTALLATION.

Figure 5.4. Notes for Bullmoose office addition plan.

Note 5 is a general CYA (Cover Your A—-) note to prevent change order requests for things like trim, luminaire clips, lenses, etc., which the contractor should have known to include in his bid.

The completed notes are shown in Figure 5.4.

We will add the notes to the drawing. Now what about details? About the only thing in our design which isn't standard construction practice is the double switching of the office luminaires, so we'll add the detail shown in Figure 4.6, and that will complete the plans. Figure 5.5 shows the completed plan.

Now we will tackle the written part of the contract documents, which will completely define the conditions, components and installation methods that are required for our project.

## Project specifications

The Construction Specifications Institute (CSI) has established a 17 division specifications standard that encompasses all of the work to be accomplished in a construction project. Within these divisions are separate sections that contain the specifications for work within the specific disciplines. Division 16000 is reserved for electrical work of all types. Section 16400 pertains to interior electrical systems, section 16500 covers lighting systems, and Section 16510 is for interior lighting systems, and is of particular interest to us. Each specification section is further broken down into three parts: Part 1 is the general conditions, which lists reference standards, special project requirements, project scope, and submittals required of the contractor; Part 2 is the products section, which lists the specific

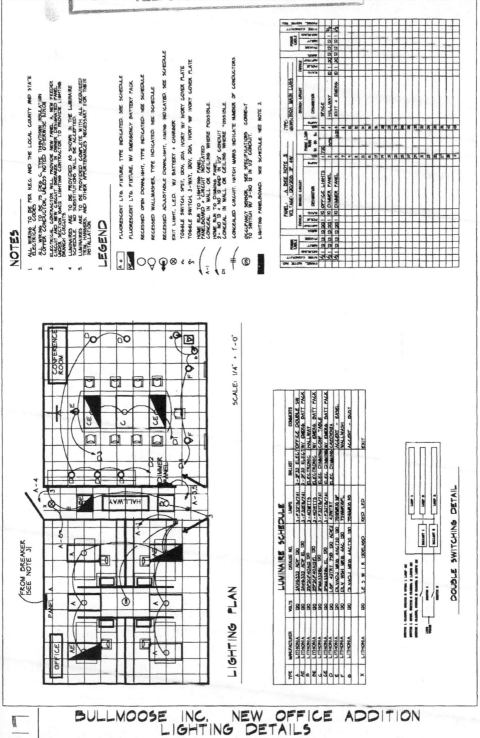

Figure 5.5. Bullmoose Office addition completed plan.

standards that the project components must meet; and Part 3 is the execution section, which spells out how the contractor is to install the components. What follows is a Section 16510 specification for our Bullmoose, Inc. project, with explanatory notes in italics.

## SECTION 16510 – INTERIOR LIGHTING

### Part 1. General

#### 1.1. References

The publications listed below form a part of this specification to the extent referenced. The publications are referred to in the text by the basic designation only.

AMERICAN NATIONAL STANDARDS INSTITUTE (ANSI)

ANSI C82.1    (1992) Ballasts for Fluorescent Lamps
ANSI C82.2    (1989) Fluorescent Lamp Ballasts – Methods of Measurement
ANSI C82.11   (1993) High-frequency Fluorescent Lamp Ballasts

AMERICAN SOCIETY FOR TESTING AND MATERIALS (ASTM)

ASTM A 641    (1992) Zinc-coated (galvanized) Carbon Steel Wire

ILLUMINATING ENGINEERING SOCIETY OF NORTH AMERICA (IES)

LHBK          (2000) Lighting Handbook, Ninth Edition

NATIONAL FIRE PROTECTION ASSOCIATION (NFPA)

NFPA 70       (1999) National Electric Code
NFPA 101      (1994) Code for Safety to Life from Fire in Buildings and Structures

UNDERWRITERS LABORATORIES INC. (UL)

UL 924        (1995) Emergency Lighting and Power Equipment
UL 935        (1995) Fluorescent Lamp Ballasts
UL 1570       (1995) Fluorescent Lighting Fixtures
UL 1571       (1996) Incandescent Lighting Fixtures

*All of the organizations listed above have put in a lot of effort and expense into developing exacting specifications for components that we have used in our design. By referencing these published standards, we incorporate the standards into our specifications.*

#### 1.2. Related requirements

Section 16050 'Basic Electrical Materials and Methods' applies to this section, with the modifications and additions specified herein.

Materials not considered to be lighting equipment are specified in Section 16402 'Interior Distribution System'.

*The project electrical engineer normally prepares all division 16000 specifications, and references are made between sections, rather than rewriting the same clauses in every section. We will approach this section from the standpoint of the lighting designer who will integrate Section 16510 into the rest of the division 16000 sections prepared by others.*

### 1.2.1. Work Included

The work included under these specifications consists of furnishing and installing all luminaires, controllers, wiring, conduit, and other necessary appurtenances to provide a fully functional lighting system as shown on the plans. The contractor shall provide all materials, labor, tools, and equipment necessary to complete the project.

## 1.3. Definitions

### 1.3.1. Average life (lamps)

The time after which 50% will have failed, and 50% will have survived under normal conditions.

### 1.3.2. Total Harmonic Distortion (THD)

The root mean square (RMS) of all the harmonic components divided by the total fundamental current.

*The power supplies used in fluorescent electronic ballasts generate harmonics that can feed back into the electrical system and cause problems with sensitive electrical equipment. It is necessary for the lighting designer to specify a maximum allowable harmonic distortion for the ballasts.*

## 1.4. Submittals

*Other sections of the specification require the contractor to submit to the designer manufacturer's catalog sheets of the materials he proposes to use on the project. The designer then reviews the submittals, and either approves or rejects them. The contractor cannot install an item until he has an approved submittal for the item signed by the designer.*

Submit the following in accordance with the section entitled 'Submittal Procedures'. Data, drawings, and reports shall employ the terminology, classifications, and/or methods prescribed by the IES LHBK, as applicable, for the lighting system specified.

### 1.4.1. Manufacturer's catalog data

    a. Fluorescent luminaires;
    b. fluorescent electronic ballasts;

   c. fluorescent lamps;

   d. incandescent luminaires;

   e. low voltage incandescent lamps;

   f. dimmer system;

   g. exit signs;

   h. emergency lighting equipment;

   i. occupancy sensors.

*This is the data that the contractor has to submit for approval. Switches, wire, panelboards, circuit breakers, and conduit are specified in Section 16402 by the electrical engineer. The lighting designer should coordinate closely with the electrical engineer to ensure that the components specified are compatible with the lighting system design*

## 1.5. Electronic ballast warranty

Furnish the electronic ballast manufacturer's warranty. The warranty period shall not be less than 5 years from the date of manufacture of the ballast. Luminaires, including ballast assembly, shall be installed within 1 year after manufacture, leaving at least 4 years of the warranty in effect after installation.

## Part 2. Products

## 2.1. Fluorescent luminaries

Luminaires shall comply with UL 1570. Fluorescent luminaries shall have electronic ballasts.

### 2.1.1. Fluorescent lamp electronic ballasts

The electronic ballast shall, as a minimum, meet the following criteria:

   a. ballast shall comply with UL935, ANSI C82.11, and NFPA 70. Ballasts shall be designed for the number and wattage of the lamps shown on the plans. Ballasts are to operate on 120 V, AC;

   b. power factor shall be 0.95 minimum;

   c. ballast shall operate at a frequency of 20 000 Hz (minimum);

   d. ballast shall have light regulation of plus or minus 10% lumen output with a plus or minus 10% input voltage regulation. Ballast shall have a maximum of 10% flicker;

   e. ballast shall be UL listed Class P with a sound rating of 'A';

   f. ballast shall have circuit diagrams and lamp connections displayed on the ballast enclosure. Ballasts shall operate lamps in a parallel configuration that permits the operation of the remaining lamps if one lamp fails;

g. ballasts shall operate in a rapid start mode;

h. ballasts shall have a full replacement warranty as specified in paragraph 1.5;

i. ballasts shall have a total harmonic distortion (THD) of 15%.

*As a general rule of thumb, the lower the THD percentage, the larger the ballast power supply, and, as we remember from Chapter 2, the larger the current inrush when the lights are switched on. If you have a large number of luminaires with very low THD ballasts connected on a circuit, it is likely that the circuit breaker will trip when you turn on the switch; 15% THD is about as low as we need to go, except in special cases.*

### 2.1.1.1. T-8 Electronic ballast

a. Ballasts shall have input wattage of 114 W maximum when operating four *F32T8* lamps.

### 2.1.1.2. T-8 Electronic dimming ballasts

a. Ballasts shall have the capability to continuously dim the output of its connected lamps from full output to 10% of full output. Dimming shall be accomplished by electronic circuitry responding to a low voltage (24 V) signal from a dimming system.

### 2.1.1.3. T5 Long twin tube ballasts

a. Ballasts shall have input wattage of 74 W maximum when operating two CF40T5 lamps.

### 2.1.1.4. Compact fluorescent electronic dimming ballast

a. Ballasts shall be solid state, dimming type, capable of continuously dimming the output of its connected lamps from full output to 10% of full output. Dimming shall be accomplished by electronic circuitry responding to a low voltage (24 V) signal from a dimming system.

## 2.1.2. Fluorescent Lamps

### 2.1.2.1. T-8 Lamps

T8 rapid start lamps shall be rated 32 W, 2800 initial lumens minimum, CRI of 72 minimum, color temperature of 4100 K, and an average rated life of 20 000 h.

### 2.1.2.2. T-5 Lamps

T5 long twin tube fluorescent lamps shall be rated 40 W, 3150 initial lumens minimum, CRI of 72 minimum, color temperature of 4100 K, and an average rated life of 20 000 h. Lamps shall be 22.6 in. maximum length, and shall have a 2G11 type base.

### 2.1.2.3. Compact fluorescent lamps

Compact fluorescent lamps shall be rated 26 W, 1800 initial lumens minimum, CRII of 72 minimum, color temperature of 4100 K, and an average rated life of 10 000 h. Lamps shall be double twin tube, with four pin base for dimming.

## 2.2. Incandescent luminaires

1.  Incandescent luminaires are to be provided complete with transformers, and circuitry required for the operation of low voltage MR-16 lamps. The luminaires are to provide tool-less lamp adjustment in both the horizontal and vertical planes to allow aiming of the lamp. All required trim, lenses, hangers, etc., required for complete installation are to be provided with the luminaires.
2.  MR-16 lamps are to be 75 W, with an average rated life of 4000 h.
    a.  Narrow spot lamps are to have a beam spread of 10° maximum, with a center of beam candlepower rating of 10 000 candlepower. *This lamp will be used to highlight the bust of Commodore Bullmoose;*
    b.  medium beam flood lamps are to have a beam spread of 24° maximum, with a center of beam candlepower rating of 3100 candlepower. *This lamp will be used to illuminate the presentation easel;*
    c.  flood lamps are to have a beam spread of 36° maximum, with a center of beam candlepower rating of 1800 candlepower. *This lamp will be used to wall wash the portraits.*

## 2.3. Dimming system

### 2.3.1. Construction

The dimming system shall be housed in a steel frame with hinged faceplate. It shall be designed to fit in a standard five-gang backbox.

### 2.3.2. Functionality

The dimming system shall provide the following features:
   a.  single control to turn off all lighting;
   b.  master raise/lower control to raise or lower the lighting levels of all luminaires simultaneously;
   c.  channel raise/lower control to raise or lower the lighting level of luminaires connected to the individual channels;
   d.  preset controls for saving preset scenes;
   e.  adjustable fade time for preset scenes. Fade time will be adjustable from 0 to 45 seconds.

### 2.3.3. Capacities

The system shall be capable of dimming a minimum combined load of 2000 W. It shall be capable of dimming fluorescent electronic ballasts, compact fluorescent lamp ballasts, and low voltage incandescent loads. It shall provide a minimum of four separate dimming channels, and have a minimum of four preset scenes involving all channels. Each channel shall have capacity for:

a. 600 W of electronic ballast load;
b. 800 W of incandescent load.

## 2.4. Exit signs

Exit signs shall be self-powered LED type, conforming to UL924, NFPA 70, and NFPA 101.

### 2.4.1. Self-powered LED exit signs (battery backup)

Sign shall be provided with automatic transfer switch, battery, and automatic high/low trickle charger contained within the sign. Battery shall be sealed electrolyte type, and shall require no maintenance, including no additional water, for a period of not less than 5 years. The battery shall be capable of powering the sign for a minimum of 90 minutes.

## 2.5. Emergency lighting equipment

Emergency lighting equipment shall conform to UL934, NFPA 70, and NFPA 101.

### 2.5.1. Fluorescent emergency system

Shall be suitable for use with solid state ballasts. Each system shall contain an automatic transfer switch, battery, and automatic battery charger contained within an enclosure in a standard luminaire. The system shall include a test switch, operable from outside the fixture, and a pilot light, visible from outside the fixture. Battery is to be sealed electrolyte type, with capacity to provide power to one lamp at an output of 1100 lumens, for a period of 90 minutes. The battery shall require no maintenance, including no additional water, for a period of not less than 5 years.

## 2.6. Occupancy sensors

Occupancy sensors shall be UL listed. Sensor mounting type shall be ceiling surface. Detector shall be combination passive infrared/ultrasonic. Sensor shall have LED positive detection indicator, adjustable delayed off – time range between 30 seconds and 15 minutes, and sensitivity adjustment. Input shall be rated for 120 V, AC. Contacts shall be rated for 15 A minimum. Sensor shall have 'fail – on' function designed to keep the lights on if the sensor fails. Sensor shall have a manual override switch.

## 2.7. Support hangers for luminaires in suspended ceilings

### 2.7.1. Wires

Wires shall conform to ASTM A 641, class 3 soft temper, 12 gauge, zinc-coated finish.

# Part 3. Execution

## 3.1. Installation

Set luminaires square, plumb, and level with ceiling and walls. Luminaires shall be in alignment with adjacent luminaires, and secure in accordance with manufacturer's directions and approved drawings. Installation shall meet the requirements of NFPA 70. Contractor shall obtain approval of the exact mounting for luminaires before commencing installation, and after coordinating with the other trades involved.

*Lighting and HVAC (heating, ventilating, and air conditioning) diffusers compete for ceiling space in a grid ceiling. In this project, it is important that our luminaires be placed where we have shown them, and any conflicts should be worked out before the two contractors begin work.*

Recessed luminaires may be supported from suspended ceiling system support tees when the ceiling system support wires are provided at a minimum of four wires per luminaire, and located not more than 6 in. from the corner of each luminaire. *In areas having seismic requirements, it will be necessary to provide an extra two wires per luminaire.* Provide support clips securely fastened to ceiling grid members, one at each corner of the luminaire. For luminaires smaller than the ceiling grid, provide a minimum of four wires per luminaire, located at the corners of the grid in which the luminaire is mounted. *In our case, this would be the recessed downlights in the conference room.* Do not support luminaires by ceiling acoustical panels. Where luminaires smaller than the ceiling grid are mounted in the center of the acoustical panel, support the luminaire with at least two 3/4 in. metal channels spanning, and secured to, the ceiling grid.

### 3.1.1. Exit signs and emergency lighting units

Wire exit signs and emergency lighting equipment on a separate circuit breaker in the lighting panelboard. Place a sticker beside this breaker which states 'EMERGENCY LIGHTING – DO NOT TURN OFF', and provide lockout.

### 3.1.2. Occupancy sensor placement

Locate the sensors in accordance with the manufacturer's recommendations to maximize energy savings and to avoid nuisance activation.

## 3.2. Field quality control

Upon completion of the installation, conduct an operating test to show that all components of the lighting system operate in accordance with this section.

And there we have it! We can send the contract documents to the architect *and* send him our invoice. Our work here is done, masked man. . . or is it?

When the project bids come in, the owner will probably suffer what is commonly known as 'price tag shock'. The cost for the project is higher than he had expected. He will huddle with the architect and the low bid contractor and try to come up with some ways to cut costs. The first item on the chopping block will be – you guessed it – the lighting system. The contractor will say that he can cut a lot of money out of the job if he is allowed to substitute generic luminaires for all that fancy stuff that you've specified. It's up to you to stand fast, and hold to your design. You have selected the luminaires through a careful process, and the system will function exactly as the architect has specified, at the lowest energy cost possible. You can enlist the support of the architect on this point. If your design is compromised, you can bet that, a year from now, nobody will remember the few dollars that were saved on the lighting, but they will curse the inferior system that they are having to endure. Your reputation as a lighting designer and the workers who occupy the space will both suffer if you allow the substitution of inappropriate fixtures into your design.

Enough of the soap box – now let's get on to those snappy exercises that we all know and love.

## Exercises

1. What are the two uses that a contractor has for the contract documents?
2. What do the technical contract documents include?
3. Under what circumstances could the designer be required to pay for part of the project? What part would the designer be asked to pay for?
4. Why do we put a symbol legend on the drawings?

5. How do we identify wire and conduit sizes in our Bullmoose, Inc. project?
6. If you are designing a government project, funded by public money, how many manufacturers must you allow to bid on your project?
7. If you are designing a project that is funded by private money, how many manufacturers must you allow?
8. What is the purpose of the technical specifications?
9. How many divisions are there in a standard construction specification?
10. How many parts are there in each technical specification section?
11. Why do we list reference publications and materials standards in Part 1 of the technical specifications?
12. Why do we require the contractor to submit manufacturer's data on the products that he intends to use on our project?
13. What does the designer do with the submittals?
14. Why don't we specify 0% THD in our ballast specifications, since harmonics are undesirable in an electrical system?
15. Why do we require tool-less adjustment in our adjustable luminaires?
16. Why do we need a 'fail – on' function for an occupancy sensor?
17. Why does the lighting contractor need to coordinate with the HVAC contractor before starting work in the ceiling?
18. Why do we require an operational test after the installation is complete?

# The second time around – retrofitting

## The problem

Back in the 1970s, the 1980s, and even into the 1990s, open plan office spaces were all the rage, and the lighting design for these spaces was done largely by rule of thumb. The luminaire of choice was the 2 ft × 4 ft (0.6 m × 1.2 m) recessed fluorescent troffer with four F40T12/735 lamps, magnetic ballasts, and flat acrylic diffusers. These luminaires were laid out at 8 ft (2.4 m) on center throughout the space.

Given these parameters, the luminaire manufacturers could only compete on one basis – price. Manufacturers scrambled to cut manufacturing costs to the bone, and in doing so, the performance of the luminaires suffered. For the most part, this didn't matter much, because the spaces were so over-lit that even with a light loss factor (LLF) of 0.40, the design still produced plenty of light (and glare) for the workers in the spaces.

On the manufacturing side, the reflector panels in the luminaires were designed to minimize manufacturing costs, rather than to improve luminaire performance. Ballasts were built as cheaply as possible, at the expense of ballast factor. Even the acrylic diffusers suffered, by becoming thinner – some as thin as 0.10 in. (2.5 mm) – which reduced the prismatic diffusion pattern to randomly sized 'lumps' on the surface of the diffuser. All this served to give the luminaire a very poor efficiency.

As energy prices continued to rise, this luminaire inefficiency put an ever larger bite on the pocketbook of building owners, causing them, for the first time, to notice the lighting. Then, in 1992, along came EPACT 92, which outlawed the manufacture of cheap F40T12 lamps, and suddenly, building maintenance engineers couldn't find replacements for burned out lamps.

Lamp manufacturers came up with an alternative that met the requirements of EPACT 92, and which would operate on the installed 40 W magnetic ballasts. This was the F34T12 'Energy Saver' lamp, which was installed as the F40T12 lamps burned out. As we said earlier in this book, there is no free lunch. Lower wattage lamps in an already inefficient luminaire meant less light output, and in general, poor lighting. To make matters worse, the F34T12 lamps didn't provide the expected energy savings, because they operated inefficiently on the old 40 W ballasts. Many millions of square feet of existing office, hospital, and manufacturing space are presently operating under these conditions, using either aging F40T12 lamps, or the F34T12 lamps.

## The solution

One solution to this problem would be to replace the old luminaires with new, efficient luminaires that use F32T8 lamps, and the much more efficient electronic ballasts. This is a very expensive option, and one that has a payback time (the time required for the energy savings to offset the installation costs) unacceptable to most building owners. A much more attractive option is to retrofit the existing luminaires to improve their efficiency. This means:

- improving the optics of the luminaires
- upgrading the ballasts of the luminaires
- installing F32T8 lamps and lampholders.

This will allow the existing luminaire housings and wiring to remain in place, while greatly improving luminaire performance and efficiency.

Just how do we go about retrofitting a luminaire? Let's take a look.

### Improving luminaire optics

When the luminaires were built, the sheet metal that forms the housing of the luminaire was shaped to contain the ballast, and to mount the lamps, without much regard for optic performance. The resulting 'box' shape of the housing allowed light from the lamps to scatter within the housing, and most of the light from the upper half of the lamp was lost.

Reflectors are now being manufactured that redirect that light downward into the usable zone. These reflectors are available to accommodate the standard two, three, and four lamp configurations of the 2 ft × 4 ft (0.6 m × 1.2 m) troffer, and are designed to fit into existing housings with a minimum of extra hardware.

Typically, the reflectors are manufactured from material that has a 90–95% reflectivity, and that is formed to direct all available light out of the luminaire. When in place, reflectors alone can boost the efficiency of the older luminaires by 20–30%, with no increase in energy usage. It is common retrofit practice to replace four F40T12 lamps and two magnetic ballasts with two F32T8 lamps, one electronic ballast, and reflectors. This saves 100 W per luminaire in lamp energy, and about 30 W per luminaire in ballast loss.

The acrylic diffusers in the older troffers were often of poor quality, and as a result, scattered some of the usable light. Dirt accumulation and aging has reduced the transmissivity of the material, resulting in additional light loss. Some of the earlier diffusers were made of polycarbonate, which, as we saw in Chapter 2, yellows with age, and that further reduces the light output of the luminaire. Modern prismatic acrylic diffusers are manufactured with precisely formed prisms that serve to direct light downward as usable light. The diffuser is normally changed during the retrofit process. Now, what about the rest of the luminaire?

## Upgrading the ballast

This is a simple one. We simply discard the old, inefficient magnetic ballasts and replace them with a new electronic ballast, don't we?

That's exactly what needs to be done, but it may not be quite that easy. The old ballast could contain PCBs (polychlorinated biphenyls), which we can't just throw in the local landfill, because PCBs have been determined to be hazardous. We first have to contact the ballast manufacturer and determine that the old ballasts do not contain PCBs. Then we discard them. If they do, they must be disposed of in a manner prescribed by the US Environmental Protection Agency (EPA). This costs a little more, but either way, we get rid of the inefficient magnetic ballasts and replace them with an electronic ballast. This gives us a big jump in efficiency since the magnetic ballasts can only operate two lamps – and therefore two ballasts are required for a four-lamp luminaire – which produces double the ballast loss. Now, what about upgrading the lamps?

## Upgrading the lamps

Naturally, we want to go with the more efficient F32T8 lamps, but let's go one step further, and use 4100 K lamps, instead of the 3500 K 'cool white' F40T12 lamps that we're taking out. This will give us a little whiter light, and also the *perceived* impression of *more* light. This will help when we replace four F40T12 lamps in an inefficient luminaire with two F32T8 lamps in our improved luminaire. The actual footcandle level at the desktop may be the same, plus or minus a few footcandles, but the space will *seem* much better lighted. If you remember the opening paragraph of this book, psychology is an important part of lighting, and it will come into play here. As a bonus, we have reduced the amount of air conditioning required to remove the waste heat from the inefficient luminaires.

It's easy to see that we have reduced the energy consumption of the system. We have replaced four 40 W lamps with two 32 W lamps, and we have improved the ballast factor from around 0.6 to at least 0.8. Plus, we have reduced the air conditioning load. But that's not all. The maintenance costs for the retrofitted luminaires will be reduced by at least half, since we now have only two lamps and one ballast per luminaire to maintain. All of this contributes toward producing a low payback time, which makes retrofitting economically attractive to building owners. In addition, if you multiply all this by the millions of luminaires currently in service, you can see how retrofitting aging luminaires can greatly improve the national energy picture.

Let's take a concrete example, and see how it looks.

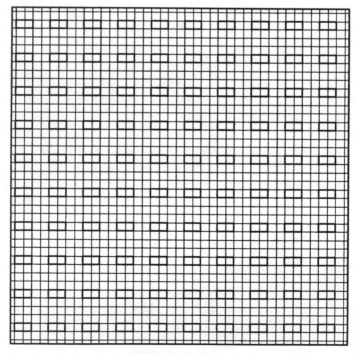

Figure 6.1. Office retrofit layout.

Let's say that we've been called in to survey a 80 ft × 80 ft open plan office space which is lighted by the typical four-lamp F40T12 luminaires, and we've been asked to show the payback for retrofitting the luminaires. The layout is shown in Figure 6.1.

We see that the luminaires are laid out on the typical 8 ft × 8 ft pattern, and that there are 100 luminaires. We can now compute the yearly energy savings for the system by calculating a 'before and

---

United Energy Associates, Inc.

PREPARED FOR: Example                    FACILITY: RETROFIT

      Hilton Head SC.

PREPARED BY:  W.J. Fielder, P.E.

      03/05/01

=== LIGHTING ENERGY AUDIT -- SAVINGS SUMMARY ===

**SAVINGS SUMMARY**

| ANNUAL CASH SAVINGS | CAPITAL RECOVERY | RETURN ON INVESTMENT |
|---|---|---|
| $4,802 | 12.18 Months | 98.48 % |

**CONVERSION SUMMARY**

EFFECTIVE COST PER KWh:     0.0750                KWd RATE:
AVERAGE WEEKLY LIGHTING HOURS:   72

| | BEFORE CONVERSION | AFTER CONVERSION |
|---|---|---|
| TOTAL FIXTURES: | 100 | 100 |
| TOTAL LAMPS: | 400 | 200 |
| TOTAL BALLASTS: | 200 | 100 |
| LIGHTING KWh PER MONTH: | 5,990 | 1,841 |
| LIGHTING KWd: | 19 | 6 |
| WATTAGE PER SQUARE FT: | 3.000 | 0.922 |

**ESTIMATED IMPROVEMENT COST AND SAVINGS**

| | | | | |
|---|---|---|---|---|
| ESTIMATED COST: | $4,876 | ANNUAL KWh SAVINGS: | | $3,735 |
| | | ANNUAL KWd SAVINGS: | N/A, | |
| NET IMPROVEMENT COST: | $4,876 | ANNUAL HVAC SAVINGS: | | $669 |
| | | ANNUAL MAINTENANCE SAVINGS: | | $398 |
| SAVINGS FOR FIVE YEARS: | $25,241 | TOTAL ANNUAL SAVINGS: | | $4,802 |
| SAVINGS FOR TEN YEARS: | $53,799 | | | |

ANNUAL ENVIRONMENTAL SAVINGS
POUNDS OF CARBON DIOXIDE:        39,836
POUNDS OF SULFUR DIOXIDE:       124,488
POUNDS OF NITRUS OXIDE:          59,754

Figure 6.2. ENLIGHTEN retrofit economic analysis (courtesy of United Energy Associates Inc.).

after' scenario, throw in the expected air conditioning savings, estimate the cost of retrofit, both labor and materials, estimate maintenance costs, run an economics analysis, and come up with an estimated payback time for the owner. If you do this by hand, it will probably take a day or two. A much easier approach is to use a packaged computer program to do the calculations for us. The program that we will use for this example is a proprietary program developed by United Energy Associates, Inc., of Winter Haven, Florida. The program is called ENLIGHTEN, and is available for purchase from United Energy Associates. Figure 6.2 shows the results of the ENLIGHTEN run for our example.

From Figure 6.2 it is seen that the capital recovery (or payback) time is 12.18 months, or a little over a year. After that, the savings to the building owner will be $4802.00 per year. This would represent a good value to the building owner, who would normally consider a payback time of under 3 years a sound investment.

For you environmentalists, you can also see that by reducing the energy consumption of the building, the amount of pollutants released into the air by generating plants has been reduced. In our case, the release of about 225 000 pounds per year of carbon dioxide, sulfur dioxide, and nitrous oxide has been avoided.

That's the retrofitting story. It's one of those rare cases where everybody wins. You as a designer should consider retrofitting as another valuable tool in your arsenal.

And that's about it for this book. Although I know you've grown to love them, you won't have any exercises to do for this chapter. Go forth and design!

# Glossary

*Accent Lighting*: Directional lighting to emphasize a particular object or to draw attention to a part of the field of view.

*Absorption*: The dissipation of light within a surface or medium.

*Accommodation*: The process by which the eye changes focus from one distance to another.

*Adaptation*: The process by which the visual system becomes accustomed to more or less light than it was exposed to during an immediately preceding period. It results in a change in the sensitivity of the eye to light.

*Air Fitting* (air bonnet, air hood, air saddle, air box): A fitting that is mounted to an air handling luminaire and connects to the primary air duct by flexible ducting. It normally contains one or two volume controls.

*Alternating Current (AC)*: Flow of electricity that cycles or alternates direction many times per second. The number of cycles per second is referred to as frequency. Most common frequency used in the US is 60 Hertz (Hz; cycles per second) in Europe, 50 Hz is the standard.

*Ambient Lighting*: General lighting, of the space (as opposed to task lighting or the lighting of the object one is looking at). It can be produced by direct lighting from recessed, surface or stem-mounted luminaires, or by indirect lighting that is wall or stem mounted, or by luminaires built into furniture or free standing.

*Amperes* (amps or A): The unit of measurement of electric current.

*Baffle*: An opaque or translucent element that serves to shield a light source from direct view at certain angles, or serves to absorb unwanted light.

*Ballast*: An auxiliary device used with fluorescent and HID lamps to provide the necessary starting voltage and to limit the current during operation.

*'Batwing' Distribution*: Candlepower distribution that serves to reduce glare and veil reflections by having its maximum output in the 30–60° zone from the vertical and with a candlepower at nadir (0°) being 65% or less than maximum candlepower. The shape is similar to a bat's wing. In fluorescent luminaires the batwing distribution is generally found only in the plane perpendicular to the lamps.

*Beam Spread*: The angle enclosed by candlepower distribution curve at the two lines that intersect the points where the candlepower is equal to 10% of its maximum.

*Branch Circuit*: An electrical circuit running from a service panel having its own overload protection device.

*Brightness (Luminance)*: The degree of apparent lightness of a surface; its brilliancy; concentration of candlepower. Brightness is produced by either a self-luminous object, by light energy transmitted through objects or by reflection. Unit of measurement of brightness is the footlambert (fl).

*Candela*: The unit of measurement of luminous intensity of a light source in a given direction.

*Candlepower*: Luminous intensity expressed in candelas.

*Candlepower Distribution Curve*: A graphic presentation of the distribution of light intensity in a given plane of a lamp or luminaire. It is determined by photometric tests. The curve is generally polar, representing the variation of luminous intensity of a lamp or luminaire in a plane through the light center.

*Capacitor*: An electric energy storage device that when built into or wired to a ballast changes it from low to high power factor.

*Cavity Ratio*: A number indicating room cavity proportions calculated from length, width and height.

*Ceiling Cavity Ratio*: A numerical relationship of the vertical distance between luminaire mounting height and ceiling height to room width and length. It is used with the zonal cavity method of calculating average illumination levels.

*Circuit Breaker*: Resettable safety device to prevent excess current flow.

*Class 'P' Ballast*: Contains a thermal protective device that deactivates the ballast when the case reaches a certain critical temperature. The device resets automatically when the case temperature drops to a lower temperature.

*Coefficient of Utilization (CU)*: A ratio representing the portion of light emitted by a luminaire in any particular installation that actually gets down to the work plane. The coefficient of utilization thus indicates the combined efficiency of the luminaire, room geometry and room finish reflectances.

*Cold Cathode Lamp*: An electric-discharge lamp whose mode of operation is that of a glow discharge.

*Color Rendering Index (CRI)*: Measure of the degree of color shift objects undergo when illuminated by the light source as compared with the color of those same objects when illuminated by a reference source of comparable color temperature.

*Color Temperature*: The absolute temperature of a black body radiator having a chromaticity equal to that of the light source.

*Cone Reflector*: Parabolic reflector that directs light downward thereby eliminating brightness at high angles.

*Contrast*: The difference in brightness (luminance) of an object and its background.

*Contrast Rendition Factor (CRF)*: The ratio of visual task contrast with a given lighting environment to the contrast with sphere illumination. Contrast measured under sphere illumination is defined as 1.00.

*Cool Beam Lamps*: Incandescent reflector lamps that use a special coating (dichronic interference filter) on the reflectorized portion of the bulb to allow infrared heat to pass out the back while reflecting only visible energy to the task, thereby providing a 'cool beam' of light.

*Cut-off Luminaires*: Outdoor luminaires that restrict all light output to below 85° from vertical.

*Cut-off Angle* (of a luminaire): The angle from the vertical at which a reflector, louver, or other shielding device cuts off direct visibility of a light source. It is the complementary angle of the shielding angle.

*Dimming Ballast*: Special fluorescent lamp ballast, which when used with a dimmer control, permits varying light output.

*Direct Current (DC)*: Flow of electricity continuously in one direction from positive to negative.

*Direct Glare*: Glare resulting from high luminances or insufficiently shielded light sources in the field of view. It usually is associated with bright areas, such as luminaires, ceilings and windows that are outside the visual task or region being viewed.

*Discharge Lamp*: A lamp in which light (or radiant energy near the visible spectrum) is produced by the passage of an electric arc through a vapor or a gas.

*Discomfort Glare*: Glare producing discomfort. It does not necessarily interfere with visual performance or visibility.

*Distribution Panel*: Panelboard containing circuit breakers that distribute power to branch circuits.

*Efficacy*: See Lamp Efficacy.

*Efficiency*: See Luminaire Efficiency.

*Equivalent Sphere Illumination (ESI)*: The level of sphere illumination that would produce task visibility equivalent to that produced by a specific lighting environment. Suppose a task at a given location and direction of view within a specific lighting system has 100 fc of illumination. Suppose this same task is now viewed under sphere lighting and the sphere lighting level is adjusted so that the task visibility is the same under the sphere lighting as it was under the lighting system. Suppose the lighting level at the task from the sphere lighting is 50 fc for equal visibility. Then the equivalent sphere illumination of the task under the lighting system would be 50 ESI fc.

*'ER' (Elliptical Reflector)*: Lamp whose reflector focuses the light about 2 ft (61 cm) ahead of the bulb, reducing light loss when used in deep baffle downlights.

*Extended Life Lamps*: Incandescent lamps that have an average rated life of 2500 or more hours and reduced light output compared to standard general service lamps of the same wattage.

*Floodlighting*: A system designed for lighting a scene or object to a luminance greater than its surroundings. It may be for utility, advertising or decorative purposes.

*Fluorescent Lamp*: A low-pressure mercury electric-discharge lamp in which a fluorescing coating (phosphor) transforms some of the ultraviolet energy generated by the discharge into light.

*Flux*: Continuous flow of luminous energy.

*Footcandle (fc)*: The unit of illuminance when the foot is taken as the unit of length. It is the illuminance on a surface one square foot in area on which there is a uniformly distributed flux of 1 lumen.

*(Raw) Footcandles*: Same as footcandles. This term is sometimes used in order to differentiate between ordinary footcandles and ESI footcandles. (Footcandles or Raw Footcandles refer only to the quantity of illumination. ESI footcandles refer to task visibility by considering both the quantity and quality of illumination.)

*Foot Lambert*: A unit of luminance of a perfectly diffusing surface emitting or reflecting light at the rate of 1 lumen per square foot.

*Fuse*: Replaceable safety device to prevent excess current flow.

*General Lighting*: See Ambient Lighting.

*General Service Lamps*: 'A' or 'PS' incandescent lamps.

*Glare*: The sensation produced by luminance within the visual field that is sufficiently greater than the luminance to which the eyes are adapted to cause annoyance, discomfort, or loss in visual performance and visibility.

*Greenfield*: Flexible metallic tubing for the protective enclosure of electric wires.

*Grounding*: Connection of electric components to earth for safety.

*Group Relamping*: Relamping of a group of luminaires at one time to reduce relamping labor costs.

*Heat Extraction*: The process of removing heat from a luminaire by passing return air through the lamp cavity.

*High Intensity Discharge (HID) Lamp*: A discharge lamp in which the light-producing arc is stabilized by wall temperature, and the arc tube has a bulb wall loading in excess of 3 W per square centimeter. HID lamps include groups of lamps known as mercury, metal halide, and high pressure sodium.

*High Output Fluorescent Lamp*: Operates at 800 or more milliamperes (mA) for higher light output than standard fluorescent lamp (430 mA).

*High Pressure Sodium (HPS) Lamp*: High intensity discharge (HID) lamp in which light is produced by radiation from sodium vapor. Includes clear and diffuse-coated lamps.

*Incandescent Lamp*: A lamp in which light is produced by a filament heated to incandescence by an electric current.

*Instant Start Fluorescent Lamp*: A fluorescent lamp designed for starting by a high voltage without preheating of the electrodes.

*Inverse Square Law*: The law stating that the illuminance at a point on a surface varies directly with the intensity of a point source, and inversely as the square of the distance between the source and the point.

*Isolux Chart*: A series of lines plotted on any appropriate set of coordinates, each line connecting all the points on a surface having the same illumination.

*Joule Heating*: Heating which occurs when an electric current is passed through a conductor having electrical resistance. The amount of heat generated is directly proportional to the resistance, and to the square of the current (Wjoule = I squared × R).

*Junction Box*: A metal box in which circuit wiring is spliced. It may also be used for mounting luminaires, switches or receptacles.

*Kilowatt-hour (kWh)*: Unit of electrical energy, or power consumed over a period of time. kWh = watts/1000 × hours used.

*Lamp*: An artificial source of light (also a portable luminaire equipped with a cord and plug).

*Lamp Efficacy*: The ratio of lumens produced by a lamp to the watts consumed, expressed as lumens per watt (LPW).

*Lamp Lumen Depreciation (LLD)*: Multiplier factor in illumination calculations for reduction in the light output of a lamp over a period of time.

*Light*: Radiant energy that is capable of exciting the retina and producing a visual sensation. The visible portion of the electro-magnetic spectrum extends from about 380–770 nm.

*Light Loss Factor (LLF)*: A factor used in calculating the level of illumination that takes into account such factors as dirt accumulation on luminaire and room surfaces, lamp depreciation, maintenance procedures and atmosphere conditions. See Maintenance Factor.

*Light Output*: Amount of light produced by a light source such as a lamp. The unit most commonly used to measure light output is the lumen.

*Lens*: Used in luminaires to redirect light into useful zones.

*Local Lighting*: Lighting designed to provide illuminance over a relatively small area or confined space without providing any significant general surrounding lighting.

*Louver*: A series of baffles used to shield a source from view at certain angles or to absorb unwanted light. The baffles are usually arranged in a geometric pattern.

*Long Life Lamps*: See Extended Life Lamps.

*Low Pressure Sodium Lamp*: A discharge lamp in which light is produced by radiation of sodium vapor at low pressure producing a single wavelength of visible energy, i.e. yellow.

*Low Voltage Lamps*: Incandescent lamps that operate at 6–12 V.

*Lumen*: The unit of luminous flux. It is the luminous flux emitted within a unit solid angle (1 steradian) by a point source having a uniform luminous intensity of 1 candela.

*Luminaire*: A complete lighting unit consisting of a lamp or lamps together with the parts designed to distribute the light, to position and protect the lamps and to connect the lamps to the power supply.

*Luminaire Dirt Depreciation (LDD)*: The multiplier to be used in illuminance calculations to relate the initial illuminance provided by clean, new luminaires to the reduced illuminance that they will provide due to dirt collection on the luminaires at the time at which it is anticipated that cleaning procedures will be instituted.

*Luminaire Efficiency*: The ratio of luminous flux (lumens) emitted by a luminaire to that emitted by the lamp or lamps used therein.

*Luminance*: See *Brightness*. The amount of light reflected or transmitted by an object.

*Lux*: The metric unit of illuminance. One lux is one lumen per square meter (lm/m2).

*Maintenance Factor (MF)*: A factor used in calculating illuminance after a given period of time and under given conditions. It takes into account temperature and voltage variations, dirt accumulation on luminaire and room surfaces, lamp depreciation, maintenance procedures and atmosphere conditions.

*Matte Surface*: A non-glossy dull surface, as opposed to a shiny (specular) surface. Light reflected from a matte surface is diffuse.

*Mercury Lamp*: A high intensity discharge (HID) lamp in which the major portion of the light is produced by radiation from mercury. Includes clear, phosphor-coated and self-ballasted lamps.

*Metal Halide Lamp*: A high intensity discharge (HID) lamp in which the major portion of the light is produced by radiation from mercury. Includes clear, phosphor-coated and self-ballasted lamps.

*Nadir*: Vertically downward directly below the luminaire or lamp; designated as 0°.

*Outlet Box*: See Junction Box.

*'PAR' Lamps*: Parabolic aluminized reflector lamps that offer excellent beam control, come in a variety of beam patterns from very narrow spot to wide flood, and can be used outdoors unprotected because they are made of 'hard' glass that can withstand adverse weather.

*Parabolic Louvers*: A grid of parabolic shaped baffles that redirects light downward and provides very low luminaire brightness.

*Penumbra*: The darkest part of a shadow.

*Plug-in Wiring*: Electrical distribution system which has quick-connect wiring connectors.

*Point Method Lighting Calculation*: A lighting design procedure for predetermining the illuminance at various points in a lighting installation by use of luminaire photometric data.

*Polarization*: The process by which the transverse vibrations of light waves are oriented in a specific plane. Polarization may be obtained by using either transmitting or reflecting media.

*Power Factor*: Ratio of: watts/volts × amperes. Power factor in lighting is primarily applicable to ballasts. Since volts and watts are usually fixed, amperes (or current) will go up as power factor goes down. This necessitates the use of larger wire sizes to carry the increased amount of current needed with low power factor (LPF) ballasts. The addition of a capacitor to a LPF ballast converts it to a HPF ballast.

*Preheat Fluorescent Lamp*: A fluorescent lamp designed for operation in a circuit requiring a manual or automatic starting switch to preheat the electrodes in order to start the arc.

*'R' Lamps*: Reflectorized lamps available in spot (clear face) and flood (frosted face).

*Rapid Start Fluorescent Lamp*: A fluorescent lamp designed for operation with a ballast that provides a low-voltage winding for preheating the electrodes and initiating the arc without a starting switch or the application of high voltage.

*Raw Footcandles*: See Footcandles.

*Reflectance*: Sometimes called reflectance factor. The ratio of reflected light to incident light (light falling on a surface). Reflectance is generally expressed in percent.

*Reflected Glare*: Glare resulting from specular reflections of high luminances in polished or glossy surfaces in the field of view. It usually is associated with reflections from within a visual task or areas in close proximity to the region being viewed.

*Reflection*: Light striking a surface is either absorbed, transmitted, or reflected. Reflected light is that which bounces off the surface, and it can be classified as specular or diffuse reflection. Specular reflection is characterized by light rays that strike and leave a surface at equal angles. Diffuse reflection leaves a surface in all directions.

*Refraction*: The process by which the direction of a ray of light changes as it passes obliquely from one medium to another in which its speed is different.

*Romex*: A 60°C cable comprised of an outer flexible plastic sheathing that contains two or more insulated wires, designated as type 'NM'. Romex is normally restricted to residential use.

*Room Cavity Ratio (RCR)*: A numerical relationship of the vertical distance between work plane height and luminaire mounting height to room width and length. It is used with the zonal cavity method of calculating average illumination levels.

*Rough Service Lamps*: Incandescent lamps designed with extra filament supports to withstand bumps, shocks and vibrations with some loss in lumen output.

*Service Entrance*: Point at which power utility wires enter a building.

*Shielding*: An arrangement of light-controlling material to prevent direct view of the light source.

*Shielding Angle* (of a luminaire): The angle from the horizontal at which a light source *first* becomes visible. It is the complementary angle of the cut-off angle. In the case of a luminaire shielded by a reflector or parabolic cell louver, it is important to ascertain also the shielding angle to the reflected image of the light source, as this is often almost as bright as the source itself.

*Silvered Bowl Lamps*: Incandescent 'A' lamps with a silver finish inside the bowl portion of the bulb. Used for indirect lighting.

*Spacing Ratio (SR)*: The ratio of the distance between luminaire centers to the height above the work plane. The maximum spacing ratio for a particular luminaire is determined from the candlepower distribution curve for that luminaire and, when multiplied by the mounting height above the work plane, gives the maximum spacing of luminaires at which even illumination will be provided.

*Spectral Energy Distribution (SED) Curves*: A plot of the level of energy at each wavelength of a light source.

*Sphere Illumination*: The illumination on a task from a source providing equal luminance in all directions about that task, such as an illuminated sphere with the task located at the center.

*Task*: That which is to be seen. The visual function to be performed.

*Task Lighting*: Lighting directed to a specific surface or area that provides illumination for visual tasks.

*Three-way Lamps*: Incandescent lamps that have two separately switched filaments permitting a choice of three levels of light.

*Transformer*: An AC device to raise or lower electric voltage.

*Transmission*: The passage of light through a material.

*Tungsten Halogen Lamp*: A gas-filled tungsten incandescent lamp containing a certain proportion of halogens.

*Veiling Reflections*: The reflections of light sources in the task that reduce the contrast between detail and background (e.g. between print and paper) thus imposing a 'veil' and decreasing task visibility.

*Vibration Service Lamps*: See Rough Service Lamps.

*Visual Comfort Probability (VCP)*: A discomfort glare calculation that predicts the percent of observers positioned at a specific location, (usually 4 ft, or 1.2 m, in front of the center of the rear wall), who would be expected to judge a lighting condition to be comfortable. VCP rates the luminaire in its environment, taking into account such factors as illumination level, room dimensions and reflectances, luminaire type, size and light distribution, number and location of luminaires, and observer location and location and line of sight. The higher the VCP, the more comfortable the lighting environment. IES has established a value of 70 as the minimum acceptable VCP.

*Visual Edge*: The line on an isolux chart which has a value equal to 10% of the maximum illumination.

*Visual Field*: The field of view that can be perceived when the head and eyes are kept fixed.

*Volt (V)*: The unit for measuring electric potential. It defines the force or pressure of electricity.

*Wall Wash Lighting*: A smooth even distribution of light over a wall.

*Watt (W)*: The unit for measuring electric power. It defines the power or energy consumed by an electrical device. The cost of operating an electrical device is determined by the watts it consumes times the hours of use. It is related to volts and amps by the following formula: Watts = Volts × Amps.

*Work Plane*: The plane at which work is done, and at which illumination is specified and measured. Unless otherwise indicated, this is assumed to be a horizontal plane 30 in. (76 cm) above the floor.

*Zonal Cavity Method Lighting Calculation*: A lighting design procedure used for predetermining the relation between the number and types of lamps or luminaires, the room characteristics, and the average illuminance on the work plane. It takes into account both direct and reflected flux.

# Index

# Teaching English, Language and Literacy

Are you looking for one book that gives a comprehensive account of primary and early years English, language and literacy teaching?

This fully revised fourth edition of *Teaching English, Language and Literacy* includes up-to-date research and updated discussion of effective teaching. Throughout the book there is guidance on England's new National Curriculum and its impact. Rooted in research evidence and multidisciplinary theory, this book is an essential introduction for anyone learning to teach English from the early years to primary school level.

The authors draw on their research, scholarship and practice to offer advice on:

- inclusion and equality, including working effectively with multilingual pupils
- speaking and listening
- developing reading, including choosing texts, and phonics teaching
- improving writing, including grammar and punctuation
- planning and assessing
- the latest thinking in educational policy and practice
- the use of multimedia
- maintaining good home–school links

All chapters include examples of good practice, coverage of key issues, analysis of research and reflections on national policy to encourage the best possible response to the exciting challenges of teaching. Each chapter also has a glossary to explain terms and gives suggestions for further reading.

This authoritative book is for all those who want to improve the teaching of English, language and literacy in schools. Designed to help inform trainee teachers and tutors, but also of great use to those teachers wanting to keep pace with the latest developments in their specialist subject, this is an indispensable guide to the theory and practice of teaching English, language and literacy

**Dominic Wyse** is Professor of Early Childhood and Primary Education at the University College London, Institute of Education. His research includes a multidisciplinary focus on the teaching of writing across the life-course, including how writing works, which is the subject of his most recent book.

**Russell Jones** was Senior Lecturer in Education and Childhood Studies at Manchester Metropolitan University.

**Helen Bradford** is part of the early years team at University College London, Institute of Education. Her research includes young children's development of literacy, focusing particularly on early writing. Her recently completed PhD looked at co-constructing writing pedagogy with 2- and 3-year-old children.

**Mary Anne Wolpert** is Affiliated Lecturer and Course Manager of the Primary PGCE course at the Faculty of Education, University of Cambridge.

# Teaching English, Language and Literacy

## Fourth Edition

Dominic Wyse, Russell Jones,
Helen Bradford,
Mary Anne Wolpert

Routledge
Taylor & Francis Group

LONDON AND NEW YORK

Fourth edition published 2018
by Routledge
2 Park Square, Milton Park, Abingdon, Oxon OX14 4RN

and by Routledge
711 Third Avenue, New York, NY 10017

*Routledge is an imprint of the Taylor & Francis Group, an informa business*

First edition published by Routledge 2001
Third edition published by Routledge 2013

*British Library Cataloguing-in-Publication Data*
A catalogue record for this book is available from the British Library

*Library of Congress Cataloging-in-Publication Data*
A catalog record for this book has been requested.

ISBN: 978-1-138-28053-3 (hbk)
ISBN: 978-1-138-28573-6 (pbk)
ISBN: 978-1-315-27200-9 (ebk)


Typeset in Bembo
by Apex CoVantage, LLC

MIX
Paper from
responsible sources
FSC
www.fsc.org    FSC® C013056

Printed and bound in Great Britain by
TJ International Ltd, Padstow, Cornwall

# Contents

# Figures

# Tables

# Preface

This fourth edition of *Teaching English, Language and Literacy* includes some significant restructuring and the introduction of new material in every chapter. More than 17 years ago, we had the idea for a comprehensive guide that wasn't available at the time. Since then, through the four editions, we have traced significant changes to teaching and policy. We have maintained a critical stance with a view to highlighting exemplary teaching on the basis of evidence, not ideology.

The work on the fourth edition was inspired by comments from reviewers of the third edition. For example, we have integrated aspects such as digital technology throughout the book (when appropriate to particular chapter topics) rather than have separate chapters. This is also true of issues to do with equality and diversity, although we have also developed a new chapter on inclusion in recognition of its changed legal status. The chapter on children's literature has been updated and relocated to be part of the section on reading.

We have also focused even more on the quality of evidence we cite to support our views on exemplary teaching and learning. This has in some cases resulted in a reduction of citations overall but also the introduction of new research of even higher quality. Another significant change was to integrate the previous chapters on assessment into one chapter, in recognition that assessment of language and literacy is part of an overall process that links the constituent parts of talking, reading and writing.

Every chapter has been updated and many have had substantial changes to reflect new research, new theory, new practice and rapidly changing government policies. England's new National Curriculum of 2014 is one obvious example that necessitated many changes to the book. The longer chapters of the book are the most obvious examples of updating (for example, the new material on children's development of spoken language) but every other chapter includes new material of some kind.

English is one of the most fascinating, controversial and challenging subjects of the curriculum. The fact that English is the language we speak also makes it a subject that is closely linked with our identities, which is one of the reasons that it often engenders passionate views. Another reason that it is important is that all teachers have to be teachers of English because learning takes place through talking, reading and writing. In the early years and primary curriculum, great stress is put on communication, language and literacy because these are essential for all other learning.

This book is a comprehensive introduction to the ideas, concepts and knowledge that are part of the study of English, language and literacy teaching and learning. It is written for trainee teachers, their tutors, for more experienced teachers and for other students of education. The partnerships between providers of teacher education and schools have maintained the need for a book that offers a comprehensive overview of the subject to enable teacher mentors to update their professional knowledge in specific areas when appropriate. It is designed as a reader that will enhance and consolidate the learning in early years and primary English programmes and as an essential guide to the teaching of English. The book's hallmarks have always been that it is:

- aimed at teachers with a view to **informing their thinking and practice**;
- **a comprehensive** account of teaching English, language *and* literacy;
- **critically evaluative** in style, e.g. in relation to government policy;
- built on an **explicit theoretical framework;**
- rooted in **research evidence** and multidisciplinary theory.

The book is divided into five parts: I. 'Introduction'; II. 'Language'; III. 'Reading'; IV. 'Writing'; and V. 'General issues'. The bulk of the book consists of short chapters that cover the variety of aspects that make up the English curriculum. All of these chapters include clear examples of practice, coverage of key issues, analysis of research and reflections on national policy. The short chapters are complemented by some longer chapters. The first of these addresses the important subject of the history of English and English teaching. The second is an update of the book's theoretical framing. The other three look at children's development in language, reading and writing, and relate this development to teaching approaches. The structure of the longer chapters allowed us to tackle some of the most important aspects of the English curriculum in depth and at a higher level. Part V is made up of issues that tend to be applicable to all three areas of language, reading and writing.

One of the important features of the book is its comprehensive scope. The subject of English is an area that boasts an impressive array of scholarship and practice. While there are many books that have addressed the modes of reading, writing and speaking and listening separately, there are very few which address the complete subject area. By doing this, we have accepted that inevitably some parts of the subject are touched on only briefly. In recognition of this, you will find more than 100 descriptions of recommended books and papers for further reading which appear in the 'annotated bibliographies' for every chapter. A novel feature of these bibliographies is a system of coding which allows you to judge the reading level and the balance between theory and practice:

*     Mainly focused on classroom practice
**   Close balance between theory and practice
*** Research- and theory-based

L1  Introductory reading
L2  Intermediate reading
L3  Advanced reading

We are fully in support of the idea that teaching should be an evidence-informed activity, and so each chapter in the book is underpinned by our reading of research. In addition to our references to papers, books and official publications, we also make reference to a range of websites. This is always a tricky business. This revision of the book took many months to complete, and in that time digital technology has continued to develop. In light of this, we have chosen sites that we hope will stand the test of time.

The most important part of reading a book like this is that it will enable you to become a better teacher. No book can offer a magic solution to becoming an effective teacher. Learning to teach – like most learning – requires practical engagement with the subject in partnership with experienced people. However, in order to establish direct and explicit links with practice, we use case studies, analysis of resources, reflections on children's work, teachers' thoughts and examples of teaching, and each chapter concludes with 'practice points', which have been written to focus attention on some of the most important practical ideas of which you should be aware.

This book covers a wide range of essential knowledge. If we consider technical vocabulary alone, there are many definitions supplied in the 'glossaries' that are a feature of every chapter. So, if you are unsure about the meaning of a particular word as you are reading, you do not need to reach for a dictionary because most of the key words are defined for you at the end of the chapter. Another aspect of knowledge that has been played down in recent years is the knowledge of issues. This is, we feel, vital to both effective teaching and success in the education profession. In order to maintain the tradition of English as a vibrant subject, we hope teachers will continue to fully engage with the issues and ideas that are explored in this book.

## Note

Throughout this book the following icons are used to assist the reader:

→  Recommends the reader looks at another chapter in the book.
☞  These words are included in the glossaries at the end of each chapter.

# Acknowledgements

We would like to record our appreciation for the outstanding reflections from the reviewers who commented on the third edition. Most of their recommendations have been acted on as we worked to refine the book. Those who were content to be named were Fiona Maine, Tom Dobson, Claire Head, Anne Bradley and Branwen Bingle – we are very grateful for your inspirational ideas.

We would like to thank all the lecturers and tutors, trainee teachers and teachers, who have read and who continue to read our book. It is only because of this continuing interest, over more than 17 years, that we have reached this fourth edition.

We would also like to thank all the Routledge staff who have been involved in making the book such a success, including Alison Foyle, who commissioned the fourth edition.

# Part I

# Introduction

# Chapter 1

# The history of English, language and literacy

One of the important aspects of historical knowledge is that it enables us to better understand the present. This chapter briefly examines three significant historical angles: the history of English as a language; the history of the teaching of English; and the history of national initiatives to improve the teaching of English. We conclude in the present by looking at the National Curriculum and the phonics screening check.

The three words 'English', 'Language' and 'Literacy' in the title of this book are significant because they are central to many of the debates that have raged about the teaching of English in primary schools. During the 1970s and 1980s, the teaching of 'Language' was the focus. The job of primary schools was to foster the development of children's language through reading, writing and, to a lesser extent, talking. This focus included the need to support multilingual children's development in English and other languages. The teachers who coordinated the subject were known as 'language coordinators'. The teaching of language in primary schools was seen as different in many respects from the teaching of English conducted in secondary schools.

With the coming of the Education Reform Act 1988, 'English' was re-established as the main focus for primary education. The subject was, however, still to be concerned with the teaching of the three language modes of reading, writing and talk. 'Speaking and Listening' became of equal importance to Reading and Writing for the first time, and this was prescribed by the National Curriculum. Coordinators were now to be called 'English' coordinators. The advent of the National Literacy Strategy (NLS) ☞ in 1997 resulted in a heavy focus on 'Literacy'. You will probably have guessed that subject leaders were renamed 'literacy coordinators'.

The first part of this chapter looks at some of the historical aspects of the subject that have shaped its development. It is important that all teachers have a historical perspective on their work; at the very least, this can give you a means to critically examine modern initiatives and to check how 'new' they really are.

We start with a brief look at some of the significant moments in the development of the English language and reflect on their continuing relevance to classroom teaching. This is followed by reflections on the history of the *teaching* of English. We conclude with an outline of some of the major national projects that have been undertaken and finish right up to date with a look at the phonics screening check.

## The English language

English, like all languages, is constantly changing. The *Oxford English Dictionary* has a large team of people who are constantly searching for new uses and new additions to the language. For example, here is an extract from the OED website:

> *June 2017 update*
> More than 600 new words, phrases, and senses have been added to the Oxford English Dictionary this quarter, including *bug chaser, chantoosie, gin daisy*, and *widdly*. You can read about other new and revised meanings in this article by Katherine Connor Martin, Head of US Dictionaries, and explore our timeline of veil words.
>
> (OED, 2017, online)

The online version of the dictionary is a spectacular resource, including as it does all known meanings for words; their grammatical function; etymology, including changes in usage and spelling over time; audio files for pronunciation by different types of speakers; sources for the examples of use of the words; etc. As well as recording language change, dictionaries play a major role in the standardisation of the language. It is interesting to note that American Standard English is represented by specific dictionaries such as those published by Merriam-Webster, but British Standard English is, for example, represented by the *Oxford English Dictionary* or *Chambers Dictionary*.

The significant influence of publishing has also resulted in standard reference works that lay down particular conventions. So if you have ever wondered how to reference properly using the 'Author – Date' method, try *The American Psychological Association (APA) Style Guide* (or for a simplified version, try *The Good Writing Guide for Education Students;* Wyse and Cowan, 2017). For teachers, the idea that language is always changing is an important one. If we place too heavy an emphasis on absolute and fixed 'rules', we may be teaching in a linguistically inaccurate or inappropriate way (→ Chapter 16). Effective teaching needs to be built on an understanding of those features of the language that are stable and those that are subject to constant change.

This process of change is by no means a recent phenomenon. Human beings' creation of alphabetic written language was a highly significant development.

All alphabets were originally derived from the Semitic syllabaries of the second millennium. The developments from both Greek script and the Roman alphabet can be seen in the use of the Latinised form of the first two letters of the Greek alphabet in the word itself, 'alphabet'. 'Alpha' was derived from the Semitic 'aleph' and 'beta' from 'beth' (Goody and Watt, 1963). Historically, the alphabet has been at the heart of some of the most enduring debates about the development of written communication, for example whether the alphabet simply emerged from logographic or pictographic forms. In Harris' (1986) examination of the origins of writing, he called this particular idea of emergence an evolutionary fallacy, arguing that the alphabet was 'the great invention' because its graphic signs have almost no limitations for human communication, unlike logos or pictographs. The continuing development of writing, for example through internet and electronic text forms, is further testament to written language's extraordinary capacity to adapt to, and be part of, cultural change.

It was during the fifth century that the Anglo-Saxons settled in England and, as always happens when people colonise, they brought changes to the language, a process that resulted in 'Old English' being established. The few texts that have survived from this period are in four main dialects ☞: West Saxon, Kentish, Mercian and Northumbrian. The last two are sometimes grouped together and called Anglian. West Saxon became the standard dialect at the time but is not the direct ancestor of modern Standard English ☞, which is mainly derived from an Anglian dialect (Barber, 1993). If you take the modern word 'cold' as an example, the Anglian 'cald' is a stronger influence than the West Saxon version, 'ceald'.

In the ninth century, the Vikings brought further changes to the language. Place names were affected: 'Grimsby' meant 'Grim's village' and 'Micklethwaite' meant 'large clearing'. The pronunciation of English speech was also affected, and it is possible to recognise some Scandinavian-influenced words because of their phonological form. It is suggested that 'awe' is a Scandinavian word and that this came from changes of pronunciation to the Old English word 'ege'. One of the most interesting things about Scandinavian loanwords ☞ is that they are so commonly used: sister, leg, neck, bag, cake, dirt, fellow, fog, knife, skill, skin, sky, window, flat, loose, call, drag and even 'they' and 'them' (Barber, 1993).

In more recent times, words from a range of countries have been borrowed. Here are a small selection of examples: French – elite, liaison, menu, plateau; Spanish and Portuguese – alligator, chocolate, cannibal, embargo, potato; Italian – concerto, balcony, casino, cartoon; Indian languages – bangle, cot, juggernaut, loot, pyjamas, shampoo; African languages – banjo, zombie, rumba, tote. However, for many of these words it is difficult to attribute them to one original country. To illustrate the complexities, consider the word 'chess':

'Chess' was borrowed from Middle French in the fourteenth century. The French word was, in turn, borrowed from Arabic, which had earlier borrowed it from Persian 'shah' 'king'. Thus the etymology ☞ of the word reaches from Persian, through Arabic and Middle French, but its ultimate source (as far back as we can trace its history) is Persian. Similarly, the etymon of 'chess', that is, the word from which it has been derived, is immediately 'esches' and ultimately 'shah'. Loanwords have, as it were, a life of their own that cuts across the boundaries between languages.

(Pyles and Algeo, 1993: 286)

The influence of loanwords is one of the factors that has resulted in some of the irregularities of English spelling. David Crystal (1997) lists some of the other major factors. Above we referred to the Anglo-Saxon period; at that time there were only 24 graphemes (letter symbols) to represent 40 phonemes (sounds). Later, 'i' and 'j', 'u' and 'v' were changed from being interchangeable to having distinct functions and 'w' was added, but many sounds still had to be signalled by combinations of letters.

After the Norman conquest, French scribes – who had responsibility for publishing texts – respelled a great deal of the language. They introduced new conventions such as 'qu' for 'cw' (queen), 'gh' for 'h' (night) and 'c' before 'e' or 'i' in words such as 'circle' and 'cell'. Once printing became better established in the West, this added further complications. William Caxton (1422–92) is often credited with the 'invention' of the printing press, but this is not accurate. During the seventh century the Chinese printed the earliest known book, *The Diamond Sutra*, using inked wooden relief blocks. By the beginning of the fifteenth century, the process had developed in Korea to the extent that printers were manufacturing bronze type sets of 100,000 pieces. In the West, Johannes Gutenberg (1390s – 1468) is credited with the development of moveable metal type in association with a hand-operated printing press.

Many of the early printers working in England were foreign (many came from Holland in particular) and they used their own spelling conventions. Also, until the sixteenth century, line justification ☞ was achieved by changing words rather than by adding spaces. Once printing became established, the written language did not keep pace with the considerable alterations to the way words were spoken, resulting in weaker links between sound and symbol.

Samuel Johnson's dictionary, published in 1755, was another important factor in relation to English spelling. His work resulted in dictionaries becoming more authoritarian and used as the basis for 'correct' usage. Noah Webster, the first person to write a major account of American English, compared Johnson's contribution to Isaac Newton's in mathematics. Johnson's dictionary was significant for a number of reasons. Unlike dictionaries of the past that tended to concentrate on 'hard words', Johnson wanted a scholarly record of the whole language. It was based on words in use and introduced a literary dimension, drawing heavily on writers such as Dryden, Milton, Addison, Bacon, Pope and Shakespeare (Crystal, 1997: 109). Shakespeare's remarkable influence on the

English language is not confined to the artistic significance of his work; many of the words and phrases of his plays are still commonly used today:

> He coined some 2,000 words – an astonishing number – and gave us countless phrases. As a phrasemaker there has never been anyone to match him. Among his inventions: one fell swoop, in my mind's eye, more in sorrow than in anger, to be in a pickle, bag and baggage, vanish into thin air, budge an inch, play fast and loose, go down the primrose path, the milk of human kindness, remembrance of things past, the sound and fury, to thine own self be true, to be or not to be, cold comfort, to beggar all description, salad days, flesh and blood, foul play, tower of strength, to be cruel to be kind, and on and on and on and on. And on. He was so wildly prolific that he could put two in one sentence, as in Hamlet's observation: 'Though I am native here and to the manner born, it is custom more honoured in the breach than the observance.' He could even mix metaphors and get away with it, as when he wrote: 'Or to take arms against a sea of troubles.'
>
> (Bryson, 1990: 57)

Crystal (2004) makes the point that although spelling is an area where there is more agreement about what is correct than in other areas of language, there's still considerable variation. Greenbaum's (1986) research looked at all the words beginning with 'A' in a medium-sized desk dictionary which were spelled in more than one way; he found 296. When extrapolating this to the dictionary as a whole, he estimated 5,000 variants altogether, which is 5.6 per cent. If this were to be done with a dictionary as complete as the *Oxford English Dictionary*, it would mean many thousands of words where the spelling has not been definitively agreed. Crystal gives some examples including: accessory/accessary; acclimatize/acclimatise; adrenalin/adrenaline; aga/agha; ageing/aging; all right/alright.

Many of Greenbaum's words were pairs but there were some triplets: for example, aerie/aery/eyrie. And there were even quadruplets: anaesthetize/ anaesthetise/anesthetize/anesthetise. Names translated from a foreign language compound the problems, particularly for music students: Tschaikovsky/ Tchaikovsky/Tschaikofsky/Tchaikofsky/Tshaikovski.

It is tempting to assume that the grammar of the English language has stabilised, but recent work indicates the scale of change that continues. In one study, more than five million books, approximately 4 per cent of all books ever published, were analysed. The units of analysis in this study were the *1-gram* and *n-gram*. The 1-gram is a meaningful sequence of characters not separated by a space that includes words, part-words (such as SCUBA), numbers, and typos (such as 'excesss'). An n-gram is a sequence of 1-grams, such as the phrases 'police station' (a 2-gram) and 'the United Kingdom' (a 3-gram). The analyses revealed significant results in relation to the ways in which the

English language continues to change. At the time the study was published the size of the language had increased by more than 70 per cent in the past 50 years, adding about 8,500 words per year. An analysis of irregular verbs showed much stability over a period of 200 years but also that 16 per cent went through change of grammatical regularisation:

> These changes occurred slowly: It took 200 years for our fastest-moving verb ('chide') to go from 10% to 90% [regular]. Otherwise, each trajectory was sui generis ☞; we observed no characteristic shape. For instance, a few verbs, such as 'spill', regularized at a constant speed, but others, such as 'thrive' and 'dig', transitioned in fits and starts (7). In some cases, the trajectory suggested a reason for the trend. For example, with 'sped/speeded' the shift in meaning from 'to move rapidly' and toward 'to exceed the legal limit' appears to have been the driving cause.
>
> (Michel *et al.*, 2010: 177)

For a more in-depth history of writing and its relationship to the teaching of writing, see *How Writing Works: From the Birth of the Alphabet to the Rise of Social Media* (Wyse, 2017).

## The teaching of English

The establishment of state education as we know it can be conveniently traced back to the 1870 Elementary Education Act. Prior to that, the education of working-class children in the United Kingdom was largely in the hands of the voluntary sector: church schools, factory schools and, in the earlier part of the nineteenth century, schools run by the oppositional Chartist and Owenite Co-operative movements. The 1870 Act led to the establishment of free educational provision in elementary schools for all children from the age of 5 up to the age of 12. Education up to the age of 10 was compulsory, but if children had met the standards required they could be exempted from schooling for the final years. State schools and voluntary sector schools existed side-by-side from that date, a distinction that is still found today. Class differences were firmly established: the elementary and voluntary schools were schools for the labouring classes and the poor. The middle and upper classes expected to pay for the education of their children; secondary education in the form of grammar and public schools was not available to the bulk of the population.

The curriculum in the voluntary schools and later in the elementary schools was extremely limited. Writing meant copying or dictation (DES, 1967: 5601). Oral work involved such things as the children learning by heart from the *Book of Common Prayer*, which included: 'To order myself lowly and reverently to all my betters' and 'to do my duty in that state of life, unto which it shall please God to call me' (Williamson, 1981: 79).

The elementary schools emerged at a time when the government exerted considerable control over the curriculum through the 'Revised Code' established in 1862, better known as 'payment by results'. This was administered through frequent tests in reading, writing and arithmetic – the three Rs. If the children failed to meet the required standards, the grant was withdrawn and the teachers did not get paid. Under such conditions curriculum development was impossible, because schools had to focus so much on the tests in order to get paid (Lawson and Silver, 1973)

Though the code was abolished in 1895, and the statutory control of the curriculum relinquished in 1902, the effects lasted well into the twentieth century, leading one inspector to comment that 30 years of 'code despotism' meant that 'teaching remained as mechanical and routine ridden as ever (Holmes, 1922: 727)' (Gordon *et al.*, 1991: 278). Despite these criticisms, however, the introduction of universal compulsory education meant that literacy rates climbed steadily.

### 'English' as a subject, 1900–39

At the start of the twentieth century, the term 'English' referred to grammar only; reading and writing were not even seen as part of the same subject. A major landmark in the development of the subject was the Newbolt Report on 'The Teaching of English in England' (Board of Education, 1921). George Sampson, a member of the Newbolt committee, writing in the same year (1921), had identified the following 'subjects' still being taught in elementary schools across the land: 'oral composition, written composition, dictation, grammar, reproduction, reading, recitation, literature, spelling, and handwriting' (Shayer, 1972: 67). The Newbolt Report sought to change that and to bring together:

> under the title of English, 'taught as a fine art', four separate concepts: the universal need for literacy as the core of the curriculum, the developmental importance of children's self-expression, a belief in the power of English literature for moral and social improvement, and a concern for 'the full development of mind and character'.
>
> (Protherough and Atkinson, 1994: 7)

This was how English became established as a subject in the secondary curriculum and was placed at the centre of the curriculum for all ages. Famously, the Newbolt Report suggested, of elementary teachers, that 'every teacher is a teacher of English because every teacher is a teacher in English' (Shayer, 1972: 70). The committee recommended that children's creative language skills be developed. They recommended the study of literature in the elementary schools. In addition, they recommended the development of children's oral work, albeit in the form of 'speech training', which they saw as the basis for written work. Finally, they challenged the nineteenth-century legacy of educational class division, placing English at the centre of an educational aim to develop the 'mind and character' of all children.

Change on the ground was slow to occur, but it was happening. The old practice of reading aloud in chorus was disappearing, silent reading was being encouraged and, in the 1920s, textbooks were published that encouraged children's free expression and that questioned the necessity for formal grammar teaching. However, even though the Newbolt Report contained evidence of the uselessness of grammar teaching, the committee had the strong feeling that self-expression could go too far, and that the best way for children to learn to write was to study grammar and to copy good models.

## The Hadow Reports

The years 1926, 1931 and 1933 saw the publication of the three Hadow Reports on secondary, primary and infant education respectively; the second (Board of Education, 1931) focused on the 7–11 age range. It had a number of specific recommendations about the curriculum in general and English in particular. Famously, it stated: 'We are of the opinion that the curriculum of the primary school is to be thought of in terms of activity and experience rather than of knowledge to be acquired and facts to be stored' (Board of Education, 1931: 139).

In English, oral work was seen as important, with an emphasis on speaking 'correctly'. 'Oral composition' – getting the child to talk on a topic of their choice or one of the teacher's – was included. 'Reproduction' involved getting the child to recount the subject matter of the lesson they had just been taught. Class libraries were encouraged and silent reading recommended, although not in school time except in the most deprived areas. And the aim? 'In the upper stage of primary education the child should gain a sense of the printed page and begin to read for pleasure and information' (ibid.: 158).

As for writing, children's written composition should build on oral composition and children should be given topics that interested them. Spelling should be related to the children's writing and reading: 'Any attempt to teach spelling otherwise than in connection with the actual practice of writing or reading is beset with obvious dangers' (ibid.: 160). The abstract study of formal grammar was rejected, though some grammar was to be taught. Bilingualism was addressed in the Welsh context, and teaching in the mother tongue was recommended. Welsh-speaking children were expected to learn English and, strikingly, English-speaking children were expected to learn Welsh.

The third Hadow Report (Board of Education, 1933) drew on ideas current at the time to suggest that formal instruction of the three Rs traditionally started too early in British schools, and recommended that for infant and nursery children: 'The child should begin to learn the 3 Rs when he [sic] wants to do so, whether he be three or six years old' (ibid.: 133).

The report noted three methods of teaching reading that were used at the time: 'look and say', 'phonics' and more contextualised meaning-centred 'sentence' methods. It recommended that teachers use a mix of the three as appropriate to the child's needs. Writing should start at the same time as reading, and children's natural desire to write in imitation of the adult writing they saw around them at home or at school should be encouraged. The child should have control over the subject matter and his or her efforts should be valued by the teacher as real attempts to communicate meaning.

The report emphasised the importance of imaginative play, and noted, 'Words mean nothing to the young child unless they are definitively associated with active experience' (ibid.: 181), and 'Oral lessons should be short and closely related to the child's practical interests' (ibid.: 182). While 'speech training' was important, drama work was recommended for the development of children's language, and nursery rhymes and game songs were encouraged alongside traditional hymns. Stories should be told and read to the children.

The Hadow Reports read as remarkably progressive documents for their time, and the principles of child-centred education that are explicit in many of their recommendations continued to inform thinking in primary language teaching for the next 50 years.

## Progressive education ☞, 1931–75

The central years of the twentieth century can perhaps be characterised as the years of progressive aspiration so far as primary language was concerned. The progressive views of the Hadow Reports began to be reflected in the Board of Education's regular guidelines, and teachers were on the whole free to follow them as they pleased. The 1944 Education Act itself offered no curriculum advice, except with regard to religious education, and central guidance on the curriculum ended in 1945. The primary curriculum in particular came to be regarded as something of a 'secret garden', to quote Lord Eccles, Tory Minister of Education in 1960 (Gordon *et al.*, 1991: 287).

The 1944 Education Act finally established primary schools in place of elementary schools, though it would be another 20 years before the last school that included all ages of children closed. At secondary level, a three-layered system of grammar, technical and secondary modern schools was established, and a new exam, the 11+, was devised to decide which children should go where. Like the scholarship exam before it, the 11+ continued to restrain the primary language curriculum, particularly with the older children, despite the fact that more progressive child-centred measures were gaining ground with younger children. With the reorganisation of secondary schools along comprehensive lines in the 1960s (encapsulated in Circular 10/65), the 11+ was abolished and the primary curriculum was technically freed from all constraint.

In retrospect, the Plowden Report on primary education (DES, 1967) can be seen as centrally representative of the progressive aspiration of 'child-centred education'. Its purpose was to report on effective primary education of the time, and it was concerned to see to what extent the Hadow recommendations had been put into effect. It functioned as much to disseminate effective practice as it did to recommend future change. The child was central: 'At the heart of the educational process lies the child' (ibid.: para. 9); and language was crucial: 'Spoken language plays a central role in learning' (ibid.: para. 54) and 'The development of language is, therefore, central to the educational process' (ibid.: para. 55).

Like its predecessors, the report emphasised the importance of talk; like its predecessors, it emphasised the fact that effective teachers of reading used a mix of approaches. Drama work and story-telling were to be encouraged; the increased importance of fiction and poetry written for children and the development of school libraries were all emphasised. The report applauded wholeheartedly the development of personal 'creative' writing (→ Chapter 14) by the children, characterising it as a dramatic revolution (1967: para. 60.1). On spelling and punctuation the committee was more reticent, noting only that when inaccuracy impeded communication should steps then be taken to remedy the deficiencies (1967: para. 60.2). Knowledge about language was seen as an interesting new area, but 'Formal study of grammar will have little place in the primary school' (1967: para. 61.2).

The Plowden Report was followed by the Bullock Report on English (DES, 1975). So far as primary age children were concerned, this spelled out in more detail much of what was already implicit in Plowden. Central to both the reports was an emphasis on the 'process' of language learning. From such a perspective, children's oral and written language would best develop in meaningful language use. A couple of quotes from the Bullock Report will illustrate the point. Of the development of oral language, it suggested: 'Language should be learned in the course of using it in, and about, the daily experiences of the classroom and the home' (ibid.: 520). Where writing was concerned: 'Competence in language comes above all through its purposeful use, not through working of exercises divorced from context' (ibid.: 528).

So far as bilingual children and children from the ethnic minorities were concerned, the Plowden Report had already recognised the contribution that such children could make to the classroom, and the Bullock committee was concerned that such children should not find school an alien place:

> No child should be expected to cast off the language and culture of the home as he [*sic*] crosses the school threshold, nor to live and act as though home and school represent two totally separate and different cultures which have to be kept firmly apart. The curriculum should reflect many elements of that part of his life which a child lives outside school.
>
> (ibid.: para. 20.5)

## Increasing political control, 1976 onwards

The ideas of progressive education remained important – despite increasingly frequent attacks – until the 1970s, when things started to change. Britain was declining in world economic importance and the oil crisis of the early 1970s was followed by an International Monetary Fund (IMF) loan which saw the Labour government of the time having to cut back on public spending. Effective child-centred education is teacher-intensive and requires small classes, and the previous decades had seen reductions in class size. That was no longer compatible with the financial constraints of the time and class sizes began to increase again. A more regulated curriculum is easier to cope with in such circumstances.

The National Curriculum itself was established by the 1988 Education Reform Act, which in the process gave the Secretary of State for Education considerable powers of direct intervention in curriculum matters. Following the Act, curriculum documents were drawn up for all the major subject areas. In line with the recommendations of the TGAT Report (DES, 1987: S227), attainment in each subject was to be measured against a ten-level scale and tested at ages 7, 11, 14 and 16. As the curriculum was introduced into schools, it became clear that each subject group had produced documents of considerable complexity. Discontent in the profession grew and a slimmed-down version was introduced in 1995. The original English document was prepared by a committee under the chairmanship of Brian Cox (DES, 1989, 1990; Cox, 1991). English was to be divided up into five 'attainment targets': Speaking and Listening, Reading, Writing, Spelling and Handwriting. These were reorganised into three in Sir Ron Dearing's 1995 rewrite, as Spelling and Handwriting were incorporated into Writing (DfE, 1995).

During the mid- to late 1980s, a number of large-scale projects were undertaken which aimed to improve the teaching and learning of English. The Schools Council, a body responsible for national curriculum development, had been replaced by the School Curriculum Development Committee (SCDC); the SCDC initiated the National Writing Project. This was in two phases: the development phase took place from 1985 to 1988 and the implementation phase from 1988 to 1989, although the Education Reform Act 1988 and the resulting National Curriculum and testing arrangements changed the focus of implementation.

One of the key problems of the time was that many children were being turned off by writing, something confirmed by some evidence from the Assessment of Performance Unit (APU). The APU found that as many as four in ten children did not find writing an enjoyable experience and 'not less than one in ten pupils [had] an active dislike of writing and endeavour[ed] to write as little as possible' (APU, 1988: 170). Somewhat later the National Writing Project gathered evidence that many children, particularly young children, tended to equate writing with transcription skills rather than composition.

The National Writing Project involved thousands of educators across the country. One of the main messages from the project was that writers needed to become involved in writing for a defined and recognisable audience, not just because the teacher said so. Connected to these ideas was the notion that writing should have a meaningful purpose. With these key concepts in place, teachers began to realise that writing tasks which were sequentially organised in school exercise books and consisting of one draft – or at best 'rough copy/neat copy' drafts – were not helping to address the audiences and purposes for writing that needed to be generated.

The National Oracy Project was also initiated by SCDC and partly overlapped with the National Writing Project. During the period from 1987 to 1991, 35 local education authorities were involved in the oracy project. The recognition that oracy, or speaking and listening, as it came to be called, needed a national initiative was in itself significant. Since the late 1960s, a number of enlightened educators had realised that talking and learning were very closely linked and that the curriculum should reflect that reality. But these people were in a minority and most educators continued to emphasise reading and, to a lesser extent, writing. The major achievement of the oracy project was to secure recognition that talk was important and that children could learn more if teachers understood the issues and planned activities to support the development of oracy. As Wells pointed out: 'The centrality of talk in education is finally being recognised. Not simply in theory – in the exhortations of progressive-minded academics – but mandated at all levels and across all subjects in a national curriculum' (Wells, 1992: 283).

The other large national project that we will touch on is the Language in the National Curriculum (LINC) project (→ Chapter 16). In 1987, a committee of inquiry was commissioned to make recommendations about the sort of knowledge about language that it would be appropriate to teach in school. The Kingman Report, as it was known (DES, 1988), disappointed right-wing politicians and sections of the press when it failed to advocate a return to traditional grammar teaching. The Cox Report (DES, 1989) ran into similar problems for the same reason, but both the 1990 and the 1995 orders for English in the National Curriculum (DES, 1990; DfE, 1995) contented themselves with general recommendations to use grammatical terms where and as the need arose. Between 1989 and 1992 most schools in England were involved with the LINC project. Its main aim was to acquaint teachers with the model of language presented in the Kingman Report. Kingman's work reaffirmed the idea that children and teachers should have sufficient 'knowledge about language' or 'KAL' if they were to become successful language users.

One of the strong features of the materials that were produced by the LINC (1991) project was that they were built on an explicit set of principles and theories:

## Principles

1    Teaching children should start positively from what they can already do.
2    The experience of using language should precede analysis.
3    Language should be explored in real purposeful situations rather than be analysed out of context.
4    An understanding of people's attitudes to language can help you understand more about values and beliefs.

## Theories

1    Humans use language for social reasons.
2    Language is constantly changing.
3    Language is a cultural phenomenon.
4    There are important connections between language and power.
5    Language is systematically organised.
6    The meanings of language depend on negotiation.

It may have been that some of these philosophies resulted in the politicians of the time refusing to publish the materials. In spite of this, the materials were photocopied and distributed widely and various publications independent of government were produced, e.g. Carter (1990).

The National Literacy Project ☞ was developed between 1996 and 1998. The project's main aim was to raise the standards of literacy in the participating schools so that they raised their achievements in line with national expectations. The project established for the first time a detailed scheme of work with term-by-term objectives that were organised into text-level, sentence-level and word-level goals. These were delivered through the use of a daily literacy hour with strict timings for the different sections. The project was supported by a national network of centres where literacy consultants were available to support project schools.

The National Literacy Project was important because it was claimed that its success was the reason that the National Literacy Strategy adopted the ideas of a Framework for Teaching and a prescribed literacy hour. However, it should be remembered that the schools which were involved in the project were schools which had identified weaknesses in their literacy teaching, and this has to be taken into account when any kind of evaluation is made about the success of the project. The other important point to bear in mind is that it was originally conceived of as a five-year project; after that time, evaluations were to be carried out. One of the features of these evaluations was that they were supposed to measure the success of the three years of the programme when schools were no longer *directly* involved in the project. In the event, the approaches of the National Literacy Project were adopted as part of the National Literacy Strategy in 1998. This occurred *before* any independent evaluation had been carried out and long before the planned five-year extent of the National Literacy Project.

The only *independent* evaluation of the project, carried out by the National Foundation for Educational Research (INFER), found that:

> The analyses of the test outcomes have indicated that, in terms of the standardised scores on reading tests, the pupils involved in Cohort 1 of the National Literacy Project have made substantial gains. All three year groups showed significant and substantial increases in scores from the beginning to end of the project.
>
> (Sainsbury *et al.*, 1998: 21)

This outcome illustrates definite progress in the fairly restricted parameters of standardised reading tests. It is not possible to conclude that the specific approach of the National Literacy Project was more beneficial than other approaches as this variable was not controlled. It is possible that the financial investment, extra support and a new initiative were the dominant factors in improved test scores rather than the particular characteristics of the recommended teaching methods. One area of concern about the findings from the evaluation was that pupils eligible for free school meals, pupils with special educational needs, pupils with English as an additional language (EAL) at the 'becoming familiar with English stage' and boys made less progress than other groups.

It seems particularly regrettable, though not surprising, that no serious attempts were made to evaluate what pupils thought of the project. Sainsbury *et al.* admitted that:

> The reading enjoyment findings are less easy to interpret. The survey showed that children do, on the whole, enjoy their reading, with substantial majorities of both age groups expressing favourable attitudes both before and after involvement in the project. These measures, however, did not change very much, indicating that the systematic introduction of different text types that was a feature of the project did not have any clearly apparent effect on children's enjoyment of reading these varied text types. In the absence of a control group, however, it is difficult to draw any more definite conclusions.
>
> (ibid.: 27)

### The National Literacy Strategy 1997–2006 and the Primary National Strategy Framework for Literacy, 2007–10

The Literacy Task Force was established on 31 May 1996 by David Blunkett, then Shadow Secretary of State for Education and Employment. It was charged with developing, in time for an incoming Labour government, a strategy to substantially raise standards of literacy in primary schools over a five- to ten-year period (Literacy Task Force, 1997: 4).

The Literary Task Force produced a final report that suggested how a National Literacy Strategy could be implemented. The recommendations heralded some of the most profound changes to English teaching. The single most important driving force behind the strategy was the introduction of target-setting: specifically that by 2002, 80 per cent of 11 year-olds should reach the standard expected for their age in English (i.e. Level 4) in the Key Stage 2 National Standard Assessment Tasks (SATs). Despite all the many changes to the curriculum since 1997, target-setting, and the associated publication of league tables, remain in place and now have an even more dominant effect on the curriculum and children's daily lives.

Earlier in this chapter we mentioned the important contribution of Brian Cox in relation to developing the guidance for the subject of English in the National Curriculum, a document that achieved a remarkable consensus in such a contentious area. Cox was extremely critical from the inception of the National Literacy Strategy: the policy on reading 'is too prescriptive, authoritarian and mechanistic', there should be 'more emphasis on motivation, on helping children to enjoy reading' (Cox, 1998: ix). Other contributors to the book were equally critical: Margaret Meek (1998: 116) criticised the 'repeated exercises in comprehension, grammar and spelling' and Bethan Marshall (1998: 109) suggested that 'the bleak spectre of utilitarianism ☞ hangs over our schools like a pall'. The words of an inspector in 1905 quoted by Marshall are another reminder of the history of the reading debates:

> A blackboard has been produced, and hieroglyphics are drawn upon it by the teacher. At a given signal every child in the class begins calling out mysterious sounds: 'Letter A, letter A' in a sing-song voice, or 'Letter A says Ah, letter A says Ah', as the case may be. To the uninitiated I may explain that No. 1 is the beginning of the spelling, and No. 2 is the beginning of word building. Hoary-headed men will spend hours discussing whether 'c-a-t' or 'ker-ar-te' are the best means of conveying the knowledge of how to read 'cat'. I must own an indifference to the point myself, and sympathise with teachers not allowed to settle it for themselves . . . 'Wake up, Johnny; it's not time to go to sleep yet. Be a good boy and watch teacher.'
>
> (Marshall, 1998: 115)

Most political education initiatives are introduced following claims that standards are falling, and the National Literacy Strategy was no exception. However, in spite of regular claims by the media, teachers, business people, politicians, etc., there was no evidence that standards of literacy had declined in England, as Beard (1999) pointed out, something that Campbell (1997) also commented upon:

> On the current moral panic over the impact of the reforms on standards of attainment in literacy and numeracy, there are two things to say. First,

> no-one can be sure about standards in literacy and numeracy because of the failure – unquestioned failure – of the national agencies (NCC, SEAC and now SCAA) to establish an effective, credible and reliable mechanism for the national monitoring of standards over time since 1989.
>
> (Campbell, 1997: 22)

One of the first attempts to evaluate the strategies was commissioned by the New Labour government. Earl *et al.*'s (2003) evaluation of the NLS and NNS (National Numeracy Strategy) included collection of data from schools as follows: a) two postal surveys (in 2000 and 2002), each to two samples of 500 schools, one for literacy and the other for numeracy. Parallel questionnaires went to head teachers and teachers; b) a postal survey to all literacy and numeracy consultants in LEAs (Local Education Authority) across England in 2002; c) repeated visits to ten selected schools (with various sizes, locations, pupil populations, levels of attainment) and their LEAs: four to six days in each school. The research team interviewed head teachers and teachers, observed literacy and mathematics lessons and analysed documents; d) interviews with literacy and numeracy managers and consultants from LEAs of the ten selected schools. The researchers also attended training sessions and staff meetings in some of those LEAs; and e) observations and interviews in 17 other schools (including special schools) and LEAs. Three of these were one-day visits to schools early in 2000, while the others were single visits as part of shadowing regional directors or HMI (Her Majesty's Inspectorate) or attending meetings locally.

Earl *et al.* (2003) found that the strategies had altered classroom practice: in particular, greater use of whole class teaching, more structured lessons and more use of objectives to plan and guide teaching. Teachers' views about the strategies were more variable than head teachers', who were more likely to be in favour. Head teachers and teachers were more supportive of the NNS than they were of the NLS. For the most part, both teachers and head teachers believed that the NNS had been easier to implement and had had greater effects on pupil learning than the NLS. Overall, Earl *et al.* reported a wide range of variation in teachers' opinion of the NLS ranging from positive to negative.

Non-government-commissioned research explored a range of issues in relation to the strategies. For example, a series of research studies all reported that the recommended pedagogy of the NLS literacy hour was resulting in rather limited teacher–pupil interaction, which was tending towards short initiation-response sequences and a consequent lack of extended discussion. Observation schedules were used in studies such as those by Hardman *et al.* (2003), English *et al.* (2002) and Mroz *et al.* (2000). Mroz *et al.* (2000) noted the limited opportunities for pupils to question or explore ideas. English *et al.* (2002) found that there was a reduction in extended teacher–pupil interactions. Hardman *et al.* (2003) found that the NLS was encouraging teachers to use more directive forms of teaching, with few opportunities for pupils to explore and elaborate on their ideas. Skidmore *et al.* (2003) used audio recordings of teacher–pupil dialogue combined with video of non-verbal communication to support their finding that teachers were dominating interaction during the guided reading segment of the literacy hour. Parker and Hurry (2007) interviewed 51 Key Stage 2 teachers in 2001 and videotaped observations of the same teachers in class literacy sessions, focusing on teacher and pupil questions and answers They found that direct teacher questioning in the form of teacher-led recitation was the dominant strategy used for reading comprehension teaching and that children were not encouraged to generate their own questions about texts. Lefstein's (2008: 731) extended case study of one primary school found that open questions were suppressed as a result of 'teacher knowledge and policy support, conditions of teacher engagement with the curricular materials, and the durability of interactional genres'.

The answer to the question of whether the NLS Framework for Teaching and its pedagogy was effective is made difficult to answer because it was not subject to rigorous large-scale experimental trial. However there is now a significant amount of evidence in general about the effectiveness of the NLS: Wyse *et al.* (2010) summarised this in their research for the *Cambridge Primary Review* by analysing studies of primary classrooms and trends in national test outcomes. Although reading showed slightly better gains than writing according to some sources, the overall trend in national test scores can be explained as modest gains from a low base as teachers learned to prepare pupils for statutory tests followed by a plateau in scores as no further gains could be achieved by test coaching. Overall, the intense focus on testing and test results in the period of the NLS resulted in a narrowing of the curriculum, driving teaching in the opposite direction to that which research indicates will improve learning and attainment.

In October 2006, the new *PNS (Primary National Strategy) Framework for Literacy* was released. The main elements that had been a feature of the NLS were still part of the PNS Framework. Teachers were offered a little more flexibility in some areas, such as in the teaching of writing, where longstanding criticisms finally and belatedly began to have an effect on policy makers. But overall the PNS Framework was little changed from the NLS. However, in 2010 a new period of radical reform began.

### Ideology or evidence?: 2010 to 2017

The election of the Conservative–Liberal Democrat coalition government in 2010 in the UK brought with it radical, and in some cases immediate, change. A new National Curriculum for primary schools that the previous New Labour government had started to implement in 2010, published on an extensive, fully functional website, was simply taken down and archived along with all the PNS resources and most other educational materials that had been developed during the New Labour terms. The ideological rationales for the rejection of previous materials were the dual promises of more freedom over the curriculum for teachers and less bureaucracy. At the same time, the new government announced yet another review of the National Curriculum (the third government-commissioned review of the primary curriculum in a period of less than ten years – for the first book-length analysis of this and other national curriculum developments in the countries of the UK, see *Creating the Curriculum*, Wyse *et al.*, 2012).

England's National Curriculum, implemented from 2014 onwards, represents the strongest grip by government on content and on pedagogy in the history of the National Curriculum. The politics and policies that led to a narrower specification of the subject English, including less attention to spoken language and more attention to elements such as formal grammar knowledge and synthetic phonics, began with a government white paper in 2010 that included the commitment to 'Review and reform the National Curriculum so that it becomes a benchmark outlining the knowledge and concepts pupils should be expected to master to take their place as educated members of society' (Department for Education, 2010: 41). The link between statutory assessment, the curriculum and school accountability was also made clear: 'The National Curriculum will continue to inform the design and content of assessment at the end of key stage two, which will apply to every child and which will provide a guide to the performance of primary schools' (op. cit.: 42). After publication of the white paper, the government commissioned a review of assessment in England led by Lord Bew. Bew's final report noted that

> there are some elements of writing – spelling, grammar, punctuation, vocabulary – where there are clear 'right' and 'wrong' answers, which lend themselves to externally-marked testing . . . Internationally a number of jurisdictions conduct externally-marked tests of spelling, punctuation and grammar . . . These are essential skills and **we recommend that externally-marked tests of spelling, punctuation, grammar and vocabulary should be developed.**
>
> (Bew, 2011: 60; bold font in original)

Some new freedoms over the curriculum in England became available to an extent, but paradoxically at the same time even more control over key and influential elements of schooling were being placed in the hands of the Secretary of State for Education and the Minister for Schools (Wyse, 2008 and 2011 feature analysis of the implications for teaching of such control). In relation to the teaching of English, Language and Literacy, a number of worrying developments took place. One was the decision to limit externally marked statutory assessment of writing for pupils aged 10–11 to the areas of spelling, grammar and punctuation; compositional aspects of writing were to be assessed by teachers. But even more troubling was the growing dominance of a single method of reading teaching called 'synthetic phonics' ☞. As you will see in Chapter 8, phonics teaching has a long history of debate. One of the trends we have identified in successive editions of our book is the growing level of control over the teaching of reading. The influence of synthetic phonics has become so pervasive that it is part of teacher training, the vetting of publishers' teaching resources, selection of advisors to government and many other areas of policy. In opposition, Nick Gibb MP was at the forefront of these trends and, as the Minister for Schools for the coalition government, intensified this work supported by Secretary of State for Education Michael Gove.

The most questionable development of all was the decision to implement a 'phonics screening check' for all 6-year-old children in England. As you will see in Chapter 21, we are not against early identification of children who struggle with reading, and have clear ideas how they can best be supported. But we are very much against the distortion of the curriculum and the effects on children, parents, teachers, schools and teacher trainers that the national testing regime in England has had.

The government response to a public consultation on whether the phonics screening check should be implemented was startling. The first problem with the consultation questionnaire design was that it did not include the question: do you think a phonics screening check should be implemented for all 6-year-old children in England? The only question in the consultation that that came close to addressing the most important issue of whether the test was desirable overall or not was this one: 'do you agree that this screening check should be focused on phonic decoding?' (DfE, 2011: 12) The response rates were: Yes: 28 per cent; No: 66 per cent; Not Sure: 6 per cent. This clearly showed a majority negative response to the main element of the phonics screening check. But in an extraordinary interpretation of the outcome of the survey, the Minister concluded that   .

> 28% of respondents agreed the check should focus on decoding using phonics. 20% respondents argued that children learn to read using a variety of strategies, including using visual and context cues, and the check should take into account these alternative strategies.
>
> (op. cit.: 4)

Therefore, 'Taking into account the consultation responses, findings from the pre-trialling and the academic evidence, we propose *to continue to develop the phonics screening check*' (op. cit.: 6, emphasis added). Surely this kind of interpretation and the resultant policy changes are undemocratic and unacceptable? Quite apart from the political issues around the phonics screening check, research continues to reveal other problems, including that the check is not a valid or reliable measure of reading (e.g. see Darnell *et al.*, 2017).

One of the most serious problems in the development of the national curriculum was the lack of attention paid to research and other evidence. When reflecting on his time as a Minister for Schools under Secretary of State for Education Michael Gove, David Laws claims that decisions were made 'not based on evidence but on hunch' (Wilby, 2017, online) and that Gove had a particular weakness for basing decisions on 'ideology and personal experience' (op. cit.). Corroboration of the lack of attention to evidence was detailed by BERA President Mary James (BERA, 2012), one of the expert group advising on the national curriculum.

Unfortunately, the government's questionable response to the outcomes of the consultation on the phonics screening check was replicated in the public consultation on the proposals for the new National Curriculum, held between February and April 2013. The consultation attracted 17,312 respondents, with 4,576 described as 'non-campaign respondents' and 12,736 described as 'campaign respondents' (i.e. organisations devoted to a particular issue. The report of the consultation made clear that campaign responses were not included in the percentages of answers to questions but were reflected in the commentaries about the answers). The question 'Do you have any comments on the content set out in the draft programmes of study?' was addressed by 3,682 respondents. With regard to the teaching of the subject English, and the teaching of grammar within that subject, 'There was recognition that the teaching of phonics, punctuation, spelling and grammar was necessary, but some felt that there was an over-emphasis on these aspects' (Department for Education, 2013: 7). It is disappointing that the number of respondents who replied about grammar was not specified in the report, as this would have provided some further evidence relevant to the strength of opinion on this issue.

There was also a follow up consultation, open from July to August 2013, on the draft legislative order, which attracted further comment about English and grammar. Although 21 respondents (11 per cent) supported the greater focus on spelling, grammar and punctuation,

> a total of 36 respondents (19%) however expressed concern in relation to the more demanding grammatical content included for years 2 and 4 . . . 52 respondents (28%) said the English primary curriculum was too prescriptive, in particular in reference to the level of specification in the appendices [where the grammatical knowledge to be learned by pupils is specified]. These respondents argued that this undermined the aims of the new national curriculum in relation to greater professional freedom and were concerned that this may have implications for the provision of a balanced and broadly based school curriculum.
>
> (Department for Education, 2013: 6)

One interpretation of these data in the second consultation is that 47 per cent of respondents were critical of the grammar specified in the National Curriculum and its appendices, but 11 per cent thought the emphasis on correct use of Standard English was commendable. An overall negative response to the proposed attention to grammar did not result in changes to this element of the national curriculum.

In addition to the emphasis in the National Curriculum programmes of study, the national statutory tests for 11-year-old pupils in England included for the first time in 2011 a separate spelling, punctuation and grammar test where formal grammar was further emphasised. In addition, the requirements for teacher assessment of writing included a strong emphasis on grammar as part of the assessment criteria. In 2016 these emphases were still in place. For example, the national statutory test for Spelling, Punctuation and Grammar included a strong emphasis on formal grammar, including questions that required knowledge of grammatical terminology (for example, '27. Underline the subordinate clause in each sentence below'; UK Government, 2016: 17). All questions in the paper attracted one mark each. Although the 2016 criteria for statutory teacher assessment of writing, produced by pupils in lessons, included aspects such as 'creating atmosphere' in their writing, there was a strong emphasis on usage according to areas of formal grammar such as 'passive and modal verbs' and 'adverbs, preposition phrases and expanded noun phrases', etc. (Standards and Testing Agency, 2015).

Reservations about the nature of the specifications for grammar teaching in the National Curriculum and its associated statutory testing continued to cause disagreement. The main government advisor for grammar in the statutory assessment system described the process of determining the curriculum for grammar as 'chaotic' and that 'We started off with the primary curriculum, which we were a bit unconfident about as none of us had much experience of primary education' (Mansell, 2017). In April 2017, a House of Commons Education Select Committee report on assessment in primary schools concluded that:

> One issue with the writing assessment is the focus on technical aspects, like grammar and spelling, over creativity and composition. We are not convinced that this leads directly to improved writing and urge the Government to reconsider this balance and make spelling, punctuation and grammar tests non-statutory at Key Stage 2.
>
> (House of Commons Education Committee, 2017: 3)

Finally, and not until 2018, government did announce some changes to the statutory assessment system that responded to the growing criticisms from educationists and from significant sections of the general public (see → Chapter 20 for these changes).

So, the paradox of policy from 2010 onwards was that some schools (particularly those that engaged with other government agendas such as becoming free schools or academies) were to have some freedom over the curriculum that was lacking during New Labour's time in government. But at the same time, the Department for Education and its Ministers were securing even greater control over early years and primary teaching as a result of the emphases in the statutory testing regimes.

Since this book's first edition in 2001 we have tracked the changes in national curriculum and assessment policy. The trajectory has been one of increasing control by government. This level of control is largely unique to England: even the other countries of the UK do not have these same circumstances (although there was some evidence that the historically democratic vision of education shared by politicians, academics and teachers alike was under threat as a result of calls for more high stakes assessment). However, in spite of this central control, there are some schools who maintain evidence-informed, progressive philosophies of primary and early years education. The teachers, supported by parents, focus rigorously on motivating their pupils and maintain professional principles in the face of 'testing times'. Professional environments like this can be exciting places to work. Future governments would be wise to build new national curricula and assessment systems that are derived from the best evidence-informed practice as a result of genuinely collaborative development of policy over more sustained periods of time.

### Practice points

- As a professional, you should evaluate all educational initiatives critically to ensure that they reflect the needs of the children that you teach.
- You need to develop a knowledge of historical developments as a vital tool for understanding educational change.
- Use the range of opportunities that are available to give your feedback about national developments in education, for example by responding to consultations, and contacting your local and national political representatives.

### Glossary

**Dialects** – regional variations of language shown by different words and grammar.

**Etymology** – the origins of words.

**Line justification** – ensuring that the beginnings and ends of lines of print are all lined up.

**Loanwords** – words adopted from other languages.

**National Literacy Project** – a three-year professional development project that was carried out with authorities and schools who wanted to raise their standards of literacy.

**National Literacy Strategy** – a national strategy for raising standards in literacy over a five- to ten-year period.

**Progressive education** – teaching approaches that rejected old-fashioned rote-learning methods in favour of methods that put the child's interests and needs first.

**Sui generis** – of its own kind; unique.

**Standard English** – the formal language of written communication in particular. Many people call this 'correct' English.

**Synthetic phonics** – an approach to teaching reading that prioritises, first and foremost, the teaching of letters and associated phonemes (sounds).

**Utilitarianism** – the idea that education and learning can be reduced to crude skills and drills.

## References

APU (Assessment of Performance Unit). (1988). *Language Performance in Schools: Review of APU Language Monitoring 1979–1983.* London: HMSO.

Barber, C. (1993). *The English Language: A Historical Introduction.* Cambridge: Cambridge University Press.

Beard, R. (1999). *National Literacy Strategy Review of Research and Other Related Evidence.* London: DfEE.

Bew, P. (2011). *Independent Review of Key Stage 2 Testing, Assessment and Accountability: Final Report.* London: Department of Education.

Board of Education. (1921). *The Teaching of English in England (The Newbolt Report).* London: HMSO.

Board of Education. (1931). *The Primary School (The Second Hadow Report).* London: HMSO.

Board of Education. (1933). *Infant and Nursery Schools (The Third Hadow Report).* London: HMSO.

British Educational Research Association (BERA). (2012). 'Background to Michael Gove's response to the report of the expert panel for the national curriculum review in England. Retrieved from www.bera.ac.uk/promoting-educational-research/issues/background-to-michael-goves-response-to-the-report-of-the-expert-panel-for-the-national-curriculum-review-in-england

Bryson, B. (1990). *Mother Tongue: The English Language.* London: Penguin.

Campbell, J. (1997). 'Towards curricular subsidiarity?' Paper presented at the School Curriculum and Assessment Authority Conference, 'Developing the Primary School Curriculum: The Next Steps', June.

Carter, R. (ed.). (1990). *Knowledge About Language and the Curriculum: The LINC Reader.* London: Hodder & Stoughton.

Cox, B. (1991). *Cox on Cox: An English Curriculum for the 1990s.* London: Hodder & Stoughton.

Cox, B. (1998). 'Foreword', in B. Cox (ed.), *Literacy Is Not Enough: Essays on the Importance of Reading.* Manchester: Manchester University Press and Book Trust.

Crystal, D. (1997). *The Cambridge Encyclopedia of Language*, 2nd edn. Cambridge: Cambridge University Press.

Crystal, D. (2004). *The Stories of English*. London: Penguin and Allen Lane.

Darnell, C., Solity, J. and Wall, H. (2017). 'Decoding the phonics screening check', *British Educational Research Journal*, 43(3): 505–527.

Department for Education (DfE). (1995). *English in the National Curriculum*. London: HMSO.

Department for Education (DfE). (2010). *The Importance of Teaching: The Schools White Paper 2010*. Norwich: The Stationery Office.

Department for Education (DfE). (2011). *Response to Public Consultation on the Year 1 Phonics Screening Check*. London: Department for Education.

Department of Education (DfE). (2013). *Reform of the National Curriculum in England: Report of the Consultation Conducted February–April 2013*. London: Department of Education.

Department of Education and Science (DES). (1967). *Children and Their Primary Schools (The Plowden Report)*. London: HMSO.

Department of Education and Science (DES). (1975). *A Language for Life (The Bullock Report)*. London: HMSO.

Department of Education and Science (DES). (1988). *Report of the Committee of Inquiry into the Teaching of English Language (The Kingman Report)*. London: HMSO.

Department of Education and Science and The Welsh Office (DES). (1987). *National Curriculum Task Group on Assessment and Testing (The TGAT Report)*. London: DES.

Department of Education and Science and The Welsh Office (DES). (1989). *English for Ages 5–16 (The Cox Report)*. York: National Curriculum Council.

Department of Education and Science and The Welsh Office (DES). (1990). *English in the National Curriculum*. London: HMSO.

Earl, L., Watson, N., Levin, B., Leithwood, K., Fullan, M., Torrance, N. *et al.* (2003). *Watching and Learning: OISE/UT Evaluation of the Implementation of the National Literacy and Numeracy Strategies*. Nottingham: DfES Publications.

English, E., Hargreaves, L. and Hislam, J. (2002). 'Pedagogical dilemmas in the National Literacy Strategy: Primary teachers' perceptions, reflections and classroom behaviour', *Cambridge Journal of Education*, 32(1): 9–26.

Goody, J. and Watt, I. (1963). 'The consequences of literacy', *Comparative Studies in Society and History*, 5(3): 304–345.

Gordon, P., Aldrich, R. and Dean, D. (1991). *Education and Policy in England in the Twentieth Century*. London: Woburn.

Greenbaum, S. (1986). 'Spelling variants in British English', *Journal of English Linguistics*, 19: 258–268.

Hardman, F., Smith, F. and Wall, K. (2003). 'Interactive whole class teaching in the National Literacy Strategy', *Cambridge Journal of Education*, 33(2): 197–215.

Harris, R. (1986). *The Origin of Writing*. London: Duckworth.

Holmes, E. A. G. (1922). 'The confessions and hopes of an ex-inspector of schools', *Hibbert Journal*, 20 (no further information in secondary source). Quoted in Gordon, P., Aldrich, R. and Dean, D. (1991). *Education and Policy in England in the Twentieth Century*. London: Woburn.

House of Commons Education Committee. (2017). *Primary Assessment: Eleventh Report of Session 2016–17: Report, Together with Formal Minutes Relating to the Report*. London: House of Commons.

Lawson, J. and Silver, H. (1973). *A Social History of Education in England*. London: Methuen.

Lefstein, A. (2008). 'Changing classroom practice through the English National Literacy Strategy: A micro-interactional perspective', *American Educational Research Journal*, 45(3): 701–737. doi: 10.3102/0002831208316256

LINC (Language in the National Curriculum). (1991). *Materials for Professional Development*. No publication details.

Literacy Task Force. (1997). *The Implementation of the National Literacy Strategy*. London: DfEE.

Mansell, W. (Producer). (2017, 9 May). 'Battle on the adverbials front: Grammar advisers raise worries about SATS tests and teaching'. Retrieved from www.theguardian.com/education/2017/may/09/fronted-adverbials-sats-grammar-test-primary

Marshall, B. (1998). 'English teachers and the third way', in B. Cox (ed.), *Literacy Is Not Enough: Essays on the Importance of Reading*. Manchester: Manchester University Press and Book Trust.

Meek, M. (1998). 'Important reading lessons', in B. Cox (ed.), *Literacy Is Not Enough: Essays on the Importance of Reading*. Manchester: Manchester University Press and Book Trust.

Michel, J.-B., Shen, Y., Aiden, A., Veres, A., Gray, M., Pickett, J., . . . Aiden, E. (2010). 'Quantitative analysis of culture using millions of digitized books', *Science*, 331(6014): 176–182.

Mroz, M., Smith, F. and Hardman, F. (2000). 'The discourse of the literacy hour', *Cambridge Journal of Education*, 30(3): 380–390.

OED. (2017). Recent updates to the OED. Retrieved from http://public.oed.com/the-oed-today/recent-updates-to-the-oed/

Parker, M. and Hurry, J. (2007). 'Teachers' use of questioning and modelling comprehension skills in primary classrooms', *Educational Review*, 59(3): 299–314.

Protherough, R. and Atkinson, J. (1994). 'Shaping the image of an English teacher', in S. Brindley (ed.), *Teaching English*. London: Routledge.

Pyles, T. and Algeo, J. (1993). *The Origins and Development of the English Language*, 4th edn. London: Harcourt Brace Jovanovich.

Sainsbury, M., Schagen, I., Whetton, C., Hagues, N. and Minnis, M. (1998). *Evaluation of the National Literacy Project Cohort 1, 1996 1998*. Slough: NFER.

Shayer, D. (1972). *The Teaching of English in Schools 1900–1970.* London: Routledge and Kegan Paul.

Skidmore, D., Perez-Parent, M. and Arnfield, D. (2003). 'Teacher–pupil dialogue in the guided reading session', *Reading Literacy and Language*, 37(2): 47–53.

Standards and Testing Agency. (2015). *2016 National Curriculum Assessments: Interim Teacher Assessment Frameworks at the End of Key Stage 2.* London: Standards and Testing Agency.

UK Government. (2016). *2016 Key Stage 2 English Grammar, Punctuation and Spelling: Paper 1: Questions.* London: UK Government.

Wells, G. (1992). 'The centrality of talk in education', in K. Norman (ed.), *Thinking Voices: The Work of the National Oracy Project.* London: Hodder & Stoughton.

Wilby, P. (2017, 1 August). 'David Laws: "The quality of education policymaking is poor"', *The Guardian.* Retrieved from www.theguardian.com/education/2017/aug/01/david-laws-education-policy-schools-minister-thinktank-epi?CMP=Share_iOSApp_Other

Williamson, B. (1981). 'Contradictions of control: Elementary education in a mining district 1870–1900', in L. Barton and S. Walker (eds.), *Schools, Teachers and Teaching.* Lewes: Falmer Press.

Wyse, D. (2008). 'Primary education: Who's in control?', *Education Review*, 21(1): 76–82.

Wyse, D. (2011). 'Control of language or the language of control? Primary teachers' knowledge in the context of policy', in S. Ellis and E. McCartney (eds.), *Applied Linguistics and the Primary School.* Cambridge: Cambridge University Press.

Wyse, D. (2017). *How Writing Works: From the Invention of the Alphabet to the Rise of Social Media.* Cambridge: Cambridge University Press.

Wyse, D., Baumfield, V., Egan, D., Gallagher, C., Hayward, L., Hulme, M., . . . Lingard, B. (2012). *Creating the curriculum.* London: Routledge.

Wyse, D. and Cowan, K. (2017). *The Good Writing Guide for Education Students*, 4th edn. London: Sage.

Wyse, D., McCreery, E. and Torrance, H. (2010). 'The trajectory and impact of national reform: Curriculum and assessment in English primary schools', in R. Alexander, C. Doddington, J. Gray, L. Hargreaves and R. Kershner (eds.), *The Cambridge Primary Review Research Surveys.* London: Routledge.

## Annotated bibliography

Cox, B. (ed.). (1998). *Literacy Is Not Enough: Essays on the Importance of Reading.* Manchester: Manchester University Press and Book Trust.
This was a powerful rejection of the concept of the Literacy Strategy. Many of its criticisms can be levelled at the PNS Framework.
L2 ★★

Goodwyn, A. and Fuller, C. (eds.). (2011). *The Great Literacy Debate: A Critical Response to the Literacy Strategy and the Framework for English.* London: Routledge.

Dubbed 'the revenge of the professors' by the publisher! An excellent account of the NLS.

**L2** ★★

Wyse, D. (2017). *How Writing Works: From the Invention of the Alphabet to the Rise of Social Media.* Cambridge: Cambridge University Press.

Exploration of writing, its processes and ultimately how it can be learned and taught better through the life course.

**L3** ★★★

National Literacy Trust www.literacytrust.org.uk/

A very useful site to find out about new initiatives and stories in the press about literacy and the teaching of English.

**L1** ★

# Chapter 2

# Thinking about learning and language

This book is based on our theories of the teaching of English, language and literacy (TELL). The chapter identifies some of the most important general theories and principles that underpin the other chapters in the book. Its interdisciplinary orientation draws on philosophical, cognitive, socio-cultural and linguistic perspectives.

Some people might question the need for theory ☞ at all. They could argue that becoming a teacher simply requires the learning of teaching techniques, that theory and research are not important, and all that needs to be learned can be learned through school practice. Of course, it is not accurate to suggest that professional practice is simply that – practice. Teachers' theories reveal themselves all the time, sometimes in turns of phrase: 'They've got no language, these kids' (deficit models); 'She's a bright girl' (nature more than nurture); 'Boys are always naughty' (gender and stereotypes); etc. A particularly well-known phrase is to call children 'able' or not. If you think about this more deeply, the idea suggests an innate level of intelligence that is not going to change. This is another deficit theory which can lead to low expectations of children. For this reason, we prefer to avoid altogether the general description of a child as being 'bright', 'clever' or 'able', and instead talk about the child's specific achievements. Theories and beliefs directly guide the practical decisions that you make all the time. They are particularly significant in guiding your decision-making in unfamiliar situations: you will encounter these throughout your career, and without previous practical experience of a given situation, you will have to make decisions based on your theories.

This chapter explores the main theoretical influences that inform the book as a whole, and our approach to the teaching of English, language and literacy. The central focus of this book is an educational one: the teaching and learning of English, language and literacy in its pedagogical and social context. But to understand this and any educational topic sufficiently requires understanding of a particular kind. In this book, evidence from education and a range of other disciplines has been fashioned through the practical considerations that are most relevant to teaching and learning.

## Thinking about education

Matters of theory and the links with application and practice are part of all learning, not just education, and are rooted in a powerful historical tradition. The ancient Greek philosopher Aristotle established the concepts of *technē* and *phronēsis* (or technical and practical reason). *Technē* is the kind of thinking required by the builder or the doctor when they make something or restore someone to good health. This is the thinking required for making things. *Phronēsis* is a different kind of practical knowledge that emerges from conduct in a public space and is more personal and experiential. Practice, in a range of occupations, was seen by Aristotle 'as something nontechnical but not, however, nonrational' (Dunne, 1993).

Although the philosophies of the ancient Greeks might seem rather distant, more recent work has related the thinking to the modern practice of teaching. In his book, Joseph Dunne argued for the modern relevance of technical and practical reason by engaging in a written 'dialogue' with five modern philosophers in consideration of the training of teachers. He concluded overall that:

> In being initiated into the practice of teaching, student-teachers need not only experience in the classroom but also the right conditions for reflecting on this experience – so that reflectiveness (which we have all the time been clarifying under the name of 'phronesis') can become more and more an abiding attitude or disposition.
>
> (ibid.: 369)

Dunne's exploration of *technē* and *phronēsis* was initially stimulated by a concern that behaviourism, most visibly applied as the objective-led teaching approach, was an inappropriate way to conceptualise and realise teaching and learning (for a critique of objective-led teaching, see Gallagher and Wyse, 2012). This kind of meaningful connection between modern educational problems and philosophy is seen at its most powerful through pragmatism. John Dewey's work is particularly illuminating. Language, or communication, was central to Dewey's philosophy more generally. Dewey regarded communication, language and discourse as a natural bridge between existence and essence (Dewey, 1925/1998). Dewey's emphasis on communication as a metaphysics of existence was ground-breaking because he started from a *process* as a philosophical origin rather than a substance as an origin (Biesta, 2013).

Unlike most philosophers, Dewey addressed education, and curriculum, explicitly in his work. Consistent with his idea about language as central to existence, Dewey regarded interaction between teachers and pupils as key to education:

> The fundamental factors in the educative process are an immature, undeveloped being; and certain social aims, meanings, values incarnate in the matured experience of the adult. The educative process is the due interaction of these forces. Such a conception of each in relation to the other as facilitates completest and freest interaction is the essence of educational theory.
>
> (Dewey, 1902: 4)

Dewey's main thesis was that good teaching is built on the educator's understanding that there should be an interaction between the child's experiences and ideas and the school's aim to inculcate learning. Less effective learning takes place if, instead of interaction, an opposition is built between experience and learning. Over-emphasis on transmission of facts to be learned from a formal syllabus is one example of such opposition. Dewey was clear that the best knowledge available to society was the appropriate material for children's learning but only through teaching that made a connection with children's experiences and thoughts. He summed up the main role of the teacher by arguing that: 'Guidance [i.e. by the teacher] is not external imposition. *It is freeing the life-process for its own most adequate fulfilment*' (op. cit.: 17; italics in original).

## Language and thinking

Philosophy has long explored the nature of thinking and its relationship to language. But this relationship has also been explored from other perspectives, including psychological ones. Lev Vygotsky's work was initially located in psychology, but one of the interesting aspects of his contribution was the fact that he viewed psychology as a tool or method rather than as a subject of investigation. Vygotsky's subjects of investigation were culture and consciousness, as the editor's introduction to one of his most well-known books explains:

> Vygotsky argued that psychology cannot limit itself to direct evidence, be it observable behaviour or accounts of introspection. Psychological inquiry is *investigation*, and like the criminal investigator, the psychologist must take into account indirect evidence and circumstantial clues – which in practice means that works of art, philosophical arguments, and anthropological data are no less important for psychology than direct evidence.
>
> (Vygotsky, 1986: xvi)

Vygotsky's contribution to understanding of the development of higher psychological processes in humans includes his important concept of 'indirect (mediated) activity', consisting of an interplay between *sign* and *tool*. In explaining mediation, Vygotsky drew on Karl Marx's idea of tool as an instrument of labour. The tool, as a physical entity, is used by humans to 'make other substances subservient to his aims' (Marx, 1887: 128): implements for food preparation and weapons being two examples from the dawn of human beings. Vygotsky and Marx cited Hegel's broader general idea that the exercise of reason (the mental capacity) is a mediational tool that results in objects acting and reacting without direct interference in the process by the person doing the reasoning. The tool is externally oriented activity that leads to changes in objects and hence allows humans to control nature, whereas the sign is internally oriented, allowing humans to control themselves through their behaviour as a means of mastering oneself. As Vygotsky said:

> The very essence of human memory consists in the fact that human beings actively remember with the help of signs. It may be said that the basic characteristic of human behaviour in general is that humans personally influence their relations with the environment and through that environment personally change their behaviour, subjugating it to their control. It has been remarked that the very essence of civilization consists of purposely building monuments so as not to forget. In both the knot [as an aid to memory] and the monument we have manifestations of the most fundamental and characteristic feature distinguishing human from animal memory.
>
> (Vygotsky, 1978: 51)

Although tool and sign are different, their use is linked as mediated activity. For Vygotsky, writing as a *sign* was seen first in the marks made by early human beings, for example carvings on sticks to represent meaning. One example he gave was of early humans in Borneo developing a particular kind of stick for digging or hoeing. The bottom part of the stick for Vygotsky represented tool (literally and metaphorically). The top part of the stick had an additional small stick attached that made a sound that acted as a call to work.

> This intertwining of sign and tool which found its concrete symbolic expression in a primitive hoeing stick shows how early the sign (and later, its highest form, the word) begins to participate in the use of tools by man, and how early it begins to fulfil a highly specific function, to be compared with nothing else in the general structure of these operations that stand at the very beginning of the development of human labour. This stick is fundamentally different from that used by apes, although without doubt they are related to each other genetically.
>
> (Vygotsky and Luria, 1994: 56)

The function of the small stick was therefore a sign because it acted as a form of communication.

Another of Vygotsky's best-known ideas was the 'zone of proximal development'. He recognised that most psychological experiments assessed the level of mental development of children by asking them to solve problems in standardised tests. He showed that a problem with this was that this testing measured only a summative aspect of development. In the course of his experiments, Vygotsky discovered that a child who had a mental age of 8 as measured on a standardised test was able to solve a test for a 12-year-old child if they were given 'the first step in a solution, a leading question, or some other form of help' (Vygotsky, 1986: 187). He suggested that the difference between the child's level of problem-solving while working alone and the child's level with some assistance should be called the zone of proximal development (ZPD). He found that those children who had a greater zone of proximal development did better at school.

There are a number of practical consequences to ZPD. Vygotsky's ideas point to the importance of appropriate interaction, collaboration and cooperation. He suggested that, given minimal support, the children scored much higher on the tests. All teachers must make decisions about the kind of interventions that they make. Although the tests showed the influence of appropriate support, they also remind us that collaboration is an important way of learning and that in the right context, there is much that children can do *without* direct instruction.

If we accept the idea of ZPD, it leaves a number of questions about how teacher interaction can best support pupils' learning within the ZPD. The term 'scaffolding' has become common in discussions about literacy teaching. For example, it is often said that teachers should 'model' and 'scaffold' aspects of writing. Unfortunately, the didactic context for these recommendations is not the same as the original concept of scaffolding. The term 'scaffolding' emerged as early as 1976 in the work of David Wood, Jerome Bruner and Gail Ross, who argued that scaffolding requires the following functions of a 'tutor': recruitment (of the learner's interest); reduction in degrees of freedom (to simplify the task); direction maintenance; marking critical features; demonstration; and frustration control, avoiding the 'major risk [of] creating too much dependency on the tutor' (Wood, Bruner and Ross, 1976: 98). Wood continued work on this in his research on the teaching techniques that mothers used with their 3–4-year-old children. The mothers, who were able to help their children complete a task that could normally only be completed by children older than 7, scaffolded their children's learning in specific ways:

- They simplified problems that the child encountered; they removed potential distractions from the central task.
- They pointed things out that the child had missed.
- The less successful parent tutors showed the child how to do the task without letting them have a go themselves, or they gave verbal instructions too frequently.

Overall, Wood (1998) identified two particularly important aspects. When a child was struggling, immediate help was offered. Then, when help had been given, the mothers gradually removed support and encouraged the child's independence. 'We termed this aspect of tutoring "contingent" instruction. Such contingent support helps to ensure that the child is never left alone when he is in difficulty, nor is he "held back" by teaching that is too directive and intrusive' (ibid.: 100).

The vital point here is that scaffolding happens in the context of meaningful interaction that is not too didactic. This idea of scaffolding is not typically what is happening when a teacher is demonstrating some aspect of the writing process. Although demonstration has a useful purpose, it should not be referred to as scaffolding and given dubious theoretical authenticity by inaccurate reference to Vygotsky. Much more thought needs to be given to the encouragement of children's independence as part of the teaching of English.

Jerome Bruner built on Vygotsky's thinking in his articulation of the 'spiral curriculum' where 'an "intuitive" grasp of an idea precedes its more formal comprehension as part of a structured set of conceptual relationships' (Bruner, 1975: 25). The idea of a spiral curriculum is important in that it suggests that knowledge and concepts need to be revisited a number of times at increasingly higher levels of sophistication. It is also important because it calls into question the notion that learning is a simple sequence, where knowledge and concepts are addressed on only one occasion. In explaining older children's apparent abandonment of reliance on *signs*, Vygotsky uses the example of memorisation and concludes that the abandonment is illusory because: 'Development, as often happens, proceeds here not in a circle but in a spiral, passing through the same point at each new revolution while advancing to a higher level' (Vygotsky, 1986: 56).

Bruner saw a close relationship between language and the spiral curriculum. He suggested that the spiral curriculum was supported by some essential elements in the learning process. Language learning occurs in the context of 'use and interaction – use implying an operation of the child upon objects' (ibid.: 25). In other words, it is important that children have first-hand experience of relevant 'objects' (including their local environment) to support their learning. In terms of English, this suggests that the writing of texts should be supported by real purposes and that the reading of texts should first and foremost be about experiencing whole texts and, second, about analysis. There are many teaching strategies that encourage the direct use of objects and the environment to stimulate talking, reading and writing. Bruner argued that this kind of language learning was 'contextualised' and should be supported by people who were expert, like the teacher.

## From language and linguistics to principles for teaching

One of the most important changes to the context for language and literacy in the curriculum of different countries has been the global growth in the use of English as a language. The growth of English is played out in the contexts of continent, country, state, district, city, town, school and classroom. This global phenomenon may seem a somewhat distant idea in relation to the daily lives of pupils and teachers in the UK. But if you pause to consider the language backgrounds and experiences of the pupils in any class, you will find that *multi-vernacularism* is part of all children's lives and so should be built on in schools (for research evidence on the related idea of *transcultural meanings*, see Wyse *et al.*, 2011). Another reason why consideration of the broader aspects of language is important is that effective teaching is informed by appropriate knowledge. Part of this knowledge is as full an understanding of language as possible. This includes the international dimensions, coupled with appropriate linguistic principles to guide practice.

In all cities in the UK, and in many rural areas, there are populations of students who are multilingual. The range is broadest in London, contrasting with the larger homogenous communities in other cities who have, for example, British Asian origins. If we look globally, in Africa every country has a different socio-political engagement with English. For example, the 13 languages enshrined in South African law contrast with the twin focus on Kiswahili and English in Tanzania. In primary schools in Tanzania, Kiswahili is the medium used for teaching; in secondary schools, English is used. In other African countries, the influence of French and Dutch colonialists provides different contexts for English. In Scandinavian countries, the success of English language teaching supported by the use of European language frameworks is tempered by popular concerns that national languages may be endangered (Simensen, 2010). China and the Chinese language are also of particular interest, in their own right and in relation to the spread of English. Bolton (2003) reminds us that the early seventeenth century was the beginning of contact between speakers of English and speakers of Chinese.

Bolton (op. cit.) also raises the issues of total numbers of speakers of Chinese and English. Arriving at accurate figures is complex (Crystal, 2010). The first problem is the world population growth of about 1.2 per cent per annum, which means that figures for less-developed nations change rapidly. Even in more stable populations, acquiring information is difficult when most census questionnaires do not include questions about linguistic background. Even if you can ask people about their language(s), there are difficulties in relation to measuring proficiency of language use, the way that people name their languages versus other forms of language such as dialect and political pressure to conform to particular ideas about the place of certain languages. Accepting these caveats, Crystal (op. cit.) estimated that the figures for numbers of speakers in 2010 were as follows:

- English, in countries where people are regularly exposed to English, including learning English in school, 2,902,853,000.
- English, second language speakers, 1,800,000,000.
- Chinese, mother tongue, all languages, 1,071,000,000.
- Mandarin Chinese, mother tongue, 726,000,000.
- English, mother tongue, 427,000,000.

A further problem for estimating the growing use of English is that there are no figures available for people who have learned English as a foreign language in countries where English does not have special status (including in China, where anecdotal accounts suggest that numbers of people learning English has grown dramatically). Halliday (2003) recognised English as a global language and made a distinction between this global spread and international variants of English that result from it. Crystal (2000) saw the possibility of an international Standard English with local and national variants. The spread of English, as is the case with any language, is an organic process that is impervious to attempts to divert it.

One of the most important implications of language change is how we should understand the multilingualism that all pupils will experience and/or encounter in their lives. As part of his *developmental interdependence* hypothesis and other work, Cummins (1979) proposed that the particular features of school discourse, such as the emphasis on particular forms of literacy learning, are part of what can make things difficult for bilingual pupils. Cummins' main conclusion was that support for bilingual pupils is necessary and there is a need to include assessments of pupils' language and literacy understanding in order to provide the most appropriate pedagogy. Reese *et al.*'s study (2000), conducted in the US with Spanish/English-speaking pupils, provided support for some elements of Cummins' theories. Its most significant finding was that even if parents were not able to speak the first language (and therefore used a second language), their engagement with books and reading was beneficial for their children's learning to read in the first language and their education more generally. Research clearly shows that support for home languages benefits the learning of another language, and that impeding the use of home languages is damaging. This is supported by recent neuroscience work. For example, Kovelman *et al.*'s (2008) research showed that bilinguals have differentiated representations in the brain of their two languages. They also found no evidence to suggest that exposure to two languages might be a source of fundamental and persistent language confusion.

A linguistically informed view of TELL prompts further thinking about pedagogy. The New London Group (NLG, 1996) argued that traditional pedagogy represents 'page-bound, official, standard . . . formalized, monolingual, monocultural, and rule governed forms of language' (p. 61). Instead, the NLG promoted the idea of *multiple literacies* that involve: a) a necessary emphasis on cultural and linguistic diversity; b) the recognition of local diversity interacting with global connectedness; and c) the idea of *situated practice* that emphasised the important concept of a community of learners. The authors were critical of what they called 'mere literacy' that involves a limited focus on language only, and is based on rules and correct usage that leads to 'more or less authoritarian pedagogy' (p. 64).

Street proposed that *new literacy studies* recognised the idea of multiple literacies and that there was an important distinction to be made between *autonomous* and *ideological* models of literacy. Autonomous models assume that the improvement of literacy, for example in economically disadvantaged groups of people, will automatically lead to societal and material advancement. Ideological models require consideration of the social and economic conditions that led to disadvantage in the first place. A helpful distinction is made by Street (2003) between literacy *events* and literacy *practices*. A literacy event takes place when written language is integral to the participants' interactions. Street's literacy practices account for literacy events but add the social models of literacy that are brought to bear by participants. Street quite rightly argues that:

> The next stage of work in this area is to move beyond these theoretical critiques and to develop positive proposals for interventions in teaching, curriculum, measurement criteria, and teacher education in both the formal and informal sectors, based upon these principles.
>
> (p. 82)

We have addressed the English language in its global context and in relation to bilingualism and pedagogy, but further selection of linguistic theory relevant to a consideration of TELL requires principles that can delimit and focus the areas of most relevance to teaching. The following principles were originally derived from analysis of the linguistic and pedagogic aspects addressed in the chapters of two publications, one aimed at researchers and policy makers (Wyse *et al.*, 2010) and one aimed at researchers (Wyse, 2011). Work on this edition was supported by major new work on written language (Wyse, 2017) which led to the following:

- Efficiency of communication, minimisation of ambiguity and constant growth in speed and range of communication drive the development of language and literacy in an organic way.
- Accounts based on analysis of language in use more powerfully, accurately and usefully explain language than prescriptive accounts.
- Oral language and written language share linguistic elements but are significantly different forms.
- Composition of meaning, in writing, is linked holistically with form, structure and orthographical elements.
- Experiencing and reflecting on the processes of reading and writing are an important resource to enhance teaching and learning.
- Language and social status (or power) are inextricably linked.

These principles underpin the book to varying degrees depending on the subject of the chapters. But for this theoretical chapter, we will restrict ourselves to one or two illuminative examples of the implications for teachers' knowledge.

Since communication of understandable meaning is the driving force of language, the aims and objectives of teaching will ensure that meaning is a constant point of reference and a purpose for language activities and interaction. Exercises that unduly focus on small components of language and literacy at the expense of whole 'texts' (in the very broadest sense) and understanding of meaning are inappropriate. An example of a more appropriate approach that reflects the centrality of meaning can be seen in relation to the teaching of writing. The teaching of writing is most effective when the emphasis is on the process of writing, including the aim to communicate a range of meanings through chosen genres of writing, coupled with structured teaching such as strategy instruction (Wyse et al., 2010).

An example of the second principle, the importance of the analysis of language in use, can be seen in relation to grammar. Teachers' knowledge of language is informed by the idea that accurate linguistic understanding draws on corpus data (e.g. Carter and McCarthy, 2006; Sealey, 2011). Data corpora are collections of real people's language in use that provide evidence for linguists. This approach to language can form the basis of enquiries in the classroom. For a simple example, take the teaching of punctuation. Custom and practice has often resulted in schools requiring pupils to use double speech marks to demarcate direct speech. But a cursory analysis of (for example) modern children's literature reveals that direct speech is often punctuated using single speech marks. Analysis of language in use enhances teacher knowledge and leads to more accurate teaching.

These examples also serve to remind us that although some conventions of language are relatively stable (spelling, for example), others are subject to much more change. Vocabulary is a particularly noticeable example of language change, but new ways of combining words and phrases are also part of the evolving linguistic landscape. Perhaps one of the most dramatic features of this language change is the way that English progresses as a world language, and what the linguistic consequences of this are for teachers around the globe.

Teachers are experienced readers and writers relative to their pupils and have a personal linguistic expertise and resource to draw upon. Their experience needs to be enhanced by the opportunity to reflect upon their own processes of reading and writing, as a means to better understand the processes that their pupils might experience. Greater knowledge of the processes of writing can come from psychological accounts (e.g. Hayes, 2006), but knowledge of the craft of writing acquired by professional writers, which is increasingly accessible in published forms, is also relevant, notwithstanding the differences between expert and novice writers (see Wyse, 2017).

The final principle reflects the idea that linguistic knowledge needs to include understandings about the language backgrounds of people in school communities that teachers are likely to encounter. To some degree, this is related to the idea of multiple literacies established by the NLG. But the central concern here is not only the way in which Standard English is established but also the implications that conceptions of Standard English, and its deployment, have for citizens. A particularly unfortunate example of (mis-)conception of Standard English is the way in which trainee teachers who use non-standard spoken varieties of English are sometimes criticised by tutors on the basis that their language does not represent a good model of the language – even though their meaning is entirely clear to their pupils. This brings us back to our first and most crucial principle: if communication of understandable meaning is the driving force of language, then the question is not whether a teacher uses Standard English or not, but rather whether they can be understood, and whether their communication helps their pupils learn.

## Practice points

- Be receptive to theories of learning and consciously use the ones that you think are important to inform your teaching.
- Remember that to help children learn, you need to take account of social factors (such as motivation) as well as cognitive ones.
- Develop the confidence to explore different approaches to teaching on the basis of theories.

## Glossary

**Theory** – a general idea, principle or set of principles that can form the basis for action.

## References

Biesta, G. (2013). *The Beautiful Risk of Education*. Boulder, CO: Paradigm Publishers.

Bolton, E. (2003). 'Chinese Englishes: From Canton jargon to global English', *World Englishes*, 21(2): 181–199.

Bruner, J. S. (1975). *Entry into Early Language: A Spiral Curriculum*. Swansea: University College of Swansea.

Carter, R. and McCarthy, M. (2006). *Cambridge Grammar of English*. Cambridge: Cambridge University Press.

Crystal, D. (2000). 'English: Which way now?', *Spotlight* (April 2000): 54–58.

Crystal, D. (2010). *The Cambridge Encyclopedia of Language*, 3rd edn. Cambridge: Cambridge University Press.

Cummins, J. (1979). 'Linguistic interdependence and the educational development of bilingual children', *Review of Educational Research*, 49(2): 222–251.

Dewey, J. (1902). *The Child and the Curriculum*. Chicago: The University of Chicago Press.

Dewey, J. (1998). 'Nature, communication and meaning: From experience and nature 1925', in L. Hickman and T. Alexander (eds.), *The Essential Dewey, Vol. 2: Ethics, Logic, Psychology*. Bloomington: Indiana University Press.

Dunne, J. (1993). *Back to the Rough Ground: Practical Judgement and the Lure of Technique*. Notre Dame, IN: University of Notre Dame Press.

Gallagher, C. and Wyse, D. (2012). 'Aims and objectives', in D. Wyse, V. Baumfield, D. Egan, C. Gallagher, L. Hayward, M. Hulme, K. Livingston, I. Menter and B. Lingard (eds.), *Creating the Curriculum*. London: Routledge.

Halliday, M. (2003). 'Written language, standard language, global language', *World Englishes*, 22(4): 405–418.

Hayes, J. R. (2006). 'New directions in writing theory', in C. MacArthur, S. Graham and J. Fitzgerald (eds.), *Handbook of Writing Research*. New York: The Guilford Press.

Kovelman, I., Baker, S. A. and Petitto, L. (2008). 'Bilingual and monolingual brains compared: A functional magnetic resonance imaging investigation of syntactic processing and a possible "neural signature" of bilingualism', *Journal of Cognitive Neuroscience*, 20(1): 153–169.

Marx, C. (1887). *Capital: A Critique of Political Economy*. Translated by Samuel Moore and Edward Aveling. Moscow, USSR: Progress Publishers, p. 128.

The New London Group. (1996). 'A pedagogy of multiliteracies: Designing social futures', *Harvard Educational Review*, 66(1): 60–93.

Reese, L., Garnier, H., Gallimore, R. and Goldenberg, C. (2000). 'Longitudinal analysis of the antecedents of emergent Spanish literacy and middle-school English reading achievement of Spanish-speaking students', *American Educational Research Journal*, 37(3): 622–633.

Sealey, A. (2011). 'The use of corpus-based approaches in children's knowledge about language', in S. Ellis and E. McCartney (eds.), *Applied Linguistics and the Primary School* (pp. 93–106). Cambridge: Cambridge University Press.

Simensen, A. M. (2010). 'English in Scandinavia: A success story', in D. Wyse, R. Andrews and J. Hoffman (eds.), *The Routledge International Handbook of English, Language and Literacy Teaching* (pp. 472–483). London: Routledge.

Street, B. (2003). 'What's "new" in new literacy studies? Critical approaches to literacy in theory and practice', *Current Issues in Comparative Education*, 5(2): 1–114.

Vygotsky, L. (1978). *Mind in Society: The Development of Higher Psychological Processes*. Cambridge, MA: Harvard University Press.

Vygotsky, L. S. (1986). *Thought and Language*. Cambridge, MA: Harvard University Press.

Vygotsky, L. and Luria, A. (1994). 'Tool and symbol in child development', in R. Van der Veer and J. Valsiner (eds.), *The Vygotsky Reader* (p. 56). Hoboken, NJ: Wiley.

Wood, D. (1998). *How Children Think and Learn*, 2nd edn. Oxford: Blackwell.

Wood, D., Bruner, J. and Ross, G. (1976). 'The role of tutoring in problem solving', *Journal of Child Psychology and Psychiatry*, 17: 89–100.

Wyse, D. (ed.). (2011). *Literacy Teaching and Education: SAGE Library of Educational Thought and Practice*. London: Sage.

Wyse, D. (2017). *How Writing Works: From the Birth of the Alphabet to the Rise of Social Media*. Cambridge: Cambridge University Press.

Wyse, D., Andrews, R. and Hoffman, J. (eds.). (2010). *The Routledge International Handbook of English, Language, and Literacy Teaching*. London: Routledge.

Wyse, D., Nikolajeva, M., Charlton, E., Cliff Hodges, G., Pointon, P. and Taylor, L. (2011). 'Place-related identity, texts, and transcultural meanings', *British Educational Research Journal, 38(6): 1019–1039.* doi: 10.1080/01411926.2011.608251

## Annotated bibliography

Olson, D. (2017). *The Mind on Paper: Reading, Consciousness and Rationality*. Cambridge: Cambridge University Press.

Argues that written language resulted in metacognitive thinking by humans. Includes attention to Vygotsky and Dewey.

L3 ★★★

Vygotsky, L. S. (1986). *Thought and Language*. Cambridge, MA: Harvard University Press.

Probably Vygotsky's best-known book. Includes the description of the zone of proximal development.

L3 ★★★

Biesta, G. (2013). *The Beautiful Risk of Education*. Boulder, CO: Paradigm Publishers.

A view from the philosophy of education. Important reminder of how Dewey's pragmatism included a central focus on communication and language.

L3 ★★★

# Chapter 3

# Inclusion and equality

> This chapter focuses on inclusive practice in teaching English, language and literacy and the fundamental importance of celebrating and valuing all learners. We start by considering categories that are commonly used to characterise pupils and then explore some questions and strategies to consider in including all children when teaching speaking and listening, reading and writing, taking particular consideration of gender and linguistic diversity.

Providing *all* children in early years and primary settings with equal access to education and high quality teaching of English, language and literacy is a philosophy that underpins the whole of this book. The examples of issues and classroom practices in other chapters of the book reflect to varying degrees our commitment to inclusion. However, this early chapter gives us an opportunity to emphasise some general considerations, and to give more attention to two aspects in particular: gender and linguistic diversity.

The National Curriculum (DfE, 2013: 9) provides six statements under the heading of inclusion. The first of these is:

> Teachers should set high expectations for every pupil. They should plan stretching work for pupils whose attainment is significantly above the expected standard. They have an even greater obligation to plan lessons for pupils who have low levels of prior attainment or come from disadvantaged backgrounds. Teachers should use appropriate assessment to set targets which are deliberately ambitious.

The National Curriculum also requires teachers to take account of their duties under equal opportunities legislation. The Equality Act 2010 established seven 'protected characteristics' and four kinds of unlawful behaviour in relation to these characteristics: direct discrimination; indirect discrimination;

harassment; and victimisation. It is unlawful for a school to discriminate against a pupil or prospective pupil by treating them less favourably because of their:

- sex
- race
- disability
- religion or belief
- sexual orientation
- gender reassignment
- pregnancy or maternity

(Department for Education (DfE), 2014, p. 8)

Legal issues around the Equalities Act are further explored by Wyse *et al.* (2016), and the full book's website has links to podcasts that support this chapter: https://study.sagepub.com/wyseandrogers/student-resources/chapter-16/podcast

The Special Educational Needs and Disability Code of Practice: 0 to 25 years (Department for Education and the Department of Health, 2015) provides extensive information about duties, particularly for children with SEN. One key element in relation to children who have English as an additional language is the importance of teachers reflecting carefully on children's language development to make an assessment of whether a child has SEN or whether any delay in development is a normal feature of developing more than one language. The Teachers' Standards (DfE, 2011: 12) require all teachers to 'adapt teaching to respond to the strengths and needs of all pupils' and to

> have a clear understanding of the needs of all pupils, including those with special educational needs; those of high ability; those with English as an additional language; those with disabilities; and be able to use and evaluate distinctive teaching approaches to engage and support them.
>
> (op. cit.)

Whilst we would not argue with the general principle that all children's needs should be supported, this does raise questions about the desirability and accuracy of categorising children in relation to their needs. Many children can be described in relation to more than one 'category'. We explore this further below, with particular reference to gender and linguistic diversity.

## Gender

Exploration of issues around gender is fraught with difficulties for teachers, particularly because stereotypes and preconceptions are rife in society. The danger of preconceptions is that they can lead to low expectations of children's capabilities. There are no simple solutions to addressing gender and attainment. First and foremost, teachers need to take an active approach to evidence about all pupils' learning and opinions. Involving children in discussion about their views is both a necessary part of education and a good way to start to address your understanding of how gender can affect English, language and literacy learning in the classroom. Teachers need to use this information to provide a curriculum that is relevant to children with differing needs and interests, whatever their gender and preferences. Giving pupils opportunities to make choices, for example over reading material and writing topics, as well as speaking and listening roles, are also important strategies to support the preferences of all children. Offering children choices also gives the sensitive teacher a deeper understanding of the pupils that they teach because it reveals aspects of children's interests and characteristics that often remain hidden when classroom practice is teacher-dominated.

One example of a supposed gender characteristic is preference for reading fiction as opposed to non-fiction. In general, it has been argued that girls tend to prefer fiction whereas boys prefer non-fiction. However, there are many qualifications that need to be made to such an idea. The characteristic does not mean that *all* boys prefer non-fiction or *all* girls prefer fiction. Therefore, in every class of children, while it may be true that the characteristic is generally accurate, there will still be boys and girls who do not conform to the stereotype. There is also a developmental angle. The youngest children tend to have a particularly strong relationship with story, nursery rhymes and songs. There is less of a distinction for them between fiction and non-fiction. As pupils get older, non-fiction forms tend to take on more importance as part of school work, but the texts that children select themselves to read at home will not necessarily follow this pattern. It is probably also true to say that most people read a wide variety of texts, hence it may be difficult to classify someone as mainly a fiction reader or mainly a non-fiction reader. So, the statement about boys' and girls' reading habits needs to be carefully qualified by the factors raised above. Before a judgement could be made about your class, you would have to analyse the reading habits of all the children. If you involve the children and parents in such a survey, it could provide helpful information for teaching and provide an opportunity for children to learn more about gender. This strategy of opening up investigation and discussion is one very good way to practically address gender issues in the classroom. As we explore in Chapter 8, developing your knowledge of all children's reading practices will help you broaden your understanding of how to develop reading for pleasure and life-long reading skills.

Even if there is evidence to show gender differences, there is still the question of how the knowledge should affect our teaching. For example, if we accept that girls prefer fiction, then we could look at this in at least two ways: (1) there is a potential problem in that they are not accessing non-fiction texts, which become increasingly important in secondary schooling and later life, and therefore we should take steps to encourage them to access non-fiction; or (2) we want to relate our teaching to girls' interests and want to motivate them, so we will expose them to even more fiction. We think that a combination of (1) and (2) is probably a sensible way forward.

There has been concern, too, about boys' attainment in literacy for some years. Raising boys' attainment, particularly that of white, working-class boys, is an important goal (Moss, 2010). Many boys do less well than girls in reading, and the difference is even more marked in writing. Strategies to consider to encourage boys' reading include:

- a holistic approach to the reading curriculum which assimilates opportunities for reading, writing, speaking and listening into an integrated whole;
- concentration on the need to encourage boys to become successful and motivated readers as well as decoders of texts (the difference between learning to read and choosing to read);
- offering a wide range of texts in a variety of media to stimulate and sustain interest;
- giving children space to talk and reflect on their reading;
- offering different gender role models who show that reading is a pleasurable activity.

A DfE study conducted in 2004 found that that gains can be made in primary literacy, particularly in the levels achieved by apparently under-achieving boys, when:

- a variety of interactive classroom activities are adopted with a 'fitness for purpose', so that both short, specific, focused activities and more sustained, ongoing activities are used, as and when appropriate;
- acknowledgement is given to the central importance of talk, to speaking and listening, as a means of supporting writing;
- the advantages to be gained through companionable writing with response partners and through group work are recognised;
- teachers are prepared to risk-take to bring more creativity and variety to literacy;
- more integrated use is made of ICT so that quality presentation can be more easily achieved and drafts amended with more ease.

(Younger *et al.*, 2004: 10)

However as stressed above, it is important to consider *which* individual children we are talking about when looking at disparities in motivation and achievement between genders.

Schools are also required under the 2010 Equality Act to consider the needs of children who have undergone or are in the process of undergoing gender reassignment. Stonewall is an organisation campaigning for the rights of lesbian, gay, bisexual and trans people and supports many schools around the country in promoting a safe and inclusive learning environment. Amongst their recommendations are:

- using opportunities to talk about difference and different families;
- reading books that challenge gender stereotypes and which feature different families;
- talking about difference when reading traditional fairy tales and sharing alternative versions that challenge traditional gender roles;
- encouraging all children to participate in different role-play activities.

## Linguistic diversity

One of the many reasons that teaching is an exhilarating experience is because of the tremendous diversity of languages within many classrooms. Over 350 languages are spoken by children in British schools, and the number of young bilingual and multilingual speakers continues to rise. Language is inextricably linked to an individual's sense of identity, their home, culture and self-esteem, and therefore all children's language should be viewed as of intrinsic value and also as a powerful resource in the classroom.

Bilingual learners ☞ in England face two main tasks in school: they need to learn English and they need to learn the content of the curriculum, tasks which should proceed hand in hand. Nearly all children learn their first language successfully at home. Children learning an additional language in a classroom are learning under conditions that are significantly different. Learning a language is more than just learning vocabulary, grammar and pronunciation. Children need to use all these appropriately for a whole range of real purposes or functions, such as questioning, analysing and hypothesising, which are clearly linked to thinking and learning skills.

Cummins (1996) adapted the metaphor of an iceberg to distinguish between basic interpersonal communicative skills and cognitive and academic language proficiency (BICS and CALP respectively for short). All children develop communicative skills first, in face-to-face, highly contextualised situations, but take longer to develop the cognitive and academic language proficiency that contribute to educational success. Cummins acknowledged that some interpersonal communication can impose considerable cognitive demands on a speaker and that academic situations may also require social communication skills. Generally speaking, children learning an additional language can become conversationally fluent in the new language in two to three years but may take five years or longer to catch up with monolingual peers on the development of CALP. A large-scale study of emergent bilingual pupils in the US clearly showed the importance of supporting all pupils' languages:

> Non-English speaking student success in learning to read in English does not rest exclusively on primary language input and development, nor is it solely the result of rapid acquisition of English. Both apparently contribute

to students' subsequent English reading achievement . . . early literacy experiences support subsequent literacy development, regardless of language; time spent on literacy activity in the native language – whether it takes place at home or at school – is not time lost with respect to English reading acquisition.

(Reese *et al.*, 2000: 633)

Bilingual and multilingual children are experts in handling language because they become adept at 'code switching' (switching between languages). This is not a sign of confusion, but has cognitive, metalinguistic and communicative advantages. Indeed, bilingual children outperform monolinguals in cognitive flexibility tasks (Bialystok, 2007) because they work constantly with two linguistic systems. These children have considerable language skills on which the teacher can build, and they are likely to have much to offer others, particularly with regard to the subject of language study. Having said this, even if the child may be skilled in language use, they may still need particular support and guidance to develop greater proficiency in the use of English at school, especially in terms of written English.

It is of course important for teachers to be aware of what children know with regard to spoken and written language. In 2016, the DfE introduced a mandatory five-point proficiency scale across listening and understanding, speaking, reading and writing to be included in the annual school census. The bell foundation, working with researchers at King's College London and the University of Cambridge, has produced an EAL Assessment Framework to support teachers with implementing this and with comprehensive tools to monitor EAL learners at each stage of language development across the four modes as well as with practical strategies to support EAL learners. The programme's objective is to improve the educational outcomes of children with English as an Additional Language.

As we argue in Chapter 23, children's home literacies, including their language, need to be valued. This means that the use of mother tongues/community languages should be positively encouraged in the classroom, both for children who are new to English and also as they become more fluent. Welcoming spoken and written forms of the languages children and their parents/ carers bring to the classroom is a fundamental and very practical strategy to building an inclusive classroom.

There is evidence to show that the least successful way to deliver English teaching to a bilingual or multilingual child at the early stages of English acquisition is to remove them from the classroom setting and provide short sharp bursts of tuition in isolation, or to group them with children who are struggling, or with SEN. More effective is the practice of immersing children in a rich spoken language with more fluent speakers of English and resourcing the classroom with a range of texts that represent different languages and cultures, including dual language texts and stories (which can also be created and recorded by children and parents) in other languages. Displays and labels in different languages will also convey the message that all languages are valued.

Cross-curricular and thematic approaches in primary classrooms often offer opportunities to acknowledge multicultural dimensions to study. Teachers can acknowledge consciously the monocultural way in which many primary school themes are conceived and open this planning up to more accurately reflect a multicultural society.

Story telling can be another particularly successful method of encouraging the multilingual child to negotiate more than one language. All cultures have their own histories, myths, legends and stories which are passed on through generations of children. These stories cross cultural boundaries; some are recognisably similar with subtle shades of difference, while others will be particular within a specific cultural context. In either case, the story itself becomes a powerful, shared experience and the telling, the retelling, the writing and reading of the range of possible stories open a rich vein of language study for the teacher to exploit. Again, parents should be seen as a valuable and authoritative resource in this area.

Here is a summary of our suggestions for supporting bilingual and multilingual learners ☞ new to English in your classroom:

- Include the child in activities/lessons right from the start.
- Make a point of speaking to the child every session even if they do not respond at first.
- Set up a buddy system.
- Give children longer to respond. Repeat what was said if necessary. Only re-phrase if this does not appear to work.
- Use pictures, photographs and picture dictionaries.
- Organise scribing of writing in first language.
- Picture books are an ideal support for language learning.
- A collection of independent 'respite' activities can be helpful for the child to consolidate and work at a less demanding pace.
- Gradually include the child in group discussion, starting with answers which require single words that the child knows.
- Be aware of the difficulties of idiomatic expressions.

(adapted from Haslam *et al.*, 2005)

Diversity and difference in a class of children is to be welcomed and valued. However, as Florian (2014: 5) notes, it is important to be wary of assigning categories or labels to children:

although knowledge about human differences is important (a student who is an English language learner is different from a student who has been diagnosed as having autism; a six year old is different form a 10 year old and so on), whatever can be known about a particular category of learners will be limited in the educational purposes it can serve, because the variations between members of a group make it difficult to predict or evaluate provision for each of the individuals within a group.

(Florian, 2014: 5)

Being an inclusive teacher of English, language and literacy requires knowing about the individual learners in your class and having the subject and pedagogical subject knowledge to use a range of teaching approaches to support them in speaking and listening, reading and writing. Assessment of the efficacy of these approaches will then help you adapt your planning and teaching where necessary.

### Practice points

- Use evidence about your class as the basis for decisions in terms of the needs of girls, boys and gender preferences.
- Organise your classroom for flexible grouping, including gender groups.
- Give opportunities for pupils to make choices as part of their English, language and literacy learning.
- Consider and plan how you can help bilingual and multilingual children to access learning at every stage of teaching.
- Acknowledge that you are unlikely to know everything about every child's culture, but in so doing acknowledge also that it is your responsibility to understand the lives of *all* the children in your class, not just those who share your own cultural background.

### Glossary

**Bilingual learners** – children who speak two languages.
**Multilingual learners** – children who speak more than two languages.

### References

Bialystok, E. (2007). 'Cognitive effects of bilingualism: How linguistic experience leads to cognitive change', *International Journal of Bilingual Education and Bilingualism*, 10(3): 210–223.

Cummins, J. (1996). *Negotiating Identities: Education for Empowerment in a Diverse Society*. Los Angeles: California Association for Bilingual Education (CABE).

Department for Education. (2011). *Teachers' Standards: Guidance for School Leaders, School Staff and Governing Bodies*. London: Department for Education.

Department for Education. (2013). *The National Curriculum in England: Framework Document: December 2014.* London: Department for Education.

Department for Education. (2014). *The Equality Act 2010 and Schools: Departmental Advice for School Leaders, School Staff, Governing Bodies and Local Authorities.* London: Department for Education.

Department for Education and the Department of Health. (2015). *Special Educational Needs and Disability Code of Practice: 0 to 25 Years: Statutory Guidance for Organisations Which Work with and Support Children and Young People Who Have Special Educational Needs or Disabilities.* London: DfE & DoH.

Florian, L. (ed.). (2014). *The SAGE Handbook of Special Education.* London: Sage.

Haslam, L., Wilkin, Y., and Kellet, E. (2005). *English as an Additional Language: Meeting the Challenge in the Classroom.* London: David Fulton.

Moss, G. (2010). 'Gender and the teaching of English', in D. Wyse, R. Andrews and J. Hoffman (eds.), *The Routledge International Handbook of English, Language and Literacy Teaching.* London: Routledge.

Reese, L., Garnier, H., Gallimore, R. and Goldenberg, C. (2000). 'Longitudinal analysis of the antecedents of emergent Spanish literacy and middle-school English reading achievement of Spanish-speaking students', *American Educational Research Journal,* 37(3): 622–633.

Wyse, D., Ford, S., Hale, C. and Parker, C. (2016). 'Legal issues', in D. Wyse and S. Rogers (eds.), *A Guide to Early Years and Primary Teaching* (pp. 301–320). London: Sage.

Younger, M., Warrington, M., Gray, J., Ruddock, J., McClellan, R., Bearne, E. *et al.* (2004). *Raising Boys' Achievement.* London: Department for Education and Skills.

## Annotated bibliography

Adoniou, M. (2013). 'Drawing to support writing development in English language learners', *Language and Education,* 27(3): 261–277.
In this article, drawing is presented as an effective strategy for teaching writing, especially when teaching English language learners.
L2 ★★

www.bell-foundation.org.uk/eal-programme/teaching-resources/eal-assessment-framework/
The bell foundation EAL Assessment Framework for supporting the assessment of children with English and an Additional Language.
L1 ★★

Wyse, D. and Rogers, S. (eds.). (2016). *A Guide to Early Years and Primary Teaching.* London: Sage.
This general guide includes chapters on 'Diversity and Inclusion' and 'Legal Issues' which provide a much wider range of topics relevant to inclusion.
L2 ★★

# Part II

# Language

# Chapter 4

# The development of language

Theories and stages of language acquisition are addressed at the beginning of this chapter. We then explore how to maximise children's capacity to learn, achieve and participate through dialogic approaches to teaching. A brief account of educational policy for language in national, statutory government curricula is presented. The chapter reveals how understanding of the centrality of language to a child's development has grown over the last 30 years.

There are two important elements of language: communication and representation. Communication is the transmission of meanings, and we know that babies engage with communication from birth. But language is also a representational system that emerges with children's cognitive skills, enabling them to understand and organise the world. Evidence from neuroscience increasingly demonstrates the importance of *connectionism* that describes the distributed nature of connections in the brain that enable complex human attributes such as language (Goswami, 2015).

Language plays a vital role in cognitive development, and its development is supported by appropriate interaction including, when children are babies, infant-directed speech by children's carers (op. cit.). Language comprises different elements that are important for effective understanding and communication: phonology, vocabulary, syntax and morphology, and pragmatics are the four basic strands which mutually support and influence each other's development. To communicate effectively, children need to develop receptive language skills in order to become increasingly able to understand the language they hear. They also need to develop expressive language skills to convey their own thoughts, feelings and desires. Thus, from a very early age children are learning through language, learning to use language and learning about language.

## Language acquisition

Infants learn language with remarkable speed: by the age of 5, provided they do not have language difficulties, all children have acquired the grammar for the main constructions of their native language (Messer, 2006) (see Figure 4.1 and Table 4.1). This is true across all cultures and in all languages (Kuhl, 2004). The term 'acquired' in this context is important because linguists make a distinction between emergent language constructions and ones which are acquired fully.

The main stages of children's oral language development begin with sounds, then the first meaningful words and then move on to phrases with two or more words. After this, children's syntax develops rapidly and on many fronts. Negative sentences such as 'I am not walking' and the use of complex sentence types will be areas that develop during the nursery phase (age 3 to 4). The ability to ask more sophisticated questions is another aspect of syntax that develops at this time.

The word morphological comes from morpheme ☞. A morpheme is the smallest unit of language that can change meaning. For example, if we take the singular 'apple' and turn it into the plural 'apples', then the letter 's' is a morpheme because it changes the meaning from singular to plural. Morphemes that can stand alone, such as 'apple', are called 'free' morphemes, and those which cannot, such as -s in apples, are called 'bound' morphemes. Children's development of morphological understanding can be seen in their capacity to invent words, such as 'carsiz' (cars).

Vocabulary development is not something that has a particular end point because we continue to add vocabulary throughout our lives. One of the features of children's development is over-extension. An example of this is where children call all meats 'chicken' because they are familiar with that word but not others, such as 'beef', 'pork', etc. The correct use of categories of objects develops in the nursery phase and beyond. Another feature of oral language development is learning about the way that the meanings of words relate to each other. Synonyms such as 'happy/joyful' and antonyms such as 'happy/sad' are part of this. This means that children can learn about vocabulary from words that they know without having to directly experience the concept of the word in question.

Phonological development ☞ has been much studied, partly because of its link with learning to read. As far as talk is concerned, there are some understandings and skills that have to be acquired before those which are beneficial for literacy. For example, the young child learns to control their vocal chords. The sound/airflow which passes from the vocal chords is altered in various ways in order to form sounds which eventually become words. The place of articulation involves use of the teeth, lips, tongue, mouth and glottis. The manner of articulation involves obstructing the airflow to varying degrees such as completely stopping it or allowing some to pass through the nose.

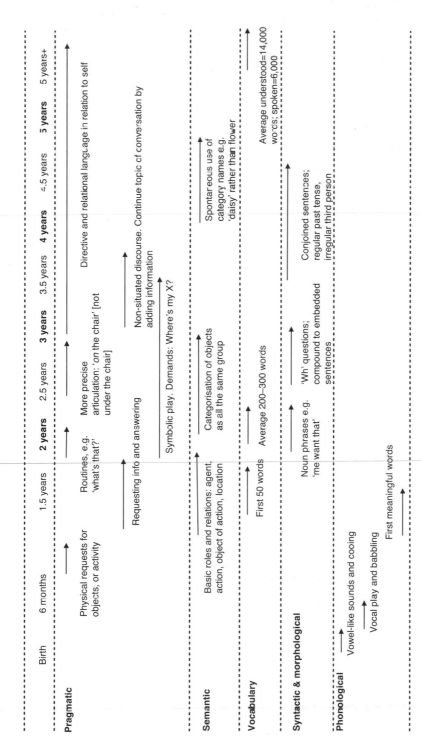

*Figure 4.1* Summary of main elements of children's oral language development: birth to age 5.

The figure content, read across the age scale (Birth, 6 months, 1.5 years, 2 years, 2.5 years, 3 years, 3.5 years, 4 years, 4.5 years, 5 years, 5 years+):

**Pragmatic**
- Physical requests for objects, or activity
- Routines, e.g. 'what's that?'
- Requesting info and answering
- More precise articulation: 'on the chair' [not under the chair]
- Symbolic play. Demands: Where's my X?
- Non-situated discourse. Continue topic of conversation by adding information
- Directive and relational language in relation to self

**Semantic**
- Basic roles and relations: agent, action, object of action, location
- Categorisation of objects as all the same group
- Spontaneous use of category names e.g. 'daisy' rather than flower

**Vocabulary**
- First 50 words
- Average 200–300 words
- Average understood=14,000 words; spoken=6,000

**Syntactic & morphological**
- Noun phrases e.g. 'me want that'
- 'Wh' questions; compound to embedded sentences
- Conjoined sentences; regular past tense, irregular third person

**Phonological**
- Vowel-like sounds and cooing
- Vocal play and babbling
- First meaningful words

Table 4.1 Main elements of children's oral language development: birth to age 5

| Age | Phonological | Age | Semantic | Age | Vocabulary | Age | Syntactic and morphological | Age | Pragmatic | Age | Pragmatic |
|---|---|---|---|---|---|---|---|---|---|---|---|
| **Birth to two months** | Vowel-like sounds such as crying and grunting | **One year** | 1.1 to 2.0 Basic semantic roles and relations (Stage I) Agent, action, object of action, location Possessive, attributive, demonstrative | **One year and six months to one year and eight months** | First 50 words acquired | **Two years** | 2.0–2.5 Grammatical morphemes (Early mastered, Stage II) In, on, -in, noun plural Begin to put words together in sentences. Noun phrases with premodification of the noun: e.g. more biscuit. Pronouns appear: e.g. 'me want that' Past tense inflection appears | **Eight months** | Requesting objects or activity, refusing, commenting, communicative games (continues to 15 months) | **Three years** | Can cope with non-situated discourse. Learning to continue topic of conversation through adding new information. Three to seven years: Directive: language is used to direct or monitor one's own or another's actions (frequency decreases) Interpretative: language is used to report on present and past experiences, and to reason (frequency increases) Projectile: language is used to predict events, to express empathy, and to create imaginary activities and roles (frequency increases) Relational: language is used to maintain one's own status and to interact |

| Two to four months | | One year and four months | | Two years | | Two years and six months | | Ten months to one year | | Two years four months to three years eight months | |
|---|---|---|---|---|---|---|---|---|---|---|---|
| Cooing | | Middle one-word period (16–20 months): agent, action, object of action (18–20 months): location, possession. Semantic development of early negation: rejection, disappearance, non-existence, denial | | Average vocabulary = 200–300 words | | 2.5–3.0 Syntactic modulation of sentences (Stage III) Auxiliary use in negation, yes–no and WH questions Starting to acquire rules for inflecting nouns and verbs: e.g. 'breaked it' or 'mouses' Multiple premodification of nouns: e.g. that red ball | | 25 vocalizations/10 minutes of free play | | Embedded requests: 'Can I have big boy shoes?' | |
| **Four to six months** | | **Two years** | | **Six years old** | | **Two years eight months** | | **One year** | | **Three years eight months to four years** | |
| Vocal play including rudimentary syllables such as /da/ or /gɔɔ/ | | Semantic aspects of WH questions (2–5 years) Objects and people (what, who) Basic event questions (who what-doing where) Event explanation (why) Event manner (how) Event time (when) | | Average vocabulary understood = 14,000 words Average spoken vocabulary = 6,000 words | | Compound sentences: e.g. 'The dog bit the cat and then he ran away' | | 50 vocalizations/10 minutes of free play (13–18 months) | | Conversation consists largely of initiation/response (I/R) exchanges | |

*(Continued)*

Table 4.1 (Continued)

| Age | Phonological | Age | Semantic | Age | Vocabulary | Age | Syntactic and morphological | Age | Pragmatic | Age | Pragmatic |
|---|---|---|---|---|---|---|---|---|---|---|---|
| Six months | Babbling such as /ba/ba/ba/ or /ga/ba/da/do | Two to three years | Refer to all members of category as the same: e.g. all flowers as flower | | | Three years | 3.0 to 3.5 Emergence of embedded sentence forms (Stage IV) and mastery of mid-difficulty grammatical morphemes Irregular past, articles the and a, be as an uncontracted main verb Post-modified phrases: e.g. 'the picture of Lego town'. Complex sentences | One year and four months | requesting information, answering questions, acknowledging response, repeating (continues to 23 months) | Four years | Elaborate oblique strategies: 'We haven't had any sweets for a long time' |
| One year | First meaningful words | Four to five | Spontaneously use category names: e.g. rose or daisy | | | Three years eight months to four years | 3.5–4.5 Emergence of conjoined sentences (Stage V) Grammatical morphemes (Late mastered, Stage V+) Regular past tense, third person present, irregular third person, contractible copula be, auxiliary be | One year and six months to one year and eight months | 75 vocalizations/10 minutes of free play (19–21 months) | Four years to four years seven months | Acquisition of auxiliary verbs (might, may, could) and negation Child's response itself increasingly becomes R/I |

|  |  |
| --- | --- |
| **Four years six months** | Coordination with ellipsis: e.g. 'The dog bit the cat and ran away' |
| **One year three months to two years** | Telegraphic directives: e.g. 'that mine', 'gimme' |
| **Two years** | Symbolic play, evoking absent objects and events, misrepresenting reality<br>Learning to continue topic of conversation through repetition |
| **Two years to two years four months** | Limited routines: 'Where's my X?', 'What's that?' |
| **Two years nine months** | Greater precision of articulation in self-repairs, increased volume and use of contrastive stress: e.g. 'It was *on* the chair!' (not under it) |
| **Three years eight months to five years seven months** | |
| **Four years seven months to four years ten months** | Greater ability to encode justifications and causal relationships allows for longer exchanges<br>Advanced embedding: 'Don't forget to buy sweets' |

*Source:* Based on Messer (2006) and Chapman (2000).

Adults model (in an unplanned way) the conventions of language, providing feedback on the effectiveness of a child's ability to communicate by responding to them. They scaffold (→ Chapter 2) the child's language learning and enable the child to test their current hypotheses about how language works. The ability of adults to take into account the limited abilities of the child and adjust their language accordingly (so that the child can make sense of them) is intuitive for most.

As can be seen in Figure 4.1 and Table 4.1, the development of language has multiple dimensions that progress in specific ways, but all are driven by the need for communication.

## Language and the bilingual child

Kuhl (2004) argues that young children usually learn their mother tongue rapidly and effortlessly, following the same developmental path regardless of culture. Bilingual children are hearing two languages – or two distinct systems – which they have to internalise and respond to. At an early age, neither language is likely to interfere with the other so young children can learn two languages easily. Reese *et al.*'s (2000) research showed that bilingual pupils' success in learning to read in English does not rest exclusively on primary language input and development. The most significant finding was that parents' engagement with reading using the second language is beneficial both for their children's reading in the first language and their education more generally. Time spent on literacy activity in a child's native language – whether at home or at school – is therefore not time lost with respect to English reading acquisition.

The social and cultural aspects of language development are important as children learn, through talk, to place themselves within a specific social context; in this way, the development of language and identity are closely linked. The quality of social experience and interaction will vary greatly between children, and, during the early years, teachers need to be aware that some children will arrive at school appearing to be confident, articulate users of the English language, whereas others seem less comfortable language users. However, teachers should beware deficit models and remember that it is too easy to label a child's spoken language as 'poor', or even to say that they have 'no language', without sufficient thought. To illustrate some of the issues with labelling, Bearne analysed a transcription of a discussion including Sonnyboy, a 6-year-old boy from a Traveller community, demonstrating his ability to 'translate' language for other children:

*Emily:*      I loves them little things.
*Sonnyboy:*  Yeah . . . I loves the little sand things – that tiny wee spade . . . And this little bucket . . .
*Teacher:*    Do you think it would be a good idea to ask Cathy to get some? *(Cathy runs a playgroup for the Traveller children on their site).*

| | |
|---|---|
| *Emily:* | What for? |
| *Teacher:* | So that you'd have some at home. |
| *Sonnyboy:* | And who'd pay for them? Would Cathy pay? |
| *Teacher:* | No, it would be part of the kit. |
| *Emily:* | I don't know what you mean. Kit – who's Kit? Me Da's called Kit – would me Da have to pay? |
| *Sonnyboy:* | Not your Da – it's not that sort of kit, Emily. It's the sort a box with thing in it that you play with . . . like toys and things for the little ones. |

(Bearne, 1998: 154)

It is important, then, that teachers understand about language diversity and the ways in which judgements are made about speakers in the classroom. From this perspective, it is equally important that teachers recognise their own histories and status as language users and resist the temptation to impose their own social criteria on the child's ongoing language development. As Bearne goes on to point out:

> Language diversity is . . . deeply involved with social and cultural judgements about what is valuable or worthy . . . Judgements are often made about intelligence, social status, trustworthiness and potential for future employment on the basis of how people speak – not the content of what they say, but their pronunciation, choice of vocabulary and tone of voice. Such attitudes can have an impact on later learning.

(ibid.: 155)

The following ideas can help to support bilingual children in the mainstream classroom:

- Encourage pupils' use of their first language in the classroom. If your knowledge of the language is poor, learn simple key phrases such as 'hello' and 'goodbye', 'please' and 'thank you'.
- Create a focus for speaking and listening activities.
- Include the child in activities and lessons right from the start; build bilingual learners' needs into the overall language and literacy objectives for the whole class.
- Integrate language learning within the lesson content of subjects other than literacy.
- Model speaking the English language. It is worth noting that some monolingual learners may also benefit from this strategy.
- Consider the use of visual aids such as pictures, photographs and real objects to support language learning.
- Involve parents in their children's learning if you can.

## Exploratory and collaborative talk

Barnes (1976) and Mercer (2000) argued that *exploratory talk* ☞ is the kind of talk that teachers should aim to develop. Exploratory talk is a way of interacting which emphasises reasoning, the sharing of relevant knowledge and a commitment to collaborative endeavour. This contrasts with *disputational talk* ☞, in which speakers are competitive rather than cooperative, each sticking to their own point of view, and *cumulative talk* ☞, in which speakers share ideas and agree with each other, but there is no critical evaluation of ideas. When children engage in exploratory talk, they are almost certain to be working in a small group with their peers. They will be sharing a problem and constructing meaning together: exchanging ideas and opinions, considering and evaluating each other's ideas, building up shared knowledge and understanding.

Collaborative learning in group work occurs when knowledge and understanding are developed through pupils talking and working together relatively autonomously (Blatchford *et al.*, 2003; Mercer and Littleton, 2007). Mercer and Littleton (2007: 23) define children as being engaged in collaborative learning 'when they are engaged in a coordinated, continuing attempt to solve a problem or in some way construct common knowledge'. The role of talk and knowledge and understanding of speaking and listening skills is therefore crucial to this process.

For successful classroom interaction to occur, a collaborative climate must be established where children feel part of a learning community in which problems are solved and understandings are developed through collective cognitive action; simply grouping children and asking them to talk together will not necessarily help them to develop talking skills. Children need to understand what is meant by 'discussion' and have the skills to engage one another in speaking and listening in order to gain value from the talk activity. Children need to be taught how to talk to one another; they have to understand and share the aims for their talk. They need to recognise that if all the group can agree on a set of rules, 'ground rules for talk', then talk can proceed in a way which will make the whole group more likely to achieve success and develop new ways of thinking.

## Dialogic teaching

Classroom dialogue contributes to children's intellectual development and their educational attainment (Mercer and Littleton, 2007). Research has further shown that both interaction with adults and collaboration with peers can provide opportunities for children's learning and for their cognitive development (Alexander, 2010). Barnes (1971) found that language is a major means of learning and that pupils' uses of language for learning are strongly influenced by the teacher's language, which prescribes them their roles as learners. Barnes suggested that pupils have the potential to learn not only by listening passively to the teacher, but by verbalising, talking, discussing and arguing. Mercer and Hodgkinson (2008) built on the work of Douglas Barnes to further explore the centrality of dialogue in the learning process.

Alexander (2015: 48) argues that 'talk in learning is not a one-way linear communication but a reciprocal process in which ideas are bounced back and forth and on that basis take children's learning forward'. During dialogue, participant children (and their teachers) are equal partners striving to reach an agreed outcome, trying out and developing what Mercer (2000) has described as the joint construction of knowledge or 'interthinking'. For younger children, this can be related to Sustained Shared Thinking (Siraj and Asani, 2014). Interthinking can be achieved through dialogue with pupils, but pupils can interthink with each other in a process of joint enquiry. Dialogic approaches to teaching are therefore based on two main premises: (1) children as active participants in learning; (2) children using language to learn.

Whole-class interactive teaching has been shown to increase pupil achievement. The key word here is 'interactive', where pupils are allowed time for talk within a framework of effective direct teaching approaches. In order to be effective at direct teaching, teachers need to understand the complexities. Direct teaching does not mean simply one-way lecturing or 'traditional' teaching: it is interactive, it can occur between pupils and the teacher and/or between pupils and pupils, and it can involve several elements:

- clear, sequenced, structured presentations;
- effective pacing and timing;
- effective demonstrations and modelling of a particular skill or procedure;
- effective interactive structured questioning and discussion;
- relaying information that pupils do not know;
- introducing and modelling technical language/key vocabulary;
- relating to and building upon existing knowledge or understanding (for example refreshing pupils' memories of previous work);
- clarifying a sequence of cognitive or practical steps appropriate to learners;
- paired discussion work between pupils;
- pupil response and feedback;
- effective summarising;
- effective consolidation.

An independent evaluation of dialogic teaching in England found that the children in schools whose teachers had received training in dialogic teaching made two additional months of progress in English and science (Jay et al., 2017).

Whole-class interactive teaching also requires skilful questioning, including the careful selection and use of questions that are asked to check understanding. The teacher might ask for examples, pursue an issue in greater depth with a particular pupil or check understanding of a process as well as the product or the single right answer. Underlying whole-class interactive teaching enables the teacher to have more communicative contact with pupils, which is itself a critical factor in effective learning.

Questioning is a powerful tool for teaching because it allows for supporting, enhancing and extending children's learning. There are essentially two types of questions that teachers can use to elicit children's understanding: lower-order and higher-order questions. Lower-order questions are sometimes called 'closed' or 'literal' questions. They do not go beyond simple recall, and children's answers are either 'right' or 'wrong'. Higher-order questions require children to apply, reorganise, extend, evaluate and analyse information in some way. Both types of question have their place within an effective pedagogy; the type of question asked and the form in which it is posed will vary in relation to its purpose.

In addition, questions need to be formulated to match children's learning needs. It is possible to differentiate questions for different abilities and different children. Different questioning *techniques* can be used in order to support children's learning more thoroughly, such as prompting, probing and redirecting. Prompting may be necessary to elicit an initial answer to support a child in correcting his or her response, for example simplifying the framing of the question, taking them back to known material, giving hints or clues, accepting what is right and prompting for a more complete answer. Probing questions are designed to help children give fuller answers, to clarify their thinking, to take their thinking further, or to direct problem-solving activities, for example, 'Could you give us an example?' Questions can also be redirected to other children – for example, 'Can anyone else help?'

In dialogic talk, the questions asked by children are as important as the questions asked by the teacher, as are the answers given. The teacher is not using questions solely for the purpose of testing pupils' knowledge, but also to enable them to reflect, develop and extend their thinking. Wragg and Brown (2001) suggest several types of response that can be made to pupils' answers and comments. Teachers can:

- ignore the response, moving on to another pupil, topic or question;
- acknowledge the response, building it into the subsequent discussion;
- repeat the response verbatim to reinforce the point or to bring it to the attention of those that might not have heard it;
- repeat part of the response, to emphasise a particular element of it;
- paraphrase the response for clarity and emphasis, and so that it can be built into the ongoing and subsequent discussion;
- praise the response (either directly or by implication in extending and building on it for the subsequent part of the discussion);
- correct the response;
- prompt the pupils for further information or clarification;
- probe the pupils to develop relevant points.

These features indicate the type of response that can be made to pupils' utterances. It is easy for the teacher to miss important clues to children's understanding when they are too concerned with leading children towards a predetermined answer, so it is important to give children time to respond and, wherever possible, build further questions from their contributions. There are other matters to consider, for example allowing thinking time (particularly for complex responses), affording pupils the opportunity to correct, clarify and crystallise their responses once uttered, i.e. not 'jumping onto' a response before a pupil has had time to finish it; building a pupil's contribution into the teacher's own plans for the sequence of the discussion; and using a pupil's contribution to introduce another question to be put to another pupil. Galton and Hargreaves (2002) found that on average a classroom teacher waits only two seconds before either repeating a question, rephrasing, it, directing it to another child or extending it themselves. Their research showed that increasing wait time from just three to seven seconds results in an increase in the following:

- the length of pupil responses;
- the number of unsolicited responses;
- the frequency of pupil questions;
- the number of responses from less capable children;
- pupil-to-pupil interactions;
- the incidence of speculative responses.

Additionally, it is important to think about pace in relation to purpose – a series of closed questions may be appropriate, but at other times we want pupils to give more thoughtful and considered responses. To summarise: discovering what pupils know and what their misconceptions are requires good communication skills, language skills and empathy. Unlike questions from teachers which elicit only brief responses from pupils, we can see that dialogic talk is a type of interaction where teachers and pupils make substantial and significant contributions.

## Language in policy and practice

Prior to the 1960s, the idea that talk should be an important part of the English curriculum would have been greeted with some scepticism. However, educational researchers became increasingly interested in the idea that learning could be enhanced by careful consideration of the role of talk. Andrew Wilkinson's work resulted in him coining the new word 'oracy' as a measure of how important he thought talk was, a fact confirmed by the *Oxford English Dictionary,* which lists Wilkinson's text historically as the first time the word was used in print:

> **1965** A. WILKINSON *Spoken Eng.* 14 The term we suggest for general ability in the oral skills is *oracy,* one who has those skills is *orate,* one without them *inorate.*

The coining of a new word is perhaps the most fitting sign of Wilkinson's legacy. The work of Wilkinson and other educationists resulted in speaking and listening becoming part of the National Curriculum programmes of study for the subject English and, since the 1980s, the recognition of oracy as part of the early years and primary curriculum had been growing. However, as you will read later in the chapter, England's national curriculum of 2014 can be seen as a retrograde step for oracy.

Recent reviews of curricula in England further support a central emphasis on oracy. The *Cambridge Primary Review* concluded that oracy must have its proper place in the language curriculum. Indeed, spoken language is central to learning, culture and life, and is much more prominent in the curricula of other countries (Alexander, 2010). In addition, an increased understanding and focus on the importance of supporting children to develop early language skills has emerged (Sylva *et al.*, 2010). This recent work suggests, amongst other key findings, the importance of what happens in relation to a child's language experiences (both at home and in early years settings) during their formative years in relation to later educational outcomes.

Whilst few would now argue that speaking and listening are not important features of early years and primary teaching and learning, there are still a number of questions that need to be asked. One of the key questions concerns the balance between speaking and listening, reading and writing. To answer this question, there is a need to separate the curriculum content to be covered from considerations of teaching style. It seems to us that most of the debates about oracy and the recent considerations of talk in teaching and learning may have more to do with teaching style than a careful consideration of programmes of study. If national curricula are present, as they are in many countries, then it is appropriate that they should specify the content of the curriculum. This can apply to communication and language/speaking and listening just as it can apply to reading and writing and other subjects in the curriculum. However, there is a need for clear thinking about what this content should be. We would argue that if teachers' practice more routinely encouraged elements such as exploratory talk and dialogic teaching, then it might not be essential for the programmes of study for oral language, reading and writing to be absolutely equal in amount of content. The ways in which reading and writing are interconnected with oral language need better articulation in curriculum policies.

## 'Speaking and Listening' in the national curriculum

The National Curriculum in England originally divided the subject of English into Speaking and Listening, Reading, and Writing. Four main areas of speaking and listening were addressed: children should learn how to speak fluently and confidently; listen carefully and with due respect for others; become effective members of a collaborative group; and participate in a range of drama activities. There was further emphasis on the importance of using spoken Standard English and some thought given to language variation. However, the emphasis of language variation lay more on the functional linguistic emphasis of language in different contexts than learning centred on topics such as accent and dialect, language and identity, language and culture, etc.

In 2003, the Qualifications and Curriculum Authority (QCA, 2003) published a resource called *Speaking, Listening, Learning: Working with Children in Key Stages 1 and 2*. The pack was designed to support the teaching of speaking and listening in primary schools and consisted of a set of materials reflecting National Curriculum requirements in English. There were several premises supported by research cited in this chapter upon which these materials were built, the first of which emphasised the fact that children need to be taught speaking and listening skills, acknowledging that those skills develop over time and as children mature. It put forward an argument as to why speaking and listening are so important, linking them with children's personal and social development. The materials described the value of talk in helping children to organise their thoughts and ideas, pointing out that speaking and listening should not be seen as part of English as a subject alone, but as extending to all curriculum areas, acknowledging that different types of talk will be appropriate in different subject areas. The interdependency of speaking and listening, reading and writing was discussed and finally, approaches to assessment were addressed (assessment of oral language is looked at in → Chapter 20).

There have been a series of changes to England's National Curriculum over the decades since its inception, and the 2014 version brought some fairly radical developments. Most notable of all was the radical de-prioritisation of oracy to barely one page of requirements for 'Spoken language' (compared to hundreds of requirements for each of reading and writing). Oracy was no longer portrayed as important in its own right and worthy of equal attention to reading and writing. Instead, oracy appears as peripheral and only relevant to underpinning reading and writing. The other main driver for spoken language, consistent with all previous versions of the National Curriculum, was in relation to Standard English. The introduction to the National Curriculum sets out a framework that includes overarching principles about 'Language and literacy'. One of the opening statements in this section is, 'Pupils should be taught to speak clearly and convey ideas confidently using Standard English' (DfE, 2013: 11). An alternative to this that represents a better fit with evidence from linguistics might be, 'Pupils should be encouraged to speak using language forms appropriate for the context of interaction'. This statement accounts for the fact that levels of formality require different kinds of language use. A classic example is the difference in language required for many job interviews versus an informal conversation with a friend.

## Language in the early years

England's *Statutory Framework for the Early Years Foundation Stage* (EYFS Department for Education, 2017) puts the development and use of communication and language at the heart of young children's learning as one of three 'prime areas' of the seven 'areas of learning and development'. However, 'literacy' is a separate area of learning and development, and one of four 'specific areas'. In addition to the lack of clarity in the terms 'prime' and 'specific', communication and language and literacy are separated: communication and language are regarded as prime but literacy as specific. The lack of rigour in this separation is a product of the longstanding political problems around topics such as phonics for reading and the place of spoken language in curricula.

The identification of 'encouraging children to link sounds and letters' before the need to 'ignite their interest' through a 'wide range of reading materials' in the EYFS document is further indication of ideology conflicting with robust evidence-based policy making. What's more, the specification of the need for children to 'listen attentively', 'respond appropriately' and 'follow instructions' etc. without mention of the need to use language to create, question, wonder, play, engage, enjoy, is a concern because it is a distortion of children's natural language development and does not represent optimal guidance based on evidence of language development. Similarly, for literacy the emphasis is not first and foremost on reading for pleasure or linking writing with its fundamental purpose to communicate meanings (i.e. communication and language); instead, the emphasis is on 'simple sentences' and 'phonic knowledge'. Although these things are vital parts of language that children should be supported to learn, the presentation of them in the EYFS document does not match the evidence about how they are likely to be most effectively learned.

The raison d'être of language is the communication of meaning, and this is done through the inextricably linked modes of talk and written text. As such, any curriculum statement should rigorously reflect this in its emphases and ordering of curriculum elements. The most optimal learning happens when whole aspects of language, such as the interaction that happens through play or the experiencing of a book, are a constant reference point for any developmentally appropriate experiences and teaching that focus on smaller elements of language, such as letters, words and sentences.

Practitioners who understand the development of communication and language are better placed to create a language-rich environment in which talk has high status. In the early years setting, children do the following:

- Develop their knowledge and understanding about how language works.
- Develop an increasingly broader range and variety of vocabulary to use.
- Develop awareness of their audience – the people they are speaking to (there is some evidence to suggest that by the age of 4, children have learned to adjust their speech according to different audiences).

- Think about the appropriate language to use according to the circumstances of the situation.
- Learn to speak coherently and with clarity to make themselves understood.
- Learn to speak with confidence.

As children develop their language, they also build the foundations for literacy, for making sense of visual and verbal signs and ultimately for reading and writing. Children need varied opportunities to interact with others and to use a wide variety of resources for expressing their understanding, including mark-making, drawing, modelling, reading and writing.

Purposeful situations must be planned in order for children to practise their language skills and become aware of what is appropriate or suitable for a specific context. Children need to learn to take turns, negotiate, share resources, listen to and appreciate another person's point of view and function in a small group situation. Opportunities for purposeful language situations are many: in role-play areas, for example, or round a talk table. Collaborative interaction can be encouraged round the water and sand trays. If there are two chairs by the computer, one child can discuss with another the applications they are using, and children can also learn to wait for their turn (the use of an egg timer to make the waiting time fair can help). The practitioner can skilfully draw children into various activities and discussions in the setting, both indoors and outdoors.

Children need to know that the setting is a place where emotions can be expressed but that there may be undesirable consequences for expressing emotions in particular ways. Being able to manage some of these emotions through talk is the challenge for both the individual child and the practitioner. For example, young children experience an intense sense of injustice if they feel they have been wronged. Consider the scenario where one child hits another, who immediately responds by hitting back. The practitioner should aim to support the child to use language as a tool for thinking by, for example, prompting the child who was hit to think about these kinds of questions: why did they hit me? Did I do anything to provoke or upset them? Why am I upset? How should I respond to being hit? What should I do if this happens again? A strong early years setting will provide guidelines for children to follow or appropriate support systems if they find themselves in this kind of situation.

Non-verbal language such as facial expressions, effective eye contact, posture, gesture and interpersonal distance or space is usually interpreted by others as a reliable reflection of how we are feeling (Nowicki and Duke, 2000). Mehrabian (1971) devised a series of experiments dealing with the communication of feelings and attitudes, such as like–dislike. The experiments were designed to compare the influence of verbal and non-verbal cues in face-to-face interactions, leading Mehrabian to conclude that there are three elements in any face-to-face communication: visual clues, tone of voice and actual words. Through Mehrabian's experiments it was found that 55 per cent of the emotional meaning of a message is expressed through visual clues, 38 per cent through tone of voice and only 7 per cent from actual words. For communication to be effective and meaningful, these three parts of the message must support each other in meaning; ambiguity occurs when the words spoken are inconsistent with, say, the tone of voice or body language of the speaker.

Young children are naturally physically expressive, such as when they are tired, upset or happy, yet they do not always understand straightaway the full meaning another child is conveying. In a situation of conflict, for example, it can be useful when practitioners point out the expression on a 'wronged' child's face to highlight the consequences of someone else's actions. Conversely, if a child is kind to another child and that child stops crying or starts to smile, then this too can be highlighted.

Similarly, the practitioner needs to be aware of the messages they are sending out to a child via their use of non-verbal language. It is important to remember that whenever we are around others, we are communicating non-verbally, intentionally or not, and children need to feel comfortable in the presence of the adults around them. According to Chaplain (2003: 69), 'children are able to interpret the meaningfulness of posture from an early age'. Even locations and positions when talking can be important. For example, it is beneficial when speaking with a young child to converse at their physical level, sitting, kneeling or dropping down on one's haunches alongside them. This creates a respectful and friendly demeanour and communicates genuine interest in the child and what they are doing.

### Practice points

- Talk with children so that they feel that you respect them, are interested in them and value their ideas.
- Give children your full attention as you talk with them; use direct eye contact to show that you are really listening.
- Find ways of encouraging children to talk in a range of contexts.
- Using specific positive praise, such as 'I really liked the way that you waited patiently for your turn on the computer'.
- Smile!

### Glossary

**Cumulative talk** – classroom interaction characterised by sharing of ideas, agreement and lack of critical evaluation of ideas.

**Disputational talk** – classroom interaction characterised by maintaining a pint of view and competing with others; views.

**Exploratory talk** – classroom interaction characterised by equal sharing of ideas and commitment to collaboration.

**Morpheme** – a minimal unit of language, with grammatical meaning, that cannot be further divided; e.g. the three morphemes in the word incoming: in come ing.

**Phonological development** – development of understanding of sounds (phonemes) and ability to use phonemes as part of speech or recognise them when reading.

## References

Alexander, R. J. (2010). *Children, Their World, Their Education: Final Report and Recommendations of the Cambridge Primary Review*. London: Routledge.

Alexander, R. J. (2015). *Towards Dialogic Teaching: Rethinking Classroom Talk*, 4th edn. York: Dialogos.

Barnes, D. (1971). 'Language and learning in the classroom', *Journal of Curriculum Studies,* 3(1): 27–38.

Barnes, D. (1976). *From Communication to Curriculum*. Harmondsworth: Penguin.

Bearne, E. (1998). *Making Progress in English*. London: Routledge.

Blatchford, P., Kutnick, P., Baines, E. and Galton, M. (2003). 'Toward a social pedagogy of classroom group work', *International Journal of Educational Research*, 39: 153–172.

Chaplain, R. (2003). *Teaching without Disruption in the Primary School*. London: RoutledgeFalmer.

Chapman, R. (2000). 'Children's language learning: An interactionist perspective', *Journal of Child Psychology and Psychiatry*, 41(1): 33–54.

Department for Education. (2013). *The National Curriculum in England: Framework Document: December 2014*. London: Department for Education.

Department for Education. (2017). *Statutory Framework for the Early Years Foundation Stage: Setting the Standards for Learning, Development and Care for Children from Birth to Five*. London: Department for Education.

Galton, M. and Hargreaves, L. (2002). *Transfer from the Primary School: 20 Years on*. London: Routledge.

Goswami, U. (2015). *Children's Cognitive Development and Learning*. York: Cambridge Primary Review Trust.

Jay, T., Willis, B., Thomas, P., Taylor, R., Moore, N., Burnett, C., . . . Stevens, A. (2017). *Dialogic Teaching: Evaluation Report and Executive Summary: July 2017*. London: Education Endowment Foundation.

Kuhl, P. K. (2004*).* 'Early language acquisition: Cracking the speech code', *Nature Reviews Neuroscience*, 5(11): 831–843.

Mehrabian, A. (1971). *Silent Messages*. Belmont, CA: Wadsworth.

Mercer, N. (2000). *Words and Minds: How We Use Language to Think Together.* London: Routledge.

Mercer, N. and Hodgkinson, S. (2008). *Exploring Talk in School*. London: Sage.

Mercer, N. and Littleton, K. (2007). *Dialogue and the Development of Thinking: A Sociocultural Approach*. New York: Routledge.

Messer, D. (2006). 'Current perspectives on language acquisition', in J. S. Peccei (ed.), *Child Language: A Resource Book for Students*. London: Routledge.

Nowicki, S. and Duke, M. (2000). *Helping the Child Who Doesn't Fit In*. Atlanta, GA: Peachtree.

Qualification and Curriculum Authority (QCA) and Department for Education and Skills (DfES). (2003). *Speaking, Listening, Learning: Working with Children in Key Stages 1 and 2*. London: DfES Publications.

Reese, L., Garnier, H., Gallimore, K. and Goldenburg, C. (2000). 'Longitudinal analysis of the antecedents of emergent Spanish literacy and middle-school English reading achievement of Spanish-speaking students', *American Educational Research Journal*, 37(3): 633–662.

Siraj, I. and Asani, R. (2014). 'The role of sustained shared thinking, play and metacognition in young children's learning', in S. Robson and S. Flannery Quinn (eds.), *The Routledge International Handbook of Young Children's Thinking and Understanding*. London: Routledge.

Sylva, K., Melhuish, E., Sammons, P., Siraj-Blatchford, I. and Taggart, B. (2010). *Early Childhood Matters: Evidence from the Effective Pre-School and Primary Education Project*. Oxford: Routledge.

Wragg, E. C. and Brown, G. (2001). *Questioning in the Primary School*. London: RoutledgeFalmer.

### Annotated bibliography

Alexander, R. J. (2017). 'Dialogic Teaching in Brief', www.robinalexander.org. uk/wp-content/uploads/2012/10/Dialogc-teaching-in-brief-170622.pdf Succinct summary of the research and theory that has informed a range of related approaches to classroom dialogue.
L2 ★★★

Mercer, N. and Hodgkinson, S. (2008). *Exploring Talk in School*. London: Sage. In addition to the valuable advice in this book, a website developed from Neil Mercer's research can be found at: http://thinkingtogether.educ.cam. ac.uk/. The website looks at research in the area of talk and provides some downloadable materials for teachers, with links to book, research projects and other websites.
L1 ★★

https://educationendowmentfoundation.org.uk/our-work/projects/ dialogic-teaching/ Provides information about the independent evaluation of an approach to dialogic teaching.
L3 ★★★

# Chapter 5

# Accent, dialect and Standard English

The emphasis in this chapter is on accent and dialect ☞ as rich resources of the English language. A discussion on Standard English ☞ flags up the political factors that are at work. We conclude with some thoughts on language and identity.

> The Jay makes answer as the Magpie chatters;
>   And all the air is filled with pleasant noise of waters.
>
> (Wordsworth, 1807: 270)

William Wordsworth's regional accent meant that water would have been pronounced 'watter' in the extract above; 'chatter' and 'water' represent a natural rhyme. Many poets have embraced the wonderful variation and authenticity that come from accent and dialect. The study of accent and dialect is an important part of knowledge about language (→ Chapter 16).

One of the reasons why the English language is considered to be so rich is because of the many intriguing and fascinating variations it has to offer. These variations reveal themselves in many ways, including through accent and dialect. While there are many people in society who regard accents and dialects as a rich source of language, there is sometimes a tendency to treat them differently in schools. Some teachers feel that they are obliged to correct children's 'mispronunciations' because of the National Curriculum's insistence on the use of Standard English. It is not difficult to become confused about the differences between accent, dialect, Standard English, the Queen's English, etc. The whole business of the child's language can seem like a linguistic minefield. A strong understanding of some of the terms can help you to know when it is appropriate to correct a child and when it may be inappropriate.

Accent is the more straightforward term because it refers only to differences in pronunciation. Accents are associated with regional and social characteristics. Issues of language and power are inseparable from consideration of accent. For example, consider the small number of speakers of accents other than *received pronunciation* ☞ that you hear in news broadcasts, or the attitudes that people have to accents from the UK cities of Liverpool, Birmingham or Glasgow. One of the engaging things about accents is trying to guess a speaker's geographical origins from their speech. However, you cannot guess this about speakers who use *received pronunciation*.

Received pronunciation (RP) is sometimes referred to as 'the Queen's English' or 'BBC English'. It is the 'posh' accent which we have come to associate with public schools, 'high society' and radio broadcasters from 50 years ago. It is different to other accents because it denies the listener any indication of the speaker's geographical origin. It is primarily a socially influenced accent rather than a geographically influenced one, and it locates the speaker in a particular social group.

All speakers of English use a dialect. Dialect refers to a specific vocabulary and grammar which is often influenced by geographical factors. It does not refer to the ways in which words are pronounced. Regional dialect includes particular words that are special to the locality. For example, a flat, circular slice of potato cooked in a fish and chip shop has a large range of names across the country: in Warrington it is a *scallop*, in South Wales it is a *patty*, in Liverpool it is a *fritter*, in West Bromwich it is a *klandike* and in Crewe it is a *smack*. Dialect also contains grammatical differences: for example, in Stoke the phrase 'Her's gone up Hanley, duck' (she's gone into Hanley) is an example of the ways in which regional dialect alters the grammatical structure of the sentence while maintaining meaning (the joke in Stoke is that this is all one word: gonnerpanley!).

It is possible to distinguish between *traditional dialects* and *mainstream dialects*. An example of a traditional dialect would be the following (see if you can work out where the speaker might come from – answer at end of chapter):

> *Ah telt thee to seh thoo hazn't to gan yem the neet*
> I told you to say you mustn't go home tonight
> (Kerswill, 2007: 41)

Mainstream dialects include non-standard dialects spoken now, such as this:

> *She come up Reading yesterday.*
> She came to Reading yesterday.
> (op. cit.)

More controversially, to the lay person, it is also more accurate to describe Standard English as a dialect. Standard English is a distinct form of the language that differs from other forms in some of its vocabulary and grammar;

hence, it is a dialect even though it does not have particular pronunciation associated with it.

A more recent phenomenon is the idea of *dialect levelling*. This is where the features of local accents and dialects are reduced in favour of new features that are adopted by speakers that cover a wider area. So-called *Estuary English* can be seen as relatively homogenised accents or dialects spoken in the South East of England (op. cit.).

In order to see the direct interaction between children's language and identity consider the example of Neil's writing (Figure 5.1).

We introduce you to the composition, spelling and punctuation of Neil's writing in later chapters, but for now can you guess where Neil lived? No? Try reading the writing out loud with a London Cockney accent! You can see how Neil's accent shows in his spelling because he attempts to convert the spoken versions of words to the written versions. Dialect also features, for example in 'two Romans *was going* to another land; the people *was the* Hojibs; and the Romans *was very* sad'. He was grappling with the conventions of Standard English writing using the linguistic experience that formed part of his identity.

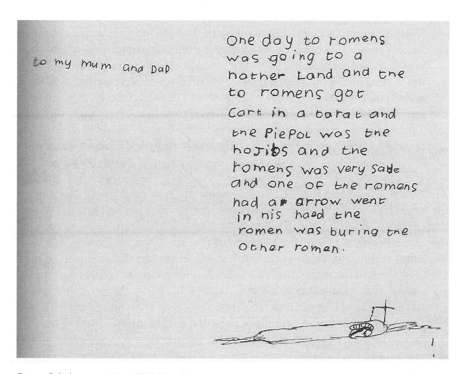

*Figure 5.1* An example of Neil's writing.

## Standard English

The question of spoken 'Standard English' is one that has also been particularly influenced by political factors. It is a complex issue that is centrally about the social context of language use, and it is bedevilled by prejudice and misunderstanding. The Cox Report contained a sensible discussion of the issue (DES, 1989), and the 1990 National Curriculum orders required only that older primary pupils should have opportunity to use spoken Standard English 'in appropriate contexts' (DES, 1990: 25) and that knowledge should rise out of the pupil's 'own linguistic competence' (DES, 1989: 6.11). The 1995 National Curriculum highlighted the issue of Standard English in a separate section which included the misleading requirement that 'To develop effective speaking and listening pupils should be taught to use the vocabulary and grammar of Standard English' (DfE, 1995: 2, 3) but otherwise left the Cox approach intact. The statement is misleading because effective speaking and listening develops in all dialects, not just Standard English.

The National Curriculum of 2015 reduced, for the first time in many years, the emphasis on spoken language. At the same time, the emphasis on a particular variation of English, described as Standard English, was increased. The changes in emphasis raise another linguistic consideration: how does the National Curriculum define 'Standard English'?

> Standard English can be recognised by the use of a very small range of forms such as those books, I did it and I wasn't doing anything (rather than their non-Standard equivalents); it is not limited to any particular accent. It is the variety of English which is used, with only minor variation, as a major world language. Some people use Standard English all the time, in all situations from the most casual to the most formal, so it covers most registers. The aim of the national curriculum is that everyone should be able to use Standard English as needed in writing and in relatively formal speaking.
>
> (DfE, 2013: 99)

The idea that Standard English is used only with 'minor variation' throughout the world is simply untenable, particularly in relation to spoken English (rather than written English, which has somewhat less variation). The suggestion that some people use Standard English 'all the time' is very dubious. Standard English is technically a dialect. In 2007, it was estimated that it was spoken by 12–15 per cent of the population in England, with use more common in speakers with higher socio-economic status (Kerswill, 2007). A more appropriate aim of the National Curriculum would be: 'children and young people should be able to use spoken and written language that communicates most effectively in relation to the context they are in'. This implies that young people should take account of the fact that language use will differ significantly according to whether that context is spoken language or written language.

## Language and identity

Trudgill (1975) offers a pyramid of social variation, at the peak of which are those who use Standard English and speak with RP, resulting in high social status. The base of the pyramid is made up of a wide range of regional accents and dialects which are afforded low social status. Given that accents and dialects are closely linked to perceptions of social status, then there are issues for teachers to consider if one form of dialect is accepted as superior in the classroom.

> There is a particular danger that, if Standard English is held up as a superior dialect which ought to replace the child's own, the child will come to resent and reject anything that has to do with Standard English – especially reading . . . there is evidence to suggest that some children at least may not learn to read because they do not want to: and that they do not want to for reasons which have to do with group identity and cultural conflict, in both of which dialect certainly plays a role.
>
> (Trudgill, 1975: 67)

As an example of identity and conflict we offer the following anecdote. When one of the authors of this book was completing his initial teacher training, he was told by a supervising tutor that he could not be a teacher until he had lost his 'working-class accent'. Over ten years later, while completing classroom-based research, the author watched a lesson given by a student teacher in a classroom where she, the teacher and the children all shared the same strong Potteries accent and dialect. Despite the fact that the lesson had progressed perfectly well, and the children had achieved the aims of the lesson, the student was failed outright by the university tutor, who 'could not understand a word that was said'. The student reported that 'I was told I can't speak. I'm common'.

These examples illustrate the close connection between accent, dialect and personal identity. The insensitive correction of regional dialects runs the risk of upsetting people, whether they are children or adults. It is for this reason that activities which encourage reflection on language in different contexts (such as role-play for different degrees of formality and looking at the differences between speech and writing) are preferable to continual correction. There have been arguments to suggest that non-standard dialects should be purged from the classroom, and teachers have resorted to all kinds of tactics to attempt to ensure this takes place:

> Generations of teachers have employed persuasion, exhortation, punishment, scorn and ridicule in attempts to prevent children from using non-standard dialects – and all of them without success. And there is no reason to suppose that they will be any more successful in the future.
>
> (Trudgill, 1975: 66)

The key linguistic issues surrounding accent, dialect and standards of language linked to deficit views of children's language are still prevalent. Stereotypes linking social class with 'poor' language are a danger not just in everyday discussions but also as part of the academic study of language, where we are usefully reminded that curiosity about, and emphasis on, language *in use* rather than idealistic linguistic models remains essential for teachers and researchers alike (Snell, 2013).

The issue for the teacher, therefore, is how to achieve the balance between using the child's own language as a motivational and cultural tool for development while at the same time illustrating that certain language forms are deemed more appropriate than others at certain times and in certain social situations. Much of this is achievable by establishing particular audiences and contexts for talk activities. However, although children have to be aware that Standard English is the norm in formal situations, they also need to be aware that there is considerable prejudice against regional accents and dialects.

### Practice points

- Regional dialects should be respected, enjoyed and seen as a rich language resource.
- Standard English is often best discussed in the context of writing or role-play.
- Helping students to understand standard conventions of writing is essential, but this always needs to be done sensitively.

### Glossary

**Dialect** – regional variations of grammar and vocabulary in language.
**Received pronunciation** – a particular accent often called 'BBC English' or 'the Queen's English'.
**Standard English** – the formal language of written communication in particular. Many people call this 'correct' English.

### References

Department for Education (DfE). (1995). *English in the National Curriculum.* London: HMSO.
Department for Education (DfE). (2013). *The National Curriculum in England: Framework document. September 2014.* London: Department for Education.
Department of Education and Science and The Welsh Office (DES). (1989). *English for Ages 5–16 (The Cox Report).* York: National Curriculum Council.

Department of Education and Science and The Welsh Office (DES). (1990). *English in the National Curriculum*. London: HMSO.

Kerswill, P. (2007). 'Standard and non-standard English', in D. Britain (ed.), *Language in the British Isles*. Cambridge: Cambridge University Press.

Snell, J. (2013). 'Dialect, interaction and class positioning at school: From deficit to difference to repertoire', *Language and Education*, 27(2): 110–128.

Trudgill, P. (1975). *Accent, Dialect and the School*. London: Open University Press.

Wordsworth, W. (1807). 'Resolution and independence', in *Oxford Library of English Poetry*, Vol. 2. Bungay: Book Club Associates.

## Annotated bibliography

The UCL Survey of English Usage is home to one of the most respected corpora of language, and analysis of grammar, internationally. It includes the *engliscious* resource to support grammar teaching.
www.ucl.ac.uk/english-usage/
**L1** ★

Crystal, D. (2004). *The Stories of English*. London: Penguin/Allen Lane.
A tremendous achievement. David Crystal shows how language and the English language in particular continue to change, driven by the need to communicate meaning. The social status of different kinds of English is clearly shown.
**L2** ★★

Grainger, K. and Jones, P. (2013). 'The "Language Deficit" argument and beyond', *Language and Education*, 27(2): 95–98. doi: 10.1080/09500782.2012.760582
The editorial for a special issue of articles all addressing the idea that certain forms of language are inferior to other forms of language.
**L3** ★★★

Britain, D. (ed.). (2007). *Language in the British Isles*. Cambridge: Cambridge University Press.
A fascinating, linguistically accurate account of language. Builds on, and is a tribute to, the work of Peter Trudgill.
**L2** ★★★

Answer to dialect question: The speaker would be from Durham in the North East of England.

# Chapter 6

# Classroom practices for talk

> The principle underlying this chapter is that spoken language is a powerful tool for learning. Teaching talk skills and strategies for the practice and consolidation of these are essential to maximise children's capacity to learn, achieve and participate. We share examples of children talking and interacting, and outline techniques for the direct teaching of talk.

As you saw in Chapter 4, 'The development of language', language is a vital part of learning and teaching. Focusing on the quality of spoken dialogue in primary classrooms can significantly improve children's educational attainment (Mercer and Littleton, 2007). The 2014 National Curriculum in England states that:

> Pupils should be taught to speak clearly and convey ideas confidently using Standard English. They should learn to justify ideas with reasons; ask questions to check understanding; develop vocabulary and build knowledge; negotiate; evaluate and build on the ideas of others; and select the appropriate register for effective communication. They should be taught to give well-structured descriptions and explanations and develop their understanding through speculating, hypothesising and exploring ideas. This will enable them to clarify their thinking as well as organise their ideas for writing.
>
> (DfE, 2013: 11)

These skills do not develop by chance but require careful planning and resourcing. Whilst for nearly all children the development of language occurs naturally, and some children talk fluently from their earliest years, *how* to use language effectively requires specific teaching. It is important for teachers to understand language development, and that oracy ☞, or speaking and listening, is taught from an early age. Opportunities for talk should be planned systematically, not only within English or literacy sessions, but across the curriculum so that children's ability to use talk for thinking and learning is given an appropriate level of attention in comparison with reading and writing skills. The teaching strategies we outline below will help broaden and enhance children's command of language by offering a range of different contexts and overt experiences of different types of talk. These include opportunities to engage in collaborative, exploratory and imaginative play; discuss, extend and justify opinions; debate and argue; listen and respond to the speech of others; and create and respond to oral performance (further ideas are illustrated in ➝ Chapter 7: Drama).

## Practices for talk in early years classrooms

Early years classrooms should provide plenty of opportunities for talk. For example, typically there will be at least one topic-linked role-play area; there may also be a role-play area outdoors. Books introduce the language of written narrative, in fiction and non-fiction. Children listen to stories, perhaps independently through headphones linked to a CD player, as well as in situations with another adult or in a group. Teachers can introduce a *talk table* where objects are arranged for children to touch, feel, look at and discuss – for example, feathers, stones or shells. Children are encouraged to talk about their experiences in the classroom, such as playing in water and sand, or what it feels like when they run their hands through gloop. Reflective early years teachers work to extend children's development as speakers and listeners and help them express themselves effectively. In the Early Years Foundation Stage (EYFS) (DfE, 2017), communication and language is one of the three prime areas crucial for every child's development. Children need to experience contexts in their Reception classrooms that offer them the opportunity to talk about everyday experiences. A key aspect of talking and listening is therefore the way in which it can encourage the development of vocabulary through opportunities for creative thinking and learning. One such approach is often referred to as sustained shared thinking (SST) where two or more individuals talk together, for example to solve a problem, clarify a concept, evaluate an activity or extend the beginning of a narrative (Siraj-Blatchford, 2004). SST can occur where children are interacting 1:1 with an adult or with a peer, or during focused group work, where language to be learned can be scaffolded and modelled. SST is effective when both parties contribute to thinking that develops and extends understanding.

The strategy of talk partners where children talk in pairs on the carpet, in response to a question from the teacher, can be used effectively (see 'think, pair, share' below). Some teachers choose to pair a stronger speaker with one who benefits from such peer support to help scaffold the child's language skills. The notion of *scaffolding* has been built upon in the findings from the *Effective Preschool and Primary Education Project* (EPPE) (Sylva et al., 2010), which show the value of SST, for example, where the importance of using open-ended questioning with children to elicit their knowledge and understanding is perceived as key to moving learning forwards.

Language skills are fundamental to communication, but the interpersonal nature of communication also calls for the development of significant social skills. This links to a second prime area in the EYFS (DfE, 2017), that of social and emotional development, and so is of importance for teachers of infant children. For example, children need to learn to take turns during group talk, to know how to ask for something appropriately, to know what it means to be a friend. It is additionally important to remember with such young children that a great deal of what we say is conveyed by non-verbal means. Tone, pitch, volume and how we project our voice are all part of the communication process. Non-verbal language, such as facial expression, effective eye contact, posture, gesture and interpersonal distance or space, is usually interpreted by others as a reliable reflection of how another person might be feeling. And confident oral communication has benefits beyond its intrinsic value; for example, studies of young writers have reported that verbal interaction with others should be an integral component of the early writing process (Kissel *et al.*, 2011).

## Practices for talk in primary classrooms

As we summarised in Chapter 4, *exploratory talk* is the optimum type of talk that teachers should aim to develop amongst pupils in their classrooms (Mercer and Hodgkinson, 2008). In other words, children are thinking together, and we can hear them thinking aloud, hypothesising and speculating. Children might use words and phrases such as 'perhaps', 'if', 'might' and 'probably'; they give reasons to support their ideas using words such as 'because' and seek support from the group. In this kind of scenario, children are listening to each other and considering their response. When children are working in this way, their reasoning becomes apparent through their talk. However, this kind of talk does not come naturally to them: they need to be guided by their teachers to understand the value of collaborative talk.

Talk is less in-depth when a teacher:

- asks a series of closed questions which necessitate single word, 'correct' answers;
- asks a series of short questions which are followed by short answers and which lead to further short questions.

In these circumstances, exploratory talk is non-existent as children are merely responding to the teacher's questioning. In situations where talk is genuinely exploratory, children:

- can generate more creative, descriptive and insightful observations without their teacher;

- cooperate as a group – they share their ideas, they listen to one another and they respond positively to new suggestions;
- look for opportunities to draw one another into the task, typically using 'tag' questions;
- are unafraid to hypothesise ☞.

These are some of the ground rules for exploratory talk:

- Everyone in the group must be encouraged to contribute.
- Contributions must be treated with respect.
- Reasons are asked for.
- Everyone must be prepared to accept challenges and justify responses.
- All relevant information should be shared and alternative outcomes need to be discussed before a group decision is taken.

The teacher should encourage the children to talk about and develop their own list of 'ground rules for talk', written in their own words. This can be placed on the wall or printed out as a reminder for small groups when they are working on their own. In order to assess the quality of group interaction, the teacher must be clear about the objectives for the task.

It is important to remember that the teacher's ability to plan exciting learning opportunities and to sometimes *leave* children to talk is an important skill in itself, although this is not to say that direct instruction and intervention does not also have their place. It is, rather, through a judicious combination of different types of talk experience that children will learn best. Guidance often stresses the need for teachers to model features of speaking and listening, but there are times when good teaching allows children to explore and interact independently. This type of talk, however, also benefits from sensitive scaffolding. In the following extract, the teacher rehearses collaborative strategies with the children:

| | |
|---|---|
| *Teacher:* | What do we need to remember in our groups? |
| *Student A:* | That everybody gets a turn to talk. |
| *Teacher:* | That everybody gets a turn to talk. |
| *Student B:* | Everybody needs to share their opinions. |
| *Teacher:* | Yeah – and are we all the same? |
| *Students:* | No. |
| *Teacher:* | Will there be someone in your group that perhaps wants to talk all the time? |
| *Students:* | Yes. |
| *Teacher:* | Will there be someone in your group who doesn't want to talk at all? |
| *Students:* | Yes. |
| *Teacher:* | How are you going to get that person who doesn't want to talk at all to say something? Shane? What do you think? How are you |

|  |  |
|---|---|
|  | going to get that person who sits there and doesn't say anything to say something in your group? Help him out, Tyber. |
| *Student C:* | Ask them. |
| *Teacher:* | Ask them – brilliant. What about that person who talks *all* the time? |
| *Student D:* | Tell them to shut up. |
| *Teacher:* | Ooh! Are you? I hope not, because that's not positive language is it? What can you do to help them out? |
| *Student A:* | Ask them and then ask somebody else and then ask the other person. |
| *Teacher:* | Brilliant. Making sure that you ask everybody in the group. Excellent. |

(Mercer and Littleton, 2007: 118)

Here the teacher has explicitly reminded children of teaching points in relation to group talk and has supported the development of ground rules for discussion.

## Strategies to promote effective talk in classrooms

When planning for talk, the teacher can draw on a repertoire of strategies to support the planning process. Strategies which can be deployed during paired work, group work and whole class teaching are outlined below.

### Talk partners

Children are put into pairs and allocated time for each to talk to the other at specific points during a teaching sequence. They might, for example, share experiences, generate ideas or reflect on what they have just learned. *Talk partners* is a strategy which means that all children are given the opportunity to think, discuss and express themselves orally. Some children may feel more confident when expressing their ideas in a paired situation rather than to the whole class. Talk partners also allows children to organise their thoughts, giving them time to think ahead of speaking out loud to a larger group.

### Group work

Group work can help children to develop the language and social skills needed for cooperation and collaboration; to use exploratory language to try out ideas; to extend their ideas as they share these with others; to support and build on each other's contributions; and to take their turns in discussion. All of these enable children to 'interthink' (Mercer and Littleton, 2007). Some examples of strategies to consider when planning for group work include the following:

## Twos to fours

The teacher sets a particular problem for a pair to discuss. After discussion, the pair meets with another pair who have been given *exactly the same task* in order to compare and elaborate on their findings.

## Think, pair, share

This is a cooperative learning strategy in which children think across questions in three distinct steps so that their thinking becomes more refined:

1    Think: children think independently about the question that has been posed, forming ideas of their own.
2    Pair: children are grouped in pairs to discuss their thoughts. This step allows them to articulate their ideas and to consider those of others.
3    Share: pairs share their ideas with a larger group (this could be the whole class).

## Envoying

When working in groups, one member of each group is allocated the role of 'envoy'. The envoy has the responsibility of gathering further information and resources as required, reporting progress to the teacher and seeking further clarification for the group. This is a particularly effective way of managing practical group activities as the teacher can focus attention on a much smaller number of children in order to maintain progress.

## Jigsawing

Children are organised into 'home' groups (of four to six children) to begin to solve a particular problem or to work on a collaborative activity. Each child in the group then has the responsibility of finding out more about one particular aspect of the problem. These children gather together in 'expert' groups in order to gather as much information as possible to then take back and share with their 'home' group. Once each child in the group has given their expert opinion to their home group, the problem-solving continues until an end point is reached.

## Assigning roles

There may be occasions when assigning roles to individuals within group work can further enhance talk. Such roles include:

•    Leader/chair – the leader organises the group, encouraging all group members to participate and to complete the task.

- Scribe – the scribe notes main points of discussion and any decisions, and checks accuracy of notes with group members.
- Reporter – the reporter works with the scribe to organise the report on findings, and by summing up and presenting ideas.
- Mentor – the mentor helps group members to carry out the task, supporting them and explaining what is needed.
- Observer – the observer makes notes on how the group works and on different contributions, then shares the observations with the group.

### *Whole class questioning*

Studies show that children learn more in classrooms where teachers use a mixture of different types of question. 'Closed' questions require relatively brief, factual answers, while 'open-ended' questions are broader, inviting a range of possible answers. In order to get the most out of whole class talk, include open-ended questions to elicit children's understanding and ideas; you may want to target individual children and specific groups in this way. Exploratory questions to promote exploratory talk might include:

- What do you think? Why?
- What makes you think . . .?
- How do you feel about . . .?
- Can you explain why . . .?
- Do you agree with . . .'s opinion?
- Is there anything that puzzles you?
- What would happen if . . .?

## Creating a learning environment which promotes talk

Conditions need to be established for effective talk to happen in classrooms. This includes the physical environment; from this perspective, the following should be considered:

- furniture organised to facilitate collaborative work in pairs and small groups;
- interactive displays;
- a collection of props and artefacts to stimulate and promote discussion and storytelling;
- a role-play area;
- posters displaying ground rules for discussion.

## Practice points

- Talk-based activities require specific planning, just as reading and writing activities do.
- Talk should be promoted within an environment that values children's contributions.
- A range of strategies should be used to promote collaborative paired, group and whole class talk.

## Glossary

**Hypothesise** – to suggest likely outcomes, reasons and/or explanations by offering opinions in relation to a subject of discussion.

**Oracy** – the capacity to express ideas clearly and fluently in spoken language.

**Tag questions** – questions that are added onto the end of a statement, e.g. '. . . isn't it?' or '. . . aren't they?'

## References

Department for Education. (2013). *The National Curriculum in England: Framework Document. September 2013.* London: Department for Education.

Department for Education. (2017). *Statutory Framework for the Early Years Foundation Stage: Setting the Standards for Learning, Development and Care for Children from Birth to Five.* London: Department for Education.

Kissel, B., Hansen, J., Tower, H. and Lawrence, J. (2011). 'The influential actions of pre-kindergarten writers', *Journal of Early Childhood Literacy*, 11(4): 425–452.

Mercer, N. and Hodgkinson, S. (2008). *Exploring Talk in School.* London: Sage.

Mercer, N. and Littleton, K. (2007). *Dialogue and the Development of Children's Thinking.* London: Routledge.

Siraj-Blatchford, I. (2004). 'Educational disadvantage in the early years: How do we overcome it? Lessons from research', *European Early Childhood Education Research Journal*, 12(2): 5–20.

Sylva, K., Melhuish, E., Sammons, P., Siraj-Blatchford, I. and Taggart, B. (2010). *Early Childhood Matters: Evidence from the Effective Preschool and Primary Education Project.* Oxford: Routledge.

### Annotated bibliography

Blatchford, P., Kutnick, P., Baines, E. and Galton, M. (2003). 'Toward a social pedagogy of classroom group work', *International Journal of Educational Research*, 39, 153–172.
An important contribution to the research on group work in classrooms, including the difference between effective and ineffective group work.
**L3** ★★★

*Dawes, L. and Sams, C. (2017). Talk Box: Activities for Teaching Oracy with Children Aged 4–8, 2nd edn. Abingdon: Routledge.*
*Practical ideas for developing talk based on theory into the benefits of exploratory talk. The techniques can be adapted for primary classrooms.*
**L1** ★

*Thinking Together.* For more information and resources on the possibilities of talk in the primary classroom, visit http://thinkingtogether.educ.cam. ac.uk/.
**L1** ★

# Chapter 7

# Drama

Reasons for teaching drama are outlined. Drama is linked to children's developing and expressive language repertoires; to play, particularly role-play in the early years; and to storytelling and writing creatively across the primary curriculum. A theoretical model of the 'building blocks of drama' is provided and some practical ideas are presented as starting points. Drama is perceived as concerning the making of meaning from a language and literacy perspective, rather than necessarily making enacting play scripts.

Since its inclusion in 1999 as a separate section in the speaking and listening requirements of the National Curriculum, over the past two decades the importance of drama in its own right as a valuable tool for learning has been formally recognised. From 2006, drama was also seen as essential for the development of literacy, with its own strand of learning objectives as part of the literacy framework for teaching. The current National Curriculum (DfE, 2013: 4) states that 'all pupils should be enabled to participate in and gain knowledge, skills and understanding associated with the artistic practice of drama', although further specification is only confined to non-statutory guidance. This chapter explains how some of this knowledge, skills and understanding can be acquired.

For children, drama incorporates the development and exploration of self-expression through the use of talk and action. It has been shown to be a powerful stimulus for writing (see for example Wyse *et al.*, 2009). And drama has the power to enable children to experience and explore a situation with their teacher and peers, and to organise their thoughts and range of vocabulary ahead of committing them to paper.

Some teachers might feel a lack of confidence about drama, claiming that it is like teaching without a 'safety net', while others will find it one of the most liberating and invigorating aspects of their pedagogical approach. Drama sessions can provide some of the most memorable, challenging, enjoyable and rigorous moments of the child's time at school. Good drama teaching builds on positive relationships and trusting interaction between teachers and learners.

## Using drama within your pedagogical repertoire

The centrality of language in children's development has included an increased recognition of the place of drama and role-play. Drama can be used across the curriculum, opening up a form of language-based study which is beneficial for the child not just educationally, but even spiritually, morally and socially. Teachers should plan accordingly, encouraging children to use the richness of the experience in a multitude of ways.

- Drama helps children to understand their world more deeply and allows them an opportunity to find ways to explore and share that understanding.
- Drama encourages self-expression and focuses the child on the art of communication.
- Although not exclusive to speaking and listening, drama includes a large oral element which allows all children to use, practise and develop their language skills, sometimes broadening their vocabulary.
- Drama offers the early years and primary teacher a route into language study and exploration that is not covered by any other form of teaching.
- Drama helps children to cooperate and collaborate with their peers through encouraging them to see themselves in a wider social context, thus promoting a greater awareness of the self and sensitivity to others. Drama can raise the self-esteem of even the most disaffected of pupils (Woolland, 2010).
- Drama can help bring a story and its characters alive through deepening perspectives and considering alternative points of view.
- Drama can invigorate children's writing (Cremin et al., 2006; Wyse et al., 2009).
- Relationships between the teacher and the child through their shared language are different in drama sessions. Expectations change, negotiated progress is a more prominent feature and there is a greater sense of active participation for the child.
- Drama, by its very nature, can be used to create direct links across the curriculum, and thus is not confined to the teaching of English.
- Drama can be highly motivating for children and highly productive for teachers as learning becomes a more dynamic process.
- Drama offers an element of negotiation and unpredictability within the curriculum.
- Drama is an art form which has played a central part in our cultural heritage.

## The early years

When children first arrive at their early years settings, they are not bound by the formal conventions of learning; for them, play is an intrinsic part of the process by which they come to know about the world and come to refine and communicate their knowledge. Early years teachers recognise the importance of play and provide a wealth of opportunities for the child to explore a variety of roles and social situations. This should never be perceived as 'mere' play. Wood and Attfield (2013), for example, describe the significance of dramatic and socio-dramatic play as a means for learning within the early years, involving complex cognitive, social and emotional processes. Dramatic play is about pretending to be someone else, role taking, imitating a person's speech and actions, using real and imagined props, first-hand and second-hand experience and the child's developing knowledge of story characters and imagined situations. Socio-dramatic play involves cooperation between two or more children, where play develops on the basis of interactions between players acting out their roles, both verbally and in terms of the acts performed. Children might act out a well-known story or a trip to the shops, for example. Role-play areas can be situated within both the indoor and outdoor learning environment, providing opportunities for dressing-up and other 'real life' and fantasy scenarios (for example, playing 'shop' or a castle for knights and princesses). Role-play areas offer children crucial opportunities to enact, to imitate, to imagine, to confront, to review and to understand the social world they inhabit. Strong early years practice builds on this understanding, acknowledging that dramatic and socio-dramatic play is part of the way in which children come to make sense of their world.

Early drama teaching builds on the child's natural inclination for play and usually develops into two areas. Firstly, there are a variety of drama games which often involve walking and clapping or mime and movement activities, often through the use of songs and music. These activities introduce some structure to drama times and establish the position of the teacher within a specific exploratory context. Secondly, and more importantly, there is a movement towards the provision of structured imaginative play, allowing the teacher to plan more carefully and encouraging the child to use their intrinsic sense of participation to explore some issues in greater detail. An example of this might be exploring characterisation within a text.

Story regularly provides a natural and productive initiation for more detailed drama sessions. Story is a familiar and important feature of early years classrooms, and there are clear links between the thematic features of children's stories (finding/losing, friends/enemies, deception, hiding, escaping, etc.) and early explorations into movement and drama. Stories also serve as a perfect medium through which the teacher can begin to introduce the speculative ideas associated with drama: What would happen if . . .?; Let's suppose that . . .; Perhaps there might be . . .? By using familiar characters and story settings, a new discourse opens through which children can explore a range of alternative possibilities. Airs and Ball (1997), for example, began the practice of using established children's stories such as *Goldilocks* and Janet and Allan Ahlberg's *Burglar Bill*, to investigate a range of dilemmas through drama, such as: should Goldilocks have entered the Three Bears' house? Why/why not? Was Burglar Bill a good man? Why/why not? How could both characters have acted differently?

## Drama with older children

The movement towards more formally planned drama work with older children needs to be capable of being both spontaneous and well structured. Woolland (2010) suggests that the building blocks for music are *pitch, melody, harmony, tempo, rhythm* and *texture*. What, then, are the building blocks of drama? Woolland offers one such possible set which begins to indicate a conceptual route for intending teachers of drama:

- **Role or character**
  Acting as if you were someone else or placing yourself as someone else in another situation.
- **Narrative**
  Ordering a sequence of events or images in such a way that their order creates meaning. Narrative involves the way in which the drama is moved forward – withholding information; sudden turn of events; surprise ending or beginning, etc.
- **Language**
  Verbal (this may include: naturalistic dialogue; a formal, heightened style of language such as a proclamation, or the beginning of a ritual; a direct address to an audience; characters talking to themselves; choral speech); or non-verbal (this may include symbols; body language; facial expression; the use of space; ritual).

Finally, Role or character, Narrative and Language all operate within a particular context:

- **Context**
  Where does the action of the drama occur?
  Is it set in a particular historical period?
  What are the relevant social/political conditions?

  (Woolland, 2010)

Teachers need to know how to use these 'building blocks' to construct meaningful and valuable drama. One method which is cited by almost all drama books is that of 'Teacher-in-Role'. At its most basic, this involves the teacher adopting the role of another person (typically historical, fictional or imagined) for the purposes of questioning and answering. Teacher-in-Role is a technique which can also be used to explore the motivation of historical figures or to generate debate about current (perhaps local) social issues. More importantly, as Bolton (1992: 32) argues, 'the main purpose of Teacher-in-Role is to do with ownership of knowledge'. While the teacher (in this simplified version) is potentially in control at all times, the nature and origin of knowledge begins to shift, so that children become instrumental not only in generating new understandings, but also (and most importantly) in understanding the process of social interaction. As children learn how to interact within this context they, of course, become capable of reversing roles and assuming the 'mantle of the expert', a term originally devised by Dorothy Heathcote in the 1980s. Mantle of the Expert is a dramatic-inquiry based approach to teaching and learning where children engage as if they are an imagined group of experts, for example police detectives solving a crime.

## Drama and texts

When drama work is linked with a text, in addition to its intrinsic merits it can deepen pupils' understanding and interpretation of that text. Here are two examples: *Little Red Riding Hood*, which is a text suitable to use with early years children in the EYFS (0–5) and Key Stage 1 (age 6 to 7), and *Goodnight Mr Tom*, suitable for children in upper Key Stage 2 (age 9 to 10).

### *Early years and early primary: drama techniques using* Little Red Riding Hood

Choose a version of the traditional tale that you and your class or group of children are familiar with.

#### *Talking partners*

Half of the class (in pairs) discuss why Little Red Riding Hood should go into the woods. Half of the class discuss why she should not. Take this to whole class discussion. Consider drawing up a 'For' and 'Against' list as part of a shared write.

#### *Mime*

This can take place either with the whole class or in small groups. It can involve making and eating cakes; walking through the wood; entering the house; Little Red Riding Hood encountering the wolf in bed and realising he is a wolf! Depending on the age of the children, a script could be developed.

*Hot seating*

Select volunteers to play Little Red Riding Hood, Little Red Riding Hood's mother and the wolf. The other children in the class need to ask each character questions. Provide opportunities for socio-dramatic play through providing a role-play area linked to the text, such as Grandma's Cottage. Use props such as a wolf mask and a red cape for children to play in role. You might also consider taking the children for a walk-in role as Little Red Riding Hood through the woods to find Grandma's Cottage in the outdoor area of your setting. What or who might the children find along the way?!

## Upper primary: drama techniques using chapter 1 of Goodnight Mr Tom

### Mime

After reading the opening extract from the book *Goodnight Mr Tom (Magorian, 1981)*, ask children to identify a character, their likely location/stance in the scene and their body language. They should back up their ideas by reading between and beyond the lines. Ask them to give a rationale for their choices.

### Freeze frame

Children work in small groups to create the opening scene, then 'freeze' the position.

### Speaking thoughts

Select children from the freeze frame and ask them to voice the thoughts of their character.

### 'Conscience alley'

Divide the class into two large groups. One half of the class should think about reasons why Mr Tom *should* take the child; the other half argue why he shouldn't. After the preparation time, 'Mr Tom' walks down the 'alley' formed by the two groups facing each other. He approaches and/or indicates children to 'voice their conscience' about the issue when Mr Tom passes them.

### Acting in pairs

Tom and Willie – the tour of Tom's house or the first meal.

*Storying*

Three large groups represent the billeting officer(s), the evacuees and the hosts. Each of the sets discusses experiences with others in a similar position. Individuals from the groups are encouraged to share their experiences, in role.

*Hot seating*

Three children play Tom, Willie and the billeting officer. Others in the class come up with questions for each character.

*Moving to film*

Draw a picture to show what the first three scenes of a film will look like. Make a list of sounds/music that will be heard in the first scene.

*Compare with the film*

Show opening and analyse differences between children's interpretations, the text and the film.

(Dominic is grateful to Ainé Sharkey for the origins of some of these ideas.)

## Drama and writing

There is evidence that drama can be a useful precursor to children's writing (e.g. Cremin *et al.*, 2006). To end the chapter, the following example shows some of the power of drama for writing. The example is taken from research carried out by Helen and Dominic that evaluated the work of museum–school partnerships to enhance writing.

Four children from a class of Year 5/6 children had been focusing on biography as a writing genre and had each written a biography of the writer Michael Morpurgo in school. In preparation for a visit to Ely Stained Glass Museum (located inside Ely Cathedral), they then started to look at the life of John the Baptist using books and internet information to stimulate class discussion.

During the museum visit, they re-enacted the Dance of Salome. The class teacher chose children to play the main characters, and the rest of the children became guests at the feast. To support the children getting into role, there were costumes to dress up in, real food to eat (grapes, dates, Turkish delight, pineapple and apricots), authentic music for Salome to dance to, and John the Baptist's head (made from papier mâché), authentically gruesome, carried in on a silver platter!

In an interview before the museum visit, the class teacher felt the drama experience that was planned as part of the trip would have a particular impact on the children's motivation to write. This was in line with the recent changes to the whole school policy that had moved towards using speaking and listening as a stimulus for writing. The school's reading levels had traditionally surpassed writing levels, so the new approach was designed as part of a whole school initiative to support raising writing levels to match those of reading. On a practical level for the school, this had meant planning a lot more role-play and drama work as part of literacy lessons, which the children enjoyed and were responding to well. These experiences were generally felt to be having a positive impact on writing outcomes. In the interview with the education officer at Ely Stained Glass Museum, she highlighted the importance of absorption and engagement leading to vivid memories during visits by school parties in order to impact on children's motivation to write.

Following the visit, the children wrote up their biography of John the Baptist, organising their work into four clear paragraphs with headings. The drama experience clearly had an impact on the children's ability to organise their material for writing. For example, the children found that they didn't need their notes as a scaffold for their writing because they had vivid memories to rely on:

*Researcher:*  Did you make any notes before you did this?
*Annie:*  Well, we made a few.
*Researcher:*  Did that help at all?
*Annie:*  I didn't really use them at all.
*Sarah:*  I used my notes and the pages from the other sheets about John the Baptist because they told me all the names of the people and where all these things happened.

Charlie was able to compare his experience of writing in the same genre pre- and post-museum experience:

*Researcher:*  You said earlier that it was better than doing the biography on Michael Morpurgo, why is that do you think?
*Charlie:*  We've had a lot more time to learn about him because we've also been to the church and learned a lot about him and how he died. He had his head chopped off and put on a silver plate and taken to Salome who gave it to her mum, who was married to Herod.

For his Michael Morpurgo biography, Charlie wrote 175 words in total. However, at 300 words, his John the Baptist biography was nearly twice the length, an impact that was common to other children in the class.

## Practice points

- You do not have to be a good actor in order to teach or use drama as part of your pedagogical practice.
- Drama does not always have to take place in the school hall; good drama can take place in five or ten minutes in the classroom without necessarily moving any of the furniture, as well as outside the school environment.
- Use children's natural creativity by giving them the chance to invent their own collaborative drama at times. In the early years and Key Stage 1 in particular, role-play areas provide excellent opportunities for collaborative language exploration.
- Use the observation of drama experiences as an opportunity to plan for new skills/subject matter.

## References

Airs, J. and Ball, C. (1997). *Key Ideas: Drama.* Dunstable: Folens.

Bolton, G. (1992). *New Perspectives on Classroom Drama.* Hemel Hempstead: Simon & Schuster.

Cremin, T., Goouch, K., Blakemore, L., Goff, E. and Macdonald, R. (2006). 'Connecting drama and writing: Seizing the moment to write', *Research in Drama Education,* 11(3): 273–291.

Department for Education. (2013). *The National Curriculum in England: Framework Document: December 2014.* London: Department for Education.

Magorian, M. (1981). *Goodnight Mr Tom.* London: Viking Press.

Wood, E. and Attfield, J. (2013). *Play, Learning and the Early Childhood Curriculum,* 3rd edn. London: Sage.

Woolland, B. (2010). *Teaching Primary Drama.* Harlow: Pearson Education Ltd.

Wyse, D., Bradford, H. and Stephenson, P. (2009). Castles, Stained Glass, and Paintings: Enhancing Children's Writing through Partnerships with Museums. Research Report (64).

## Annotated bibliography

Loizou, E., Michaelidesm, A. and Georgiou, A. (2017). 'Early childhood teacher involvement in children's socio-dramatic play: Creative drama as a scaffolding tool', *Early Childhood Development and Care.* doi: 10.1080/03004430.2017.1336165

Fascinating paper revealing the possibilities for participatory teacher roles in the development of children's socio-dramatic play.

**L2** ★★★

Taylor, T. (2016). *A Beginner's Guide to Mantle of the Expert.* Norwich: Singular Publishing.

Pedagogic approaches for enhancing the primary school curriculum through drama.
**L1** ⋆

See *www.mantleoftheexpert.com* for more information, including resources and planning ideas.
**L1** ⋆

# Part III

# Reading

# Chapter 8

# The development of reading

Helping children learn to read is one of the most important roles that early years and primary teachers carry out. In order to support children effectively, it is necessary to be aware of the ways children might develop. This chapter begins with an analysis of interaction in an early years setting. Then research on reading development across languages is surveyed. The second half of the chapter explores the most effective ways to teach reading and why policy makers frequently propose practice that lacks a sufficient research base.

How children learn to read and – of particular significance to this chapter – how they can most effectively be taught to read is a concern for researchers, teachers, policy makers and societies in general. If a child does not learn to read, they cannot play a full part in society once they reach adulthood, nor during childhood can they access the full school curriculum. In an ideal world, these different groups of people with an interest in the teaching of reading would have sufficient shared understanding of how it is best taught. A shared understanding could allow people to act in ways that complemented rather than contradicted each other in the best interests of supporting children's reading. It appears that this is indeed an ideal world, because while theory and research have much to tell us about reading pedagogy, the route from research evidence to policy and practice is one that is far from smooth.

Learning to read is not simply a matter of acquiring knowledge about written language and skills in decoding, but becoming involved in cultural practices of meaning-making (Hall, 2003; Heath, 1983; Street, 1984). Socio-cultural theory has much to offer our understanding of reading teaching, and particularly learning to read, but in addition to some socio-cultural considerations we wish to advance the theory of contextualisation. This is explained in relation to phonics teaching.

The following extract of children talking with an adult took place in an early years setting for children aged 3 to 4 in England. The transcript of dialogue was part of a project carried out in the early years centre. Taking an ethnographic approach, the research analysed the ways that the children engaged with print and texts, and what the implications were for practitioners. One of the children was working in the 'writing area' in his classroom. He had been folding a piece of sugar paper into an irregular structure, to which he added some marks with felt pen. He turned to show the adult what he had done. In seeking to ensure that the photograph of a child's writing was appropriately oriented, the adult was drawn into a pedagogic role that centred on the children's good-natured disagreement about letters and phonemes.

*Adult:*     Oh that's good, a parcel for Ben [*Mark's friend who did not attend the same early years centre*], I like that. I'll take a picture of it like that.
*Mark:*     I want to hold it like that.
*Adult:*     Do you, that makes the writing upside down, is that alright? OK. You want to hold it. Well tip it back a bit so that I can get the writing. Look at the camera, you can see it . . . That looks like a letter M.
*Michael:*  No it's a /m/ [*Michael voices the sound*].
*Adult:*     It's a /m/ is it?
*Michael:*  /m/ for mummy. It's for my mummy.
*Neil:*     No it's M for mummy.
*Adult:*     That's right, M is the name of the letter isn't it, and /m/ is the sound.
*Neil:*     No M! [*spoken very firmly*]
*Adult:*     M's the name yes. They're both right . . . Is that mummy?
*Neil:*     Yes.

Pedagogy is revealed in many ways through this short extract. The teacher had organised a writing area with a range of resources to support children's mark-making, and the children were encouraged to use the area in a similar way to the other play-based areas in the classroom. The children were able to exercise choice over the kind of mark-making that they carried out. Mark had chosen to construct a parcel for his friend Ben, so in other words he had decided who the audience for his text was. Purposeful activity determined by the children, with, in Mark's case, a real audience in mind for his writing, was affected on occasion by the interaction of more expert language users – the adults who worked in the centre, including the researcher, but also the children's peers whose development of literacy varied due to their different home backgrounds.

The adult used the term M (the letter name) rather than the sound /m/, in part because of their view that learning about the distinction between letter names and the sounds associated with letters is an important feature of early years literacy learning. The ensuing discussion about letter names and sounds arguably represents significant learning about complex ideas, through a discussion led in a spirited way by children because they were interested in the topic. As a way of resolving the dispute, and to help the children's understanding, the adult intervened to offer some information to the children. The role of the adult as a more expert language user was a facilitative one, but also one that involved a kind of direct teaching informed by ongoing formative assessment of the children's discussion.

To sum up, a series of understandings about reading and writing were addressed in the course of the interaction: a) texts can communicate meaning to specific audiences; b) text has to be oriented in a particular way; c) the letter M can be called M or /m/; d) the letter M is the first letter of the word 'mummy'; e) there is a complex relationship between letters and the sounds that they represent in English. The children's knowledge of this relationship was in the early stages, but it is likely that such conversations would support their emerging understanding. From a socio-cultural perspective, the example highlights some key features of pedagogy: a) children having some control over their learning within frameworks established by teachers; b) learning located in scenarios that are meaningful to children; c) the interaction of the adult as expert language user extending children's learning by responding to the children's interests with a clear understanding, and high expectation, of the knowledge to be developed.

The scenario above represented a snapshot in time that gives a glimpse into children's learning. The consideration of longer periods of time takes us into the realms of reading development.

## Reading development

Descriptions of stages of development for reading and writing are an important feature of educational curricula. In England, this feature is revealed in the National Curriculum expectations for different year groups, for example of grammar, punctuation and spellings to be learned. The National Literacy Strategy in England attempted to do this through the sequencing of objectives and guidance papers on progression. It is reasonable to assume that reading and writing curricula should reflect research evidence on development, but it is often unclear whether such evidence has been considered.

Tierney (1991: 180) observed that although there were a large number of longitudinal studies ☞ about children's encounters with print, too many of them focused narrowly on decoding skills. Tierney also argued that reading was represented in longitudinal work more than writing and that the writing research tended to be dominated by 'cross-sectional comparisons of students varying in age or ability rather than studies that have looked at the same children at different ages' (ibid.: 189). This kind of cross-sectional comparison can be seen in Loban's (1976) data on language development, which was acquired mainly through sentence-level and word-level tests. The concluding chart of sequences and stages strongly emphasised grammatical development consistent with the kind of tests that were used.

Later work, such as that by Harste *et al.* (1984), convincingly portrayed the active and constructive nature of children's meaning-making. Wells (1986) confirmed such findings, and additionally concluded that listening to stories was more significantly correlated with children's literacy acquisition than looking at picture books and talking about them; drawing and colouring; or playing at writing. Ferreiro and Teberosky's influential semi-longitudinal study (1982: 263) concluded that by the age of 4, most children understand the main principle that 'writing is not just lines or marks but a substitute object representing something external to the graphics themselves'.

One feature of the research evidence focusing on the development of reading and writing has been the analysis of stages of development that are common to most children. Often these syntheses are derived from case-study data. For example, the Centre for Language in Primary Education (CLPE, 1989) developed reading scales through their work trialling the primary language record with teachers in London. Bearne (1998) worked closely with a group of 80 teachers to identify their expectations for children's development at the beginning and end of each school year.

Another important strand of case-study research, and longitudinal work, has been the use of individual child case studies, a methodology which, like Brooker (2002), we consider to be significant in the goal to better understand early childhood literacy. Seminal work ☞ in the field includes White (1954) and Butler (1975). The most important study of this kind, because of the richness of the data and its close link with other research evidence, is that by Bissex (1980) of her son Paul's development as a reader and writer between the ages of 5 and 11. Dyson (1983) cited the significance of the Bissex study, arguing that it was unusual because it provided evidence of the purposes which writing serves in children's lives, something that she argued many studies had neglected because of their narrow focus.

A common criticism of individual child case studies is that their findings cannot be reliably generalised because of such small sample sizes. However, the Bissex study has shown that some generalisation is possible. Gentry (1982) carried out an analysis of the examples of Paul Bissex's spelling. Gentry identified five stages of spelling development which subsequently Ellis (1994: 156) confirmed: 'it is now generally agreed that children move through five distinct stages of spelling, viz: "precommunicative", "semi-phonetic", "phonetic", "transitional", and "correct"'.

The early 1980s saw a number of individual child case studies from Australia, the US and the UK which provided further data about children's development as readers and writers (those by Lass, 1982; Baghban, 1984; Kamler, 1984; Payton, 1984; Schmidt and Yates, 1985). In addition to the descriptions of developmental progression that these studies offered, a range of general conclusions was put forward. The most common themes of these conclusions were the need to alleviate the disjunction between home and school literacy learning, and the important part that appropriate social interaction played in literacy learning. However, one of the limitations of these studies was their lack of references to other similar case studies in order to build more reliable evidence about developmental progression. The lack of reference to other similar studies was something that continued in subsequent decades. Minns (1997) identified six key areas that the children in her study developed:

1  understanding that print carries a message;
2  the ability to predict key phrases and memorise chunks of book language;
3  familiarity with book handling and directionality;
4  understanding and use of metalanguage;
5  understanding that there was a correspondence between letters and sounds;
6  the ability to discriminate between letter shapes and recognise individual words.

However, like Fadil and Zaragoza (1997), there was no reference to other case studies, not even to the seminal work of Bissex. Campbell (1999) concluded that children's learning is supported when they are actively involved and interested and that story reading and the opportunity to choose books are beneficial. Although Campbell did make reference to other case studies, it is not clear from his analysis how previous studies informed his reflections on developmental progression. Kress (2000) similarly did not explicitly build on previous developmental studies in his interesting thesis, which pointed to the multiple meanings of children's spelling attempts.

The rich data that has been portrayed in case studies and syntheses provides useful insights into the development of reading and writing. The milestones presented in Tables 8.1, 8.2, and 8.3 were built from a synthesis of the case studies reviewed so far in this chapter and from case studies of two other children (see Wyse, 2007, for examples from the data).

Knowledge about rich pictures of children's development is important because it can influence the way we teach our children. By knowing developmental milestones, it is possible to anticipate these and provide teaching at the appropriate level. This influence means that our pedagogy ☞ is related to what we know about how children develop. However, although milestones of development can be useful for thinking about progress, the detail of reading development is rather more complicated than such linear models suggest.

Wyse and Goswami (2012) summarised thinking arising from research on reading development. A key understanding is the child's awareness of larger features of language preceding awareness of smaller features of language, in a trajectory that begins at birth. At word-level, this begins with experiencing objects and their names through talk and interaction. But even at this stage, the syllable is the primary perceptual linguistic unit, something that is true across languages. Following awareness of syllables, children become aware of intrasyllabic features such as onsets and rimes.

Research on development has looked across different languages. Differences in reading development across languages are revealed when phoneme awareness is studied. One of the main challenges for children and their teachers is that English is one of the hardest languages to learn to read and write.

*Table 8.1* Expectations for a child's reading at age 4

| *What you can expect* | *What you can do to help* |
| --- | --- |
| Understands distinction between print and features | Talk about pictures and talk about print. Encourage children to point to print or point to pictures. |
| Can recognise and understand some words and signs in the environment | Encourage children to read food packets and to play 'shop'. Read signs, logos, and labels with them. Comment on text that appears on TV. Talk about greetings cards. |
| Understands that text has specific meaning | Read stories and other books with children. When reading a text that is at the child's level, read all the words as written. Talk about what particular words and sentences mean. |
| Plays at reading | Make sure that children have easy access to a really good range of books. Encourage their playing with the books and their pretend reading. Encourage them to pretend to read to others or even to cuddly toys. |
| Uses words and phrases from written language when retelling stories | Respond to children's requests to hear favourite stories. Encourage children to predict what is coming next in a story. Suggest that they join in with repetitive phrases. Celebrate when they remember phrases from favourite stories. |
| Needs other people to help with reading aloud | Read aloud daily with children. Encourage discussion and always be looking to develop children's independence to read words. |
| Will choose favourite picture books to be read aloud | Encourage daily reading. Give easy access to books. Read children's favourites but introduce them to new books as well. |
| Uses picture cues and memory of texts | Once children are familiar with a book, encourage them to tell the story by looking at the pictures as prompts for their memory of the text. |
| Understands orientation of print and books | Talk about the front and back covers of books, where the print is and, where the print starts on a page. Comment on books that play with these conventions. |
| Salient visual cues used to remember some familiar words like own name | Give frequent opportunities to read, write and play games with words such as names. |

*Table 8.2* Expectations for a child's reading at age 7

| *What you can expect* | *What you can do to help* |
| --- | --- |
| Silent reading established | Provide time, space, opportunities and resources to encourage children to read regularly. |
| Can accurately read increasing number of unknown texts independently | Ensure that children have access to 'new' books on a regular basis. |
| Uses expression when reading aloud | Have fun with using expression when you are reading to children. Encourage children to read at a good speed and with expression when they share books with you. If children are involved in any kind of performance where they have to read out loud, such as an assembly, help them with expression and clarity. |

| What you can expect | What you can do to help |
| --- | --- |
| Uses a range of word-reading strategies appropriately | Help children to use semantic, phonological and orthographic knowledge to work out tricky words. Praise them for good guesses and supply the correct word if necessary but give them time to think. |
| Stronger individual preferences for particular texts | Encourage children to develop preferences for particular topics and types of text. Talk to them about their preferences and those of the class. |
| Likes reading longer stories in addition to returning to picture books | Provide access to books with more text and fewer pictures. |
| Sight-word reading for rapidly increasing bank of familiar words | The more children read, the more sight words they will acquire. |
| Phonological knowledge fully established. Growing awareness of irregularities of English spelling | Phonics teaching will have taken place. Help children to see that the one-letter-makes-a-sound idea is not accurate. Discuss the irregularities of English spelling. |

Table 8.3 Expectations for a child's reading at age 11

| What you can expect | What you can do to help |
| --- | --- |
| Reflective reader with strong preferences | Discuss texts that children are reading and seek to extend their understanding of the issues raised. |
| Uses different reading styles for different texts | Encourage children to become involved with things like map-reading or locating information on the internet through cross-curricular work. |
| Can follow instructional texts | Do some cooking together which requires use of a recipe book. Involve children in following instructions to assemble things, including instructions that their peers have written. |
| Can sort and classify evidence | Encourage reading and writing of a range of formats for summarising information. |
| Varies pace, pitch and expression when reading aloud and varies for performance purposes | Discuss occasions when children have to perform. Encourage involvement in dramatic activities at home and at school. |
| Can adopt alternative viewpoints | The starting point for this might be the ability to empathise with others. Encourage consideration of evidence from different sides of an argument. |
| Recognises language devices used for particular effects | Enjoy the imagination of authors who like to play with text effects. Reread texts like poetry to discover effects. |
| Can discuss different author styles | Encourage children to read a series of books by the authors that they like and to think about their style. |
| Enjoys selecting and reading appropriate adult texts | Encourage access to newspapers and magazines. |

The reason that it is easier to learn to read in some languages than others is to do with their linguistic complexity. As Goswami (2005) has shown, there are two key factors in this. The first is the way that consonants and vowels are linked together. Some languages, such as Italian, Spanish and Chinese, are based on a simple consonant–vowel syllable structure (like *panini* [pa-ni-ni] in Italian or *tapa* [ta-pa] in Spanish). This makes them less complex than English. English has very few consonant–vowel syllables. Words such as 'baby' and 'cocoa' are examples that do exist (see Table 8.4). The most frequent syllable type in English is consonant–vowel–consonant, as in words like 'dog' and 'cat'.

The second key factor is the consistency of how the written symbols represent sounds. In some languages, such as English and Dutch, one letter or one cluster of letters can have many different pronunciations. In other languages, such as Greek, Italian and Spanish, the letters and clusters are always pronounced in the same way no matter which word they appear in, which is simpler to learn. A large study of 14 European languages clearly showed the dramatic differences by measuring the reading of real words and made-up words in different languages, with English right at the bottom of the list (see Table 8.5).

*Table 8.4* Examples of words in English with consonant/vowel syllables

| Word | Syllable | Consonant | Vowel(s) | Syllable | Consonant | Vowel(s) |
|------|----------|-----------|----------|----------|-----------|----------|
| Baby | ba | b | a | by | b | y |
| Cocoa | co | C | o | coa | c | oa |

*Table 8.5* Data (% correct) from the large-scale study of reading skills at the end of grade 1 in 14 European languages

| Language | Familiar real words | Pseudo-words |
|----------|---------------------|--------------|
| Greek | 98 | 92 |
| Finnish | 98 | 95 |
| German | 98 | 94 |
| Austrian German | 97 | 92 |
| Italian | 95 | 89 |
| Spanish | 95 | 89 |
| Swedish | 95 | 88 |
| Dutch | 95 | 82 |
| Icelandic | 94 | 86 |
| Norwegian | 92 | 91 |
| French | 79 | 85 |
| Portuguese | 73 | 77 |
| Danish | 71 | 54 |
| Scottish English | 34 | 29 |

*Source*: Goswami (2005: 275).

## The teaching of reading

The history of the debates about approaches to the teaching of reading has repeatedly hinged on fundamental disagreements related to different theories of learning to read. Until the early 1800s, there was no distinction between the teaching of reading and the teaching of spelling in England. Therefore, evidence on reading teaching at this time can be found in the hundreds of different spelling textbooks that were published. Michael (1984) argued, on the basis of an analysis of the spelling books, that reading teaching was for most teachers a bottom–up approach. As early as 1610 this was clear, for example: 'Therefore let the scholler, being thus traded (i.e. schooled) from letters to syllables of one Consonant: from syllables of one Consonant, to syllables of many Consonants: from syllables of many Consonants, to words of many syllables; proceede to sentences' (1984: 57). Michael suggests that the approach was a consequence of the prevailing view that complex things could be learned by children only if they were first broken down into their component parts (a theory that still informs some current opinions of how children learn to read). In his chapter that examines the growth of whole-word methods, Michael also identifies what he sees as the first published reference to whole-word teaching. Charles Hoole, in his translation of the preface to Comenius' *Orbis Sensualium Pictus* (1659) said, 'reading cannot but be learned; and indeed too, which thing is to be noted, without using any ordinary tedious spelling, that most troublesome torture of wits, which may be wholly avoyded by this Method [the whole word method]' (1984: 60).

The seminal text in the debate of the modern era was Jean Chall's (1983) book *Learning to Read: The Great Debate*, which was first published in the 1960s. In it she defines the differences between two models:

> The top-down models relate . . . to the meaning–emphasis approaches of beginning reading and stress the first importance of language and meaning for reading comprehension and also for word recognition . . . The reader theoretically samples the text in order to confirm and modify initial hypotheses.
>
> The bottom-up models – those that view the reading process as developing from perception of letters, spelling patterns and words to sentence and paragraph meaning – resemble the code emphasis, beginning reading approaches.
>
> (Chall, 1983: 28–29)

Chall's use of the term reading 'model' is perhaps better replaced with 'approach', because reading models generally seek to account for reading development rather than approaches to teaching. The classic example of a top-down approach to reading would be the 'real book approach' or the 'whole language approach', and the contrasting bottom–up approach would be 'phonics'.

Reading models are important because they can efficiently summarise complex areas of knowledge. However, an important consideration is the

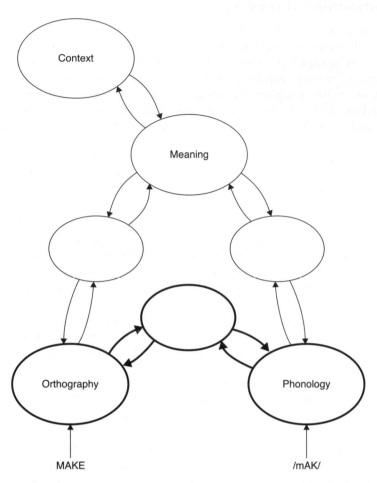

*Figure 8.1* A parallel distributed processing model of reading (Lupker, 2005: 49).

relationship between the model and approaches to the teaching of reading. For example, one model that accounts for word recognition is the parallel distributed processing model (see Figure 8.1).

Parallel distributed processing (PDP) models are based on the idea of the brain establishing 'sets of distributed, sub-symbolic codes representing the attributes of the words we know' (op. cit.). The word recognition system involves three types of mental representations: orthographic, phonological and semantic. The appropriate connections between the representations have to be learned. When presented with a word, the units at all levels begin to activate. Activation patterns are initially quite inaccurate but through experience weights between units become adjusted in order to make processing more accurate the next time.

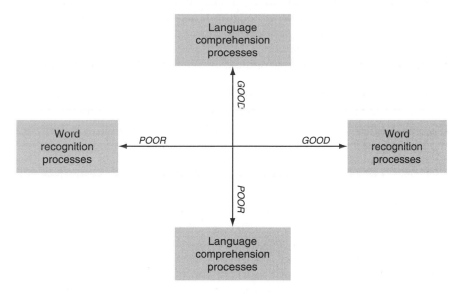

*Figure 8.2* An interpretation of the simple view of reading.

An important aspect of the model depicted in Figure 8.2 is the *interaction* between the four main elements that leads to word recognition. This idea of interaction between elements is shared by models that have in the past informed the teaching of reading. For example, the three-part model of word recognition that featured semantic, syntactic and graphophonic *cueing* that came out of the work of Goodman (1969) and in the UK from Arnold's synthesis (Arnold, 1982) of Goodman's ideas. This three element model was also similar to Clay's model (1979) and the *searchlights* model that underpinned the National Literacy Strategy in England from 1998 to 2006 (DfEE, 1998). However, by 2007 the searchlights model had been replaced by Stuart and Watson's interpretation of the *simple view of reading* (SVR) (Figure 8.2) that featured as an appendix to the controversial government-commissioned Rose Report on the teaching of early reading (Rose, 2006).

Stuart and Watson's case for the SVR was more complex, research-informed and nuanced than that of the main text of the Rose Report yet, like the Rose Report, it had some fundamental flaws. Stuart and Watson rightly saw Clay's model of reading strategies as an influence on the NLS searchlights model but were unduly critical of Clay's work as a result of failing to take into account the substantial international evidence base supporting Clay's approach. One of their key warrants was that: 'The case for change that we discuss below [i.e. further into the text of the appendix to the Rose Report] rests on the value of explicitly distinguishing between word recognition processes and language comprehension processes' (Rose, 2006: 74 – Appendix 1, Stuart and Watson).

While the authors attempted to say that they didn't necessarily mean that phonics teaching should be separate from other language elements, the clear direction of their recommendations overall was just that (for a more extensive critique of the model, see Harrison, 2010). Rose's recommendations, informed by the simple view of reading, resulted in synthetic phonics being imposed throughout England.

Stuart and Watson based their model to a certain extent on Gough and Tunmer's (1986) seminal paper outlining another *simple view of reading*: Reading = Decoding × Comprehension (R = D × C). Gough and Tunmer's model has been subject to numerous interpretations, so it is important to register that there would be almost unanimous agreement with their statement that both 'decoding and comprehension are essential to reading' (p. 9). It is also important to remember that Gough and Tunmer claimed that their model was not about reading teaching, although they did refer to the reading debates in the introduction to their seminal article and suggest that reading models and teaching are interconnected, which is true.

Gough and Tunmer accepted that 'if there is no comprehension then reading is not taking place' (p. 7), so although they also concede that 'comprehension is not sufficient, for decoding is also necessary' (p. 7), the logical opposite to their first statement is, 'if there is no decoding then reading is not taking place'. This requires a precise definition of decoding. Although their definition ranges somewhat, they seem to finalise it as 'knowledge of letter–sound correspondence rules'. This reveals a flaw in the logic. Children younger than 3 years of age can read words such as their name or words in their environment that are part of packaging, without knowledge of letter–sound correspondence rules. Similarly, the young child's ability to memorise large chunks of story texts and 'read' these in sync with the turning of the pages in a picture book involves a form of reading comprehension. Contrary to the simple view, in these cases comprehension is being demonstrated without decoding. The main limitations of the simple view are: a) the lack of fit between the model and early reading development; b) the lack of accommodation in the model for the transactions (Rosenblatt, 1985) of reading; and c) the lack of accommodation of reading of visual elements such as pictures.

Further investigation of the SVR reveals subtle but important changes in definitions from the original SVR paper in 1986 to a subsequent paper in 1990, where decoding is not defined as the phonemes and graphemes that are the focus of synthetic phonics but is actually 'efficient word recognition' (Savage *et al.*, 2015: 821). Further broadening of the original definitions is seen in the definition of linguistic comprehension as 'the full set of linguistic skills, such as parsing, bridging and discourse building' (Hoover and Gough, 1990: 131). The concept of building discourse is so broad as to bring into question the fitness for purpose of this definition in relation to establishing a model of reading showing precise causal relationships. However, Savage *et al.* (2015: 839) do present new research showing the classroom-level influences of D and listening comprehension (LC) on RC, independent of pupil-level variance in D and LC respectively. They conclude that their findings 'are consistent with broad theoretical models of reading that emphasise ecological or contextual factors as well as pupil level factors in early reading development such as the Component Model of Reading (e.g. Joshi and Aaron, 2012)'.

We would argue that a more appropriate logic for reading teaching using the presentation of logic in a philosophical style informed by socio-cultural theory is as follows. If R = Reading, U = Understanding, Mg = Meaning and → is the process of understanding written language in order to establish meaning, then

$$R = (U \rightarrow Mg)$$

Reading requires understanding of visual symbols (VS) and their relationship to phonology (P = units of sound), but VS ↔ P is only a component part of (U → Mg), as is comprehension (C) (psychologists separate linguistic comprehension and reading comprehension; in the model below, C = reading comprehension because wider concepts of comprehension are represented by U → Mg).

$$R = (U \rightarrow Mg)/(C + (VS \leftrightarrow P))$$

But reading includes visual image interpretation (VII), so

$$R = (U \rightarrow Mg)/(C +/- VII + (VS \leftrightarrow P))$$

The basis for some items and relationships in the formula above is built on psycholinguistic grain size theory (Ziegler and Goswami, 2005) with the addition of visual image interpretation (a fuller account of TELL theory where the formula was first published can be found in Wyse, 2011a). Psycholinguistic grain size theory accounts for the developmental sequence of learning across languages. In spite of differences in the phonological structure of languages being learned, for example, the transparency of the orthography, the developmental sequence is the same: shallow sensitivity of large phonological units to a deep awareness of small phonological units, i.e. syllables, onset–rime, nucleas coda, phoneme and phone.

### Research evidence and effective phonics teaching

One of the most significant contributions to debates about research evidence and the teaching of reading was the report of the US National Reading Panel (NRP) on reading instruction, carried out by the National Institute of Child Health and Human Development (National Institute of Child Health and Human Development, 2000). This extensive work addressed a number of questions, including: 'Does systematic phonics instruction help children learn

to read more effectively than non-systematic phonics instruction or instruction teaching no phonics?' (p. 3) As far as differences between analytic and
synthetic phonics are concerned, the NRP concluded that

> specific systematic phonics programs are all significantly more effective
> than non-phonics programs; however, they do not appear to differ sig-
> nificantly from each other in their effectiveness although more evidence is
> needed to verify the reliability of effect sizes for each program.
>
> (ibid.: 93)

The NRP also concluded that reading teaching should not focus too much
on the teaching of letter–sound relations at the expense of the application
of this knowledge in the context of reading texts. And that phonics should
not become the dominant component in a reading programme, so educators
'must keep the end in mind and insure that children understand the purpose
of learning letter-sounds' (2000: 2–96). The importance of the cautions about
phonics becoming a dominant component is given added weight if we consider the findings of Camilli *et al.* (2003). Camilli *et al.* replicated the meta-
analysis from the NRP phonics instruction report and found a smaller but
still significant effect for systematic phonics ($d = 0.24$) than the NRP, but also
found an effect for systematic language activities ($d = 0.29$) and an effect for
individual tutoring ($d = 0.40$). Hence, the effect for individual tutoring was
larger than the effect for systematic phonics; the effect for systematic language
activities was slightly larger but comparable with that for systematic phonics.
These findings resulted in their conclusion that 'systematic phonics instruction
when combined with language activities and individual tutoring may *triple* the
effect of phonics alone' (Camilli *et al.*, 2003).

In 2006, the UK Department for Education and Skills (DfES) commissioned a
systematic review of approaches to the teaching of reading. The methodology of
the NRP was refined to produce a meta-analysis that included only Randomised
Controlled Trials (RCTs) ☞. On the basis of their work, Torgerson *et al.* concluded,
once again in direct contrast to the Rose Report, that 'There is currently no strong
RCT evidence that any one form of systematic phonics is more effective than
any other' (2006: 49). This finding links with their pedagogical recommendation from a replication study in 2018 that 'in our view there remains insufficient
evidence to justify a "phonics only" teaching policy; indeed, since many studies
have added phonics to whole language approaches, balanced instruction is indicated' (Torgersen, *et al.* 2018: 27). One of the difficulties of forming policy
recommendations for reading pedagogy is that this balanced instruction can easily
be disrupted by policy thrusts that lack a sufficient evidence base.

This work in the US and the UK was complemented by an Australian government report recommending that

> teachers [should] provide systematic, direct and explicit phonics instruc-
> tion so that children master the essential alphabetic code-breaking skills

required for foundational reading proficiency. Equally, that teachers [should] provide an integrated approach to reading that supports the development of oral language, vocabulary, grammar, reading fluency, comprehension and the literacies of new technologies.

(Australian Government, Department of Education, Science and Training, 2005: 14)

The Australian report also appropriately cautioned that:

While the evidence indicates that some teaching strategies are more effective than others, no one approach of itself can address the complex nature of reading difficulties. An integrated approach requires that teachers have a thorough understanding of a range of effective strategies, as well as knowing when and why to apply them.

(ibid.: 14)

Early years educators have been particularly concerned about the dangers of an inappropriate curriculum being imposed on young children. The research evidence on this matter was quite clear and once again contradicted the report. The majority of evidence in favour of systematic phonics teaching referred to children aged 6 and older. Some 20 out of the 43 studies covered in the Torgerson *et al.* (2006) and NRP reviews were carried out with children aged 6 to 7. Only nine studies were carried out with children aged 5 to 6. No studies were carried out with 4-year-olds. The idea that children younger than 5 will benefit from synthetic phonics is not supported by evidence and is arguably one of the most controversial recommendations of the Rose Report. In spite of all the powerful evidence, policy makers in England have consistently, and with increasing emphasis over many years, promoted phonics teaching at the expense of other more important aspects of reading (Wyse and Styles, 2007).

Wyse and Goswami (2008) carried out a review of research internationally in response to the Rose Report. Part of their analysis included establishing categories for the effective phonics instruction pedagogy that was part of the experimental trials included in the two systematic reviews featured above (Torgerson *et al.*, 2006; National Institute of Child Health and Human Development (NICHD), 2000). One category of studies was 'contextualised phonics instruction'. Further analysis of the pedagogy of these studies provides insights into effective teaching (it is important to remember that these experimental trials were only the ones that addressed phonics teaching. If other areas of reading research were taken into account, such as reading comprehension, an even stronger case is likely).

Table 8.6 summarises the key features of contextualised phonics teaching revealed in a series of studies (further information about the methods used in the studies can be found in Wyse and Goswami, 2008). The key feature of the effective approach in the study by Berninger *et al.* (2003) was the combination of word recognition activities with comprehension teaching

Table 8.6 The pedagogy of contextualised phonics teaching

| Study | Overall teaching context | Key features of contextualised phonics teaching |
|---|---|---|
| Berninger et al. (2003) | Combination of explicit word recognition and reading comprehension was most effective. Reading comprehension training included language cueing at text level: e.g. 'Tell the plot or main events in the story so far'. | Combination of word recognition and comprehension teaching in reading lesson. |
| Blachman et al. (1999) | This study covered kindergarten through to, and including, grade 1. Overall sequence: (1) review of sound-symbol associations learned previously; (2) phoneme analysis and blending skills; (3) automatic recognition of words; (4) ten to 15 minutes of reading connected text. | High-frequency word selected from stories that the children would be reading and that are introduced as part of sessions. Reading of connected text. Writing to dictation so teacher could assess students' progress. Vocabulary development and comprehension not neglected. Children's understanding of words and their understanding of stories was supported. |
| Brown and Felton (1990) | The selection of reading programmes was based on those which were 'complete instructional programmes which emphasised both word identification and comprehension'. | Mastery of the skills taught was reinforced through the use of controlled readers and the coordination of reading and spelling. |
| Evans and Carr (1985) | General analysis of 20 classrooms. In these classrooms, reading was instructed primarily through basal readers and workbooks rather than student-generated stories, phonics drill rather than sight-word banking, and supervised practice at cloze-type prediction from context, using relatively unfamiliar reading materials. | Recognised the importance of helping pupils understand how to coordinate dual task performance such as word analysis with predictive context use. |
| Foorman et al. (1997) | Synthetic phonics group did activities including reading practice: 'Language, Alphabet, Reading, spelling decks, New Concept, Reading practice, Handwriting, Spelling, Review, Verbal expression, Listening' (p. 260). | The more successful phonics intervention is characterised as 'synthetic', but this was integrated as part of the daily lesson format which included reading practice. |

| Study | Description |
|---|---|
| Foorman et al. (1998) | Carried out during 90-minute language arts period. Direct instruction in letter–sound correspondences practised in decodable text … emphasis on phonemic awareness, phonics … and literature activities. | The more successful direct code approach mixed phonemic awareness and phonics with literature activities. The phonics rules are introduced using alliterative stories and controlled vocabulary text in order to practice. Skills in oral language comprehension and motivation for stories are developed. |
| Greaney et al. (1997) | Reading Recovery-type lessons but with more flexibility. | The reading of familiar and less familiar books is an integral part of the programme. |
| Martinussen and Kirby (1998) | Programmes included broader features such as work with shapes, matrices, and sequential analysis. The reading of picture books by instructors was part of the programmes. | Phonics teaching was embedded in the reading of texts. |
| Santa and Hoien (1999) | The Early Steps programme has a particular emphasis on story reading, writing and phonological skills. Similar to Reading Recovery. The programme evaluated is Early Steps, an intervention with one-to-one tutoring and with particular emphasis on story reading, writing, and phonological skills … 'the program emphasises real book reading' (p. 62). | The programme represents what the researcher considers to be a balanced approach and one that fitted well philosophically with programmes already in place in the district. Included reading connected text and daily writing. |
| Tunmer and Hoover (1993) | Reading Recovery approach. | Work on phonological and visual similarities of words was used in the context of lessons to learn strategies about how and when to apply such knowledge. Wherever possible, the teachers chose clear and memorable examples from texts that had been read. Children were encouraged to identify unfamiliar words in their reading, using strategies they had been taught and to help with spelling words in writing. |
| Umbach et al. (1989) | Broad approach which included teaching of text orientation, sounding-out, comprehension and problem-solving skills. | Argues for combined phonics and comprehension teaching. |
| Vickery et al. (1987) | Broader programme: multisensory teaching approach for reading, spelling and handwriting (MTARSH). | Comprehension is part of programme. Attention to broader areas of learning is included. |

including language cueing at text-level as part of the lessons, although the word recognition training was contextualised in words rather than whole texts. The combination of phonics teaching with comprehension teaching was also a feature of the studies by Umbach *et al.* (1989) and Vickery *et al.* (1987). Similarly, Blachman *et al.*'s (1999) study featured the reading of connected text as a part of the lesson. Brown and Felton (1990) found that reinforcement of skills through use of whole texts was beneficial. Evans and Carr (1985) recognised the importance of using word analysis skills in combination with other tasks, such as predicting meaning when using relatively unfamiliar reading materials. The studies by Foorman (Foorman *et al.*, 1997 and 1998) were set in the context of language arts lessons which included literature activities. Phonics teaching was embedded in the reading of texts in the Martinussen and Kirby (1998) study in addition to the less common use of global and bridging tasks.

The studies by Santa and Hoien (1999), Tunmer and Hoover (1993) and Greaney *et al.* (1997) all based the pedagogy of the effective intervention on Clay's (1979) *Reading Recovery* approach with modifications. Reading Recovery teaching is well specified by Clay and also has the benefit of a particular high number of research evaluations (U.S. Department of Education, Institute of Education Sciences, What Works Clearinghouse, 2013) which enable in-depth understanding of the pedagogy. Reading Recovery lessons begin and end with the use of whole texts. In summary, the teaching of sub-word-level features such as phonemes, and the decoding of words, appears to be effective when embedded in whole texts. It is now proven beyond reasonable doubt that Reading Recovery is one of the most effective ways to help pupils who are struggling with reading (D'Agostino and Harmey, 2016).[1] The results of multiple experimental trials, and other research, in a range of different countries have conclusively proven its impact. For that reason, it is mystifying that governments in the UK have not recently invested in Reading Recovery.

Contextualised phonics teaching is likely to be effective because new understandings can be applied in real contexts in order to consolidate what are often complex areas of learning for pupils. The use of whole texts also enables systematic comprehension teaching to be very closely linked with phonics teaching. In addition, although the focus may be the teaching of reading, writing is frequently a part of the lessons. This enables pupils to understand the links between the decoding and encoding of language, something that once again consolidates their understanding. Further evidence to support our position can be seen in guidance that yet again stresses that a balanced approach to reading, where decoding and comprehension in reading are seen as important but not sufficient on their own, is the most effective because pupils also need to be *motivated* to read for pleasure, and metacognitively aware of the strategies they can use (Education Endowment Foundation, 2016).

## Conclusions

The Rose Report resulted in unprecedented levels of direct political involvement in the work of teachers and teacher trainers. Its conclusions about synthetic phonics did not adequately reflect the evidence that was available. It missed an opportunity to take forward the debate about reading in a constructive way, opting instead for its more controversial approach and findings. The problems with a lack of attention to robust research evidence at the expense of political ideology was intensified during the creation of the programmes of study in the 2015 National Curriculum, and in the revision to the Early Years Foundation Stage, in England. Of particular concern was the use of the phonics screening check. Although the reading of pseudo-words as part of a test in a psychological research have some validity, the use of pseudo-word reading in a high stakes national test intended to hold schools to account is not a valid assessment. In addition to political controversy around high stakes assessment in England, research has cast further doubt on the phonics screening check, in particular the failure to test a sufficiently wide range of grapheme–phoneme correspondences and the lack of attention to the fact that children's ability to decode words is also dependent on vocabulary knowledge, not just phonics skills, and as a consequence bringing into question the purpose and validity of the phonics screening check (Darnell et al., 2017).

Wyse et al. (2010) summarise what they see as a pattern in the kinds of policy responses to the teaching of reading described above. First, a dubious characterisation by politicians that education is failing, followed by 'decisive' action giving a reason for greater centralisation as the government 'takes responsibility' for improving performance; then the inevitable increase in a battery of measures to gauge progress and the impact of curricular interventions ('high-stakes testing'); a gradual seeping in of pedagogical control as well as curricular control, leading to loss of teacher autonomy and agency. Such control eventually becomes unworkable and uninspiring, and ceases to provide the results it is intended to deliver; and so a reaction sets in, freeing up teachers to have more agency within what and how they teach; allowing more space for creativity across the whole English and language curriculum; and making the curriculum more closely related to the outside world. Perry et al.'s (2010) work confirms the gap between evidence and policy making, the politicisation of decisions on pedagogy, the speed of policy change, and the role of the media in exacerbating some of these problems. Moss and Huxford (2007) suggested that a focus by researchers on understanding the contexts of policy enactment, more than 'finding new content for policy to convey' (2007: 72), would be a helpful way forward, yet in spite of some researchers' direct engagement with government on policy implementation and their understanding of the contexts of policy enactment, the problems with politicisation and lack of attention to the full range of evidence remain, suggesting different problems such as increasing political control of education (see Wyse, 2011b).

## Practice points

- A truly objective and balanced approach to the teaching of reading is vital.
- Work on your observation skills and extend your understanding of children's reading development.
- Phonics teaching should be regular, brief and as enjoyable as possible.

## Glossary

**Longitudinal studies** – research that looks at development over several years rather than a year or less.

**Pedagogy** – approaches to teaching and learning.

**Randomised Controlled Trial** – a particular kind of research which analyses the effect of an intervention on experimental groups and a control group to which the participants have been randomly allocated.

**Seminal work** – classic (often old) academic work that continues to be referenced by large numbers of writers.

## Note

1 Sinead Harmey (from D'Agostino and Harmey, 2016) joined the International Literacy Centre at UCL Institute of Education in 2017. The ILC has rights to some aspects of the Reading Recovery brand.

## References

Arnold, H. (1982). *Listening to Children Reading*. London: Hodder & Stoughton.

Australian Government, Department of Education, Science and Training. (2005). *Teaching Reading. Report and Recommendations: National Enquiry into the Teaching of Literacy*. Barton, Australia: Department of Education, Science and Training.

Baghban, M. (1984). *Our Daughter Learns to Read and Write: A Case Study from Birth to Three*. Newark, DE: International Reading Association.

Bearne, E. (1998). *Making Progress in English*. London: Routledge.

Berninger, V. W., Vermeulen, K., Abott, R. D., McCutchen, D., Cotton, S., Cude, J., *et al.* (2003). 'Comparison of three approaches to supplementary reading instruction for low-achieving second-grade readers', *Language, Speech, and Hearing Services in Schools*, 34: 101–116.

Bissex, G. L. (1980). *Gnys at Wrk: A Child Learns to Write and Read*. Cambridge, MA: Harvard University Press.

Blachman, B., Tangel, D., Ball, F., Black, R. and McGraw, D. (1999). 'Developing phonological awareness and word recognition skills: A two-year intervention with low-income, inner-city children', *Reading and Writing: An Interdisciplinary Journal*, 11: 239–273.

Brooker, L. (2002). 'Five on the first of December!: What can we learn from case studies of early childhood literacy?', *Journal of Early Childhood Literacy*, 2(3): 291–313.

Brown, L. and Felton, R. (1990). 'Effects of instruction on beginning reading skills in children at risk for reading disability', *Reading and Writing: An Interdisciplinary Journal*, 2: 223–241.

Butler, D. (1975). *Cushla and Her Books*. Auckland: Hodder & Stoughton.

Camilli, G., Vargas, S. and Yurecko, M. (2003). 'Teaching children to read: The fragile link between science and federal education policy', *Education Policy Analysis Archives*, 11(15). Retrieved 1 March 2006, from http://epaa.asu.edu/epaa/v11n15/

Campbell, R. (1999). *Literacy from Home to School: Reading with Alice*. Stoke-on-Trent: Trentham Books.

Centre for Language in Primary Education (CLPE) and Inner London Education Authority (ILEA). (1989). *The Primary Language Record Handbook for Teachers*. London: Centre for Language in Primary Education.

Chall, J. S. (1983). *Learning to Read: The Great Debate*, updated edn. New York: McGraw-Hill.

Clay, M. (1979). *The Early Detection of Reading Difficulties*, 3rd edn. Auckland: Heinemann Education.

D'Agostino, J. and Harmey, S. (2016). 'An international meta-analysis of reading recovery', *Journal of Education for Students Placed at Risk (JESPAR)*, 21(1): 29–46.

Darnell, C., Solity, J. and Wall, H. (2017). 'Decoding the phonics screening check', *British Educational Research Journal*, 43(3): 505–527.

Department for Education and Employment (DfEE). (1998). *The National Literacy Strategy Framework for Teaching*. Sudbury: DfEE Publications.

Dyson, A. H. (1983). 'Individual differences in emerging writing', in M. Farr (ed.), *Advances in Writing Research, Vol. 1: Children's Early Writing Development*. Norwood, NJ: Ablex.

Education Endowment Foundation. (2016). *Improving Literacy in Key Stage One*. London: Education Endowment Foundation.

Ellis, N. C. (1994). 'Longitudinal studies of spelling development', in G. D. A. Brown and N. C. Ellis (eds.), *Handbook of Spelling: Theory, Process and Intervention*. Chichester: John Wiley and Sons.

Evans, M. and Carr, T. (1985). 'Cognitive abilities, conditions of learning, and the early development of reading skill', *Reading Research Quarterly*, 20: 327–350.

Fadil, C. and Zaragoza, N. (1997). 'Revisiting the emergence of young children's literacy: One child tells her story', *Reading*, 31(1): 29–34.

Ferreiro, E. and Teberosky, A. (1982). *Literacy Before Schooling*. Portsmouth, NH: Heinemann Educational Books Ltd.

Foorman, B. R., Francis, D. J., Fletcher, J. M., Schatschneider, C. and Mehta, P. (1998). 'The role of instruction in learning to read: Preventing reading failure in at-risk children', *Journal of Educational Psychology*, 90: 37–55.

Foorman, B. R., Francis, D. J., Winikates, D., Mehta, P., Schatschneider, C. and Fletcher, J. M. (1997). 'Early interventions for children with reading disabilities', *Scientific Studies of Reading*, 1: 255–276.

Gentry, J. R. (1982). 'An analysis of developmental spelling in *Gnys at Wrk*', *The Reading Teacher*, 36: 192–200.

Goodman, K. (1969). 'Analysis of oral reading miscues: Applied psycholinguistics', *Reading Research Quarterly*, 5(1): 9–29.

Goswami, U. (2005). 'Synthetic phonics and learning to read: A cross-language perspective', *Educational Psychology in Practice*, 21(4): 273–282.

Gough, P. B. and Tunmer, W. E. (1986). 'Decoding, reading and reading disability', *Remedial and Special Education*, 7(1): 6–10. doi: 10.1177/074193258600700104

Greaney, K., Tunmer, W. and Chapman, J. (1997). 'Effects of rime-based orthographic analogy training on the word recognition skills of children with reading disability', *Journal of Educational Psychology*, 89: 645–651.

Hall, K. (2003). *Listening to Stephen Read: Multiple Perspectives on Literacy*. Buckingham: Open University Press.

Harrison, C. (2010). 'Why do policy-makers find the "simple view of reading" so attractive, and why do I find it so morally repugnant?', in K. Hall, U. Goswami, C. Harrison, S. Ellis and J. Soler (eds.), *Interdisciplinary Perspectives on Learning to Read: Culture, Cognition and Pedagogy*. London: Routledge.

Harste, J. C., Woodward, V. A. and Burke, C. L. (1984). *Language Stories and Literacy Lessons*. Portsmouth, NH: Heinemann Educational Books Ltd.

Heath, S. B. (1983). *Ways with Words: Language, Life and Work in Communities and Classrooms*. Cambridge: Cambridge University Press.

Hoover, W. and Gough, P. (1990). 'The simple view of reading', *Reading and Writing: An Interdisciplinary Journal*, 2: 127–160.

Kamler, B. (1984). 'Ponch writes again: A child at play', *Australian Journal of Reading*, 7(2): 61–70.

Kress, G. (2000). *Early Spelling: Between Convention and Creativity*. London: Routledge.

Lass, B. (1982). 'Portrait of my son as an early reader', *The Reading Teacher*, 36(1): 20–28.

Loban, W. (1976). *The Development of Language Abilities, K-12.* Urbana, IL: National Council for Teachers of English.

Lupker, S. (2005). 'Visual word recognition: Theories and findings', in M. Snowling and C. Hulme (eds.), *The Science of Reading. A Handbook.* London: Blackwell.

Martinussen, R. and Kirby, J. (1998). 'Instruction in successive and phonological processing to improve the reading acquisition of at-risk kindergarten children', *Developmental Disabilities Bulletin*, 26: 19–39.

Michael, I. (1984). 'Early evidence for whole word methods', in G. Brooks and A. K. Pugh (eds.), *Studies in the History of Reading.* Reading: Centre for the Teaching of Reading, University of Reading.

Minns, H. (1997). *Read It to Me Now! Learning at Home and at School*, 2nd edn. Buckingham: Open University Press.

Moss, G. and Huxford, L. (2007). 'Exploring literacy policy making from the inside out', in L. Saunders (ed.), *Educational Research and Policy-Making: Exploring the Border Country between Research and Policy.* London: Routledge.

National Institute of Child Health and Human Development. (2000). *Report of the National Reading Panel: Teaching Children to Read: An Evidence-Based Assessment of the Scientific Research Literature on Reading and Its Implications for Reading Instruction: Reports of the Subgroups* (NIH Publication no. 00–4754). Washington, DC: US Government Printing Office.

Payton, S. (1984). *Developing Awareness of Print: A Young Child's First Steps Towards Literacy.* Birmingham: University of Birmingham, Educational Review.

Perry, P., Amadeo, C., Fletcher, M. and Walker, E. (2010). *Instinct or Reason: How Education Policy Is Made and How We Might Make It Better.* Reading: CfBT Education Trust.

Rose, J. (2006). *Independent Review of the Teaching of Early Reading.* Nottingham: DfES Publications.

Rosenblatt, L. (1985). 'Viewpoints: Transaction versus interaction: A terminological rescue operation', *Research in the Teaching of English*, 19(1): 96–107.

Santa, C. and Hoien, T. (1999). 'An assessment of early steps: A program for early intervention of reading problems', *Reading Research Quarterly*, 34: 54–79.

Savage, R., Burgos, G., Wood, E. and Piquette, N. (2015). 'The simple view of reading as a framework for national literacy initiatives: A hierarchical model of pupil-level and classroom-level factors', *British Educational Research Journal*, 41(5): 820–844.

Schmidt, E. and Yates, C. (1985). 'Benji learns to read naturally! Naturally Benji learns to read', *Australian Journal of Reading*, 8(3): 121–134.

Street, B. (1984). *Literacy in Theory and Practice*. Cambridge: Cambridge University Press.

Tierney, R. J. (1991). 'Studies of reading and writing growth: Longitudinal research on literacy development', in J. Flood, J. M. Jenson, D. Lapp and J. R. Squire (eds.), *Handbook of Research on Teaching the English Language Arts*. New York: Macmillan.

Torgerson, C., Brooks, G., Gascoine, G., and Higgins, S. (2018). 'Phonics: reading policy and the evidence of effectiveness from a systematic "tertiary" review', *Research Papers in Education*. doi:10.1080/02671522.2017.1420816

Torgerson, C. J., Brooks, G. and Hall, J. (2006). *A Systematic Review of the Research Literature on the Use of Phonics in the Teaching of Reading and Spelling*. London: Department for Education and Skills (DfES).

Tunmer, W. and Hoover, W. (1993). 'Phonological recoding skill and beginning reading', *Reading and Writing: An Interdisciplinary Journal*, 5: 161–179.

Umbach, B., Darch, C. and Halpin, G. (1989). 'Teaching reading to low performing first graders in rural schools: A comparison of two instructional approaches', *Journal of Instructional Psychology*, 16: 22–30.

U.S. Department of Education, Institute of Education Sciences, What Works Clearinghouse. (2013, July). *Beginning Reading Intervention Report: Reading Recovery®*. Retrieved from http://whatworks.ed.gov

Vickery, K., Reynolds, V. and Cochran, S. (1987). 'Multisensory teaching approach for reading, spelling, and handwriting, Orton-Gillingham based curriculum, in a public school setting', *Annals of Dyslexia*, 37: 189–200.

Wells, G. (1986). *The Meaning Makers: Children Learning Language and Using Language to Learn*. Sevenoaks: Hodder & Stoughton.

White, D. (1954). *Books Before Five*. Auckland: New Zealand Council for Educational Research.

Wyse, D. (2007). *How to Help Your Child Read and Write*. London: Pearson and BBC Active.

Wyse, D. (2011a). 'Towards a theory of English, language and literacy teaching', in D. Wyse (ed.), *Literacy Teaching and Education: SAGE Library of Educational Thought and Practice*. London: Sage.

Wyse, D. (2011b). 'Control of language or the language of control? Primary teachers' knowledge in the context of policy', in S. Ellis and E. McCartney (eds.), *Applied Linguistics and the Primary School*. Cambridge: Cambridge University Press.

Wyse, D. and Goswami, U. (2008). 'Synthetic phonics and the teaching of reading', *British Educational Research Journal*, 34(6): 691–710.

Wyse, D. and Goswami, U. (2012). 'The development of reading', in J. Marsh, N. Hall and J. Larson (eds.), *Handbook of Early Childhood Literacy*, 2nd edn. London: Sage.

Wyse, D., Andrews, R. and Hoffman, J. (2010). 'Introduction', in D. Wyse, R. Andrews and J. Hoffman (eds.), *The Routledge International Handbook of English, Language and Literacy Teaching*. London: Routledge.

Wyse, D. and Styles, M. (2007). 'Synthetic phonics and the teaching of reading: The debate surrounding England's "Rose Report"', *Literacy*, 47(1): 35–42.

Ziegler, J. and Goswami, U. (2005). 'Reading acquisition, developmental dyslexia and skilled reading across languages: A psycholinguistic grain size theory', *Psychological Bulletin*, 131(1): 13–29.

## Annotated bibliography

https://educationendowmentfoundation.org.uk/school-themes/literacy/
The Education Endowment Foundation have funded a significant number of studies of whether a range of interventions improve pupils' learning or not.

Hall, K., Goswami, U., Harrison, C., Ellis, S. and Soler, J. (eds.). (2010). *Interdisciplinary Perspectives on Learning to Read: Culture, Cognition and Pedagogy*. London: Routledge.
Powerful, evidence-based contribution to understanding the teaching and learning of reading.
L2 ★★★

Darnell, C., Solity, J. and Wall, H. (2017). 'Decoding the phonics screening check', *British Educational Research Journal*, 43(3), 505–527.
A robust examination of the phonics screening check.
L3 ★★★

Wyse, D. and Goswami, U. (2008). 'Synthetic phonics and the teaching of reading', *British Educational Research Journal*, 34(6): 691–710.
Put forward a new analysis of high-quality research in relation to the teaching of early reading in order to offer a strong critical response to the Rose Report. One of the most cited articles in 40 years of the British Educational Research Journal.
L3 ★★★

# Chapter 9

# Children's literature

Children's literature is the lifeblood of the teaching of English, language and literacy. In this chapter, we consider its rich variety and research around the importance of teachers' knowledge of and enthusiasm for sharing texts with children. We explore picture books, poetry, fiction, non-fiction and digital reading in a variety of forms as well as some of the issues related to their selection and use, offering some practical suggestions for working with literature in the classroom.

Margaret Meek (1988) argued powerfully that the specific texts that children experience are one of the most important aids to learning to read and that the challenges for young readers are to learn not merely how to decode, but to understand that there is meaning behind the words. The *Cambridge Primary Review*, the most comprehensive review of primary education in the past 40 years, stated that children's imaginations need to be excited in order that children can

> advance beyond present understanding, extend the boundaries of their lives, contemplate worlds possible as well as actual, understand cause and consequence, develop the capacity for empathy, and reflect on and regulate their behaviours . . . [W]e assert the need to emphasise the intrinsic value of exciting children's imagination. To experience the delights – and pains – of imagining, and of entering into the imaginative world of others, is to become a more rounded person.
>
> (Alexander, 2010: 199)

Literature offers 'mirrors, windows and sliding glass doors' (Sims Bishop, 1990) into real and imagined worlds, prompting a better understanding of ourselves and others and reflection on what it is to be human. Children's literature lies at the heart of teaching English.

It is helpful here to consider the perspective that reader-response theory ☞ brings to the act of reading. Rosenblatt (1985) emphasised the subjectivity of transactions between reader and text, formulating a continuum from 'aesthetic' (for pleasure) to 'efferent' (for meaning) styles of reading. Reader-response theory places the emphasis on reading as a creative act of meaning-making by the reader; meaning does not reside solely in the text but 'is made from the reader and the book' (Hunt, 1991: 66). Reading is, in effect, a performance in which the reader makes connections with existing knowledge and experience to make sense of the text. Reading and the choices of text to be read are therefore powerful tools for stimulating thinking in the classroom.

What is meant by the term 'children's literature'? Scholars in the field have contested the notion itself, and for readers in the twenty-first century, the very concept of text includes representations of meanings in an expanding range of forms. For the purposes of this chapter, we will take Bearne and Styles' broad definition of the term, meaning 'texts written to entertain the young' (2010: 22). More than 30 years ago, Hardy (1977) wrote that 'narrative is a primary act of mind' (1977: 13), arguing that story can play a key role in helping children structure and understand the world. From a socio-cultural perspective, reading and the texts that children read entails a shift of emphasis from the individual to the 'social and cultural context in which literacy occurs' (Hall, 2003: 134). In selecting texts to engage children in the classroom, teachers therefore need to be cognisant of the texts with which children engage outside the classroom and to support them in making connections between these and those that they encounter in school. This range needs to take into account what it means to be a reader in the twenty-first century and include not only narratives, but a broad variety of print and online texts. We argue that, in order to teach English effectively, it is essential that teachers have a knowledge of children's literature in a rich variety of forms.

In a United Kingdom Literacy Association (UKLA) project (Cremin et al., 2008), data was collected from 1,200 teachers regarding their personal reading habits, their knowledge of children's literature and the way they used literature in the classroom. When asked to list 'good' children's writers, only 46 per cent of the teachers could name six. In response to the same question about children's poets, 10 per cent listed six and 58 per cent of the respondents could only name one, two or no poets; 22 per cent named no poets at all. Well over half the sample (62 per cent) were only able to name one, two or no picture fiction creators; 24 per cent named no picture fiction authors/illustrators, while only 10 per cent named six. The authors of the Executive Summary report (2007–08) suggested that the question of teachers' knowledge of a diverse enough range of writers to enable them to make informed recommendations to young readers is a cause for concern and that teachers' knowledge of children's poetry, picture fiction and global literature needs considerable development.

In the second phase of the project, teachers were supported to improve their subject knowledge of children's literature and to develop a more inclusive reading for pleasure pedagogy. The project was undertaken in five local authorities. The sample of 27 participating schools involved one primary-level pupil referral unit, five infant, two junior and 19 primaries. Forty-three teachers were involved, 80 per cent of whom were not responsible for literacy in their schools, and three 'focus' children were identified in each class. Findings showed that children who had been identified as reluctant and disaffected readers became drawn into reading; their perceptions of their abilities as readers and self-confidence subsequently improved. They demonstrated increased pleasure in reading and began to read both more regularly and more independently. Children's talk about reading and texts also became significantly more spontaneous, informed and extended. Finally, the majority of the children's attainment showed above average increases across the year. The report's authors concluded that 'reading for pleasure urgently requires a higher profile in primary education to raise both attainment and achievement and increase children's engagement as self-motivated and socially engaged readers' (p. 4).

Surveys over the past decade, however, indicate some troubling trends in reading habits and reading for pleasure (RfP). Findings from the National Literacy Trust's literacy survey of 42,406 pupils aged 8 to 18 in 2016 showed that one child in three enjoyed reading 'quite a lot' and overall, nearly six children in ten (58.6 per cent) said that they enjoy reading either very much or quite a lot. However, this leaves four children in ten who either only enjoy reading a bit or not at all. There were marked differences by ethnic background, with fewer pupils from white ethnic backgrounds enjoying reading compared with pupils from mixed or black ethnic backgrounds, and pupils from Asian backgrounds most likely to say that they enjoy reading. Another national study (Hempel-Jorgensen *et al.*, 2017) revealed that teachers' perceptions and classroom pedagogy constrains struggling boys RfP. McGeown (2015) and Moss and Washbrook (2016) also found growing attitude problems and gender differences with 6–7-year-olds. International statistics also (OECD, 2010) point to a decline in attitudes, enjoyment and frequency of boys' RfP.

What are the implications then of these national and international reports for beginning teachers? Key findings from Cremin *et al.*'s (2014) work show that considerable knowledge of children's literature and other texts is fundamental. Building on Commeyras *et al.*'s (2003) American research, they argue that those professionals who are 'reading teachers' – both readers who teach and teachers who read – are better positioned to develop genuinely reciprocal reading communities. By sharing their own experiences of reading, these teachers make a positive impact on children's desire to read and the frequency of reading at home and at school. Other key findings from the research are that teachers need to develop reciprocal and interactive reading communities within the classroom and beyond and embrace an RfP pedagogy that encompasses:

- social reading environments;
- reading aloud;

- informal book talk, inside-text talk and recommendations;
- independent reading time.

We will discuss practical strategies for developing this further in Chapter 13.

## Selecting literature for the classroom

As a starting point, think about your memories of the texts you encountered as a child at home, in school and on the playground and which were important to you. To what extent were these shared or private experiences and how and by whom were you influenced in your choices? How might these early experiences influence the way you guide the less experienced readers in your class? How will you share your current reading of children's literature to enthuse the pupils with whom you work? It is vital, however, that you do not rely on a 'canon' of remembered favourites but explore a wide range of contemporary literature, too.

The internet has a huge range of resources, not least from publishers who are finding increasingly imaginative ways to market their books, as well as information about numerous children's book awards, for example the Kate Greenaway, Carnegie and UKLA. Websites such as the Just Imagine Story Centre (www.justimaginestorycentre.co.uk) and Booktrust (www.booktrust.org.uk) are valuable resources for teachers and will help you find out about new and established texts. These sites feature recommended books, reviews, regular interviews with authors and a range of practical resources to support teachers in selecting and working with texts in the classroom. Further recommendations for websites are included at the end of this chapter.

In deciding which texts to use in the classroom, teachers are obviously constrained by budget considerations and the resources available to them. However, your selection should be influenced by:

- children's preferences and recommendations;
- guidance from English subject leaders and other colleagues;
- support from librarians;
- long and short lists from children's book awards;
- published guides;
- your own personal preferences, based on wide reading.

Considerations of diversity must also be taken into account Although improving yearly, the majority of children's books published still tend to feature white main characters, with books about people of colour only just outpacing those about animals, trucks and other non-human characters. If the literature to which children are exposed is genuinely to offer 'sliding glass doors' (Sims Bishop, 1990), it needs to represent dimensions of race, ethnicity, gender, sexual orientation, socio-economic status, physical abilities and religious beliefs. Publishers that specialise in providing a broad range of texts include Letterbox library, Tamarind Press and Mantra Lingua. Lantana Publishing, a relatively new company, has diversity as its ethos and works with authors and illustrators from a wide range of cultures and nationalities. In any classroom, it is essential that diversity is represented and celebrated through literature and that the imposition of any narrow social and cultural judgements is resisted.

The issue of quality in relation to any text is clearly important, if contentious. How do we define quality? For teachers, one of the main criteria has to be about the learning that is likely to arise from the reading of the book. However, the genre ☞ of the text and the nature of the experience you are trying to provide will alter the type of learning. In addition, another way to determine quality is through children's preferences for and enjoyment of particular texts. But a dilemma exists here; if children have not had the opportunity to read a wide range of texts and actively discuss their quality, then their judgements may be incomplete. The same is true of teachers: if you have not read widely and analytically, it is difficult to have informed judgements about the quality of texts. Exploring recommendations through the options mentioned above will support you to develop your skill in bringing appropriate and enjoyable texts to your classroom.

Texts operate on a number of semantic levels. First and foremost, the texts should appeal directly and powerfully to children, but adults should also find aspects that engage their curiosity and analytic skills. High-quality texts, particularly fiction and poetry, are usually characterised by the different layers of meaning they contain, layers that reveal themselves only through rereading and analysis. The following list suggests things that you could be thinking about when selecting texts from the wide range of genres for your classroom/setting:

- How will the text support the children's learning?
- Will the subject and content of the text interest the children?
- Does the text link with the children's experience in a meaningful way and/or offer new perspectives?
- Is the amount and level of print appropriate for the reading stamina of the children (this will include some texts which really stretch the children's capabilities)?
- What kind of prior knowledge might children need to access the text (including context and vocabulary)?
- What kind of knowledge will children acquire by reading the text?
- How effectively is language used?
- If present, how effectively are images used?

Your most important aim should be to inspire children to read for themselves.

For infant classrooms (ages 5 to 7), you may like to have a stock of very familiar texts, ones which have been recently introduced and texts that are completely new to your class to share on a daily basis. For junior classrooms, it is often easy to feel that there simply is not time for you to read aloud to the class, but the benefits of having stories, poems and texts to share – again on a daily basis wherever possible – cannot be over-emphasised. Barrs and Cork (2001: 39) argue that when teachers and others read aloud, this enables children to 'attend more closely to the language of the text' without having to focus on decoding and allows teachers to model expression and intonation as well as reading strategies and behaviours.

In the following sections, we discuss the broad range of literature that children should encounter in the classroom, select some favourite texts of our own which have successfully been used in classrooms, and include some text-inspired activities that might take place.

## Picture books

There are many well-established classics ☞ and also contemporary picture books in which the quality of the images and text is breath-taking. Authors and illustrators now have available to them a sophisticated range of media and paper engineering to produce ever more inventive books. A wealth of research has shown that sharing picture books with children leads to highly satisfying and educationally rich experiences for children and adults alike (Arizpe and Styles, 2016; Sipe and Pantaleo, 2008). Picture books develop varied and complex skills and stimulate children's pleasure in reading and writing in the classroom, inspiring confidence. In her study of young bilingual children responding to complex contemporary picture books, Coulthard (2016: 88) highlights

> the intellectual challenge and the capacity of these books both to stimulate and provide ways of demonstrating thinking . . . the power of these texts to inspire pupils to talk in a way that pushes their language to the outer limits.

In simple terms, picture books use image as well as written text to convey meaning. Immediately, though, this statement requires clarification as there are many excellent examples of wordless picture books which provide opportunities for sophisticated reading and writing. Indeed, Lewis (2001) testifies to the difficulty of providing a single definition, as many picture books are 'subtle, indeterminate and resistant to easy categorisation' (2001: 44). Generally, however, picture books' unique nature is based on the combination of verbal and visual modes of communication, of words and image. Styles *et al.* (1996) talk about the gap that exists between the image and the words, the gap that has to be filled by the child's imagination. Often there is a disparity between word and image, so children have to work hard to fill in the gaps and create meaning. It is the complexity of this relationship that offers endless possibilities for multi-layered meanings and interpretation and makes picture books such a rich resource for use in both Key Stage 1 and 2 classrooms and beyond.

A significant feature of picture books is the way that authors and illustrators use images and words to make links to other stories and texts (a feature called intertextuality ☞). A classic example of this is Janet and Alan Ahlberg's *The Jolly Postman or Other People's Letters* (1986), where reference is made to a host of fairy-tale characters as the eponymous 'hero' delivers their letters. Children can delight in taking out the various contents of the envelopes and detecting the many intertextual allusions and puns. There is a wealth of opportunities for children to adopt the format and create their own versions of the postman's journey.

The Ahlbergs' stories can rightfully be regarded as classics because of their longevity and continuing appeal. There are many more recent postmodern picture books ☞ which also play with intertextual links in ingenious ways. Amongst the plethora of examples are Child's *Who's Afraid of the Big Bad Wolf* (2002), Browne's *Into the Forest* (2004), Grey's *The Pea and the Princess* (2004) and Woolvin's *Little Red* (2016). An important contribution of these authors and illustrators to children's literature is the manner in which they root their work in stories which increasingly are being narrated in a variety of formats, from oral tales and books to comics, television and films, that children may have encountered outside the classroom. Whilst being rooted in a Western tradition, they also enable children from a wide range of linguistic and cultural backgrounds to recognise stories and characters with which they may be familiar. Children delight in being 'detectives', spotting the clues to other texts and playing with the metafictive ☞ elements of the stories. In addition, publishers such as Letterbox library produce dual language texts of stories from a range of cultures.

In selecting texts, and visual texts in particular, it is important to think about how images of race and culture are represented. Trish Cooke's book *So Much!* was a multiple prize-winner when it was published in 1994. Although prizes do not always identify the best books, on this occasion the awards of the Smarties Book Prize, the Kurt Maschler Award and the *She*/WH Smith awards were justified. Indeed, Anthony Browne is quoted on the back of the book: 'It is always a delight to see an established artist taking risks, breaking new ground and succeeding brilliantly'. *So Much!* explores an aspect of Black British children's culture and, like many children's books, has a naturally repetitive structure:

> They weren't doing anything
> Mum and the baby
> nothing really . . .

Then,

DING DONG!

'Oooooooh!'
Mum looked at the door,
the baby looked at Mum.
It was . . .
        (Cooke, 1994: 7)

The text encourages children to predict what will happen next, thus helping to develop an important reading strategy, and recognises their enthusiasm for guessing and problem-solving. The illustrations show accurate and positive images of a British Afro-Caribbean extended family, and as each character arrives at the house they first want to do something with the baby, such as squeeze him (Auntie Bibba), kiss him (Uncle Didi), eat him (Nanny and Gran-Gran) or fight him (Cousin Kay Kay and Big Cousin Ross):

And they wrestle
and they wrestle.
He push the baby first,
the baby hit him back.
He gave the baby pinch,
the baby gave him slap.
And then they laugh
and laugh and laugh.
'Huh huh huh!'
        (ibid.: 28)

The language of the book brilliantly uses rhythms and repetitions of African-English.

Two books which celebrate the power of children's imagination rooted in their everyday experience are *Billy's Bucket* (Gray and Parsons, 2003) and *Traction Man Is Here* (Grey, 2005). *Billy's Bucket* describes a boy who wants only one thing – a bucket – for his birthday. Once filled with water, he imagines all the amazing things that could be swimming around in it and, in a humorous twist at the end, confounds his parents' scepticism about his choice and the pleasure to be derived from a mundane household object. *Traction Man* is a wonderfully evocative depiction of a boy's imagination, inspired by the Action Man figure that he receives for Christmas. Both text and illustration draw richly on many genres, including the comic, superhero and toy story. This is a wonderful book simply to share across the primary age range but can also be used as a stimulus for children's play, talk and writing. Some suggestions for starting points include:

- Discuss with the children stories/texts/films they know that feature super-heroes. What qualities do superheroes have? What sorts of adventures do superheroes have?
- Before sharing the text, have a selection of objects displayed to represent the villains of the story (pillows, dishcloth, spade, sock, broom, scissors). Ask children to discuss what part the props might play in a story about a superhero.
- Act out the story for the children, using props and sound effects.

After reading the text:

- Ask children to act out a scene from the story, firstly with props and then without. They could try miming/freeze framing the scene and then invent their own dialogue and play script.
- Have a role-play area with props, superhero outfits etc.
- Make a story box (with Action Man figure, spoons, scrubbing brush, etc.) for a book corner/role-play area.

Opportunities for writing:

- *The Further Adventures of Traction Man* – a sequel to the book in a form of your choice (adventure, mystery, sci-fi, fantasy, historical and contemporary fiction, dilemma stories, dialogue/play, myths, legends, fairy tales, fables, traditional tales).
- 'If I could be a superhero' – a list poem, a haiku, adapt a nursery rhyme.
- 'The *Traction Man* Advertising Campaign' – a radio or TV script to per-form, posters, packaging for Traction Man, letter to persuade head teacher to invite Mini Grey to World Book Day.
- 'My Christmas' – a diary entry, letter, newspaper report for Christmas Day.
- 'How to care for your Traction Man' – an instruction leaflet, a recipe for one of Granny's cakes, a map of one of the episodes in the story.
- A fact file, an alphabet book of superheroes, encyclopaedia entries of the 'baddies' in the book, a catalogue of Christmas gifts.
- other options – captions for role-play area, list of objects for Traction Man's survival kit, new endpapers ☞ for the book, packaging for Scrub-bing Brush toy, thank you letter to Granny . . .

Fortunately for children, teachers and adults alike, there are now two further sequels to Traction Man's adventures!

Many picture books will appeal to children (and adults) of all ages. In con-sidering the vast range available, it is worth stressing again that many can be used very effectively in Key Stage 2 classrooms as they offer ways of approaching complex issues. There are some excellent examples of wordless picture books which can engage children and lend themselves easily to cross-curricular work. Baker's *Window* (2002) and *Belonging* (2004) are concerned with issues of environmentalism and urban and human growth. The books are presented as a series of photographs of Baker's collages, which deserve repeated scrutiny for the many messages and interpretations they offer. Tan's *The Arrival* (2007) is a haunting wordless graphic novel ☞, depicting in a series of sepia-tinged images the journeys of migrants. Former Children's Laureate Anthony Browne's work is notable because of the way that his books often focus on important issues while maintaining genuinely interesting narratives. Examples of such issues include: sexism (*Piggybook*); self-esteem and bullying (*Willy the Champ*); one-parent families (*Gorilla*); freedom and captivity (*Zoo*); gender and sibling rivalry (*The Tunnel*). All of his books are accompanied by mesmeric illustrations and are excellent examples of picture books that can be used very successfully in Key Stage 2 classrooms.

## Poetry

Poetry is a vital resource in early years and primary classrooms. Benton (1978) argued that poetry is the most condensed form of language we have and therefore has the potential to deepen a child's knowledge of what language is and does more subtly than any other form of literature. It supports language acquisition and learning, as well as being a rich introduction to literary heritages. As W. H. Auden said, 'everything we remember, no matter how trivial: the mark on the wall, the joke at luncheon, word games, these like the dance of a stoat or the raven's gamble are equally the subject of poetry' (Izzo, 2011: 120).

Children have an innate response to rhythm and an instinct for musical language which may well start in the mother's womb and is a natural response in babies. Pleasures in patterned language are a marked, cross-cultural feature of childhood. Outside the classroom, young children are often exposed to the language, form and content of poetry through advertisements, jingles and songs. These become supplemented with playground games and chants (often rude!) and songs learnt in school. Reflect on your own experiences of poetry from childhood, including the television and radio jingles, songs and patterned rhythmic texts you experienced at home and school. Which poems and poets do you recall? Can you remember any specific examples of lines, verses or classic poems and nursery rhymes?

Nursery rhymes and action rhymes are an early introduction to many features of the English language and these first experiences of poetry are often oral. They offer the child short snippets of language to remember, repeat and share. Rhyme can be invested with emphasis and physical action, for example 'Ring-A-Ring-Of-Roses' ends with the phrase 'all fall DOWN', which the child learns to accentuate by intonation and by literally falling down. Similarly, the rhyme:

> Round and round the garden
> like a teddy bear
> one step, two step
> and TICKLE HIM UNDER THERE

teaches the child anticipation, turn-taking, humour and the joy of another shared fragment of language. There are close links here with communal songs and stories, but poetry has the particularly important feature (at this stage) of brevity: it is manageable and memorable. The most successful poets for the young have always understood that poetry should be synonymous with playfulness and daily pleasure, and that sharing rhymes in early years and infant classrooms is an essential part of a rich language curriculum. Recordings of nursery rhymes are also very useful in early years and infant classrooms, and nonsense verse and books of riddles are a rich of source of language play.

Building on children's early oral experiences, how can poetry continue to be experienced and valued in the classroom? Regular opportunities to browse through a large range of poetry, both of individual poets' collections and also of anthologies, to find favourites and return to these should be a feature of the ongoing curriculum and not merely a function of English lessons. There are many high-quality classic and modern anthologies to have on your classroom bookshelf. *The Puffin Book of Utterly Brilliant Poetry* (1999) also includes interviews with the poets represented. Anthologies of classic poetry include *The Walker Book of Classic Poetry and Poets* (2001), and there are of course individual collections of older and 'classic' poets such as Stevenson's (1885) *A Child's Garden of Verses*. An unusual anthology which brings together concrete (or shape) poetry ☞ is *A Poke in the I* (2005), and a favourite anthology of ours, which draws together poetry from a range of cultures, is *You'll Love This Stuff* (1986). Examples of single poet collections are also essential, and the more you can acquaint yourself and children with, the better. Any list, even as a starting point, is of course incomplete, but some you might like to sample are: Carol Ann Duffy's *New and Collected Poems for Children*, Roger McGough's *All the Best*, Adrian Mitchell's *Daft as a Doughnut*, Tony Mitton's *Plum*, Grace Nichols' *Everybody Got a Gift* and Benjamin Zephaniah's *Wicked World*.

When reading poetry with children as part of a specific unit of work, there is a temptation to move into analysis and to ask 'what does the poet really mean?' However, a more useful approach is to question 'what are the different ways that a poem can be read? What did this poem remind you of?' This can offer illuminating insights into children's understandings and promote dialogue in the classroom. It should also be remembered that poetry is meant to be *heard*, and that children should be given opportunities to develop the specific skills required to listen to the sounds of words in poetry to interpret meaning. The Children's Poetry Archive is an invaluable resource here. Visitors can listen to aural clips of an ever-expanding range of contemporary English language poets and poets from the past as well as hear interviews with selected poets. All the poems and poets selected have been chosen with children in mind; they include work by well-known 'children's poets' such as Ahlberg, Berry, Bloom, Dahl, Nichols, Rosen and Wright and can be searched by poet, poem or theme. Many poets also have examples of readings on their own websites and the benefits of inviting a poet in to school, so that children can hear poetry read live and work with poets, cannot be over-emphasised.

Children's everyday contact with spoken language is a useful starting point for their poetry writing. Simply asking children to collect together the language they come across builds naturally on early oral experiences of rhyme, rhythm and language and leads into experimentation with form and content based on personal experience. Children are capable of extraordinary observations and often make startling conceptual links between what they see, hear, feel, know and imagine and how they compare those understandings.

Using the constructs of particular poetic forms such as the haiku ☞ or the cinquain ☞ can be interesting starting points for writing poetry. Giving children specific numbers of lines (three in the case of haiku and five for cinquains) and then moving on to the classic syllabic patterns in each line can provide a supportive framework (in a haiku, the syllabic pattern is 5, 7, 5 and in a cinquain, 2, 4, 6, 8, 2). Regular rhythmic patterns such as the limerick also provide similar opportunities for children to work within specific poetic structures that are light-hearted and offer reasonably quick returns for their linguistic investment, although like many classic forms this takes time to master. Teachers can also use poems as a model to act as a framework for children to write their own. Examples that are often used are Kit Wright's *The Magic Box* and Miroslav Holub's *Go and Open the Door*. However, while having a framework can be a helpful tool, it should of course be remembered that these are wonderful poems in their own right and that using models for imitation can detract from children's ability to experiment with form and content in order to create new meaning.

Poetry offers the opportunity for children to find their 'voice' and draw on their language resources to express their own ideas and feelings. Poetry-rich classrooms provide children with the chance to exercise real choice over the content and form of their writing. Finding a balance between form and freedom, however, is a challenge. Whilst models and structures may be helpful to an extent, children should be encouraged to experiment both with form and, importantly, with the development of voice; children need to know about how working within poetic form requires the poet to adjust and adapt language, thought and feeling. This can come about only through extensive experience of reading poetry.

Writing poetry can be a liberating and challenging experience both for children and teachers. At its best, it can be vigorous, committed, honest and fascinating and is probably the most personal writing we ask children to do. As such, we also need to respond to their writing with care. Poetry writing should above all be about searching for things that genuinely matter to the writer. As Ted Hughes said: 'Almost everybody, at some time in their lives, can produce poetry. Perhaps not very great poetry, but still, poetry they are glad to have written' (1967: 33).

### Ideas for working with poetry

- Read poetry all the time and share a wide range of poets and forms.
- Listen to poets reading their own work as well as actors reading poems.
- Play games with language.
- Construct displays of children's and poets' work with poems of the week/month.
- Ask children to review and then collate their own anthologies.
- Provide opportunities for children to write their own collections of poetry.
- Provide opportunities for children to learn poems by heart and for choral speech and poetry performances.
- Organise school poetry assemblies and festivals.
- Encourage children to take part in poetry competitions.
- Invite poets into the classroom.

Finally, it is also worthwhile to be mindful of Mitchell's exhortation at the beginning of his collection *Daft as a Doughnut*:

> GOOD LUCK, TEACHERS!
> Please don't use these poems or any
> Of my other work in exams or tests.
> But I'm happy if people choose to read
> Them aloud, learn them sing, dance
> Or act them in or out of school.
> (Mitchell and Ross, 2004)

## Longer fiction

National curricula often require the teaching of a range of fiction and poetry (including modern, classic and texts drawn from a variety of cultures), play scripts, myths, legends and traditional folk and fairy stories. Traditional stories and folk tales are a key feature of the infant curriculum. They have a particularly important role to play in children's narrative experience because of their origin in oral traditions and speech patterning (Fox, 1993). Fairy tales also lend themselves easily to oral retelling and role-play. Modern retellings and parodies are available in poetic, short story, film, picture book and app versions, offering many opportunities for younger and older children to engage and play with traditional stories in dramatic and written formats. Little Red Riding Hood, for example, has been recast in the Roberts' *Little Red* picture book (2005) as a boy; in *Goldilocks and Just the One Bear* (Hodgkinson, 2012) Little Bear has the opportunity to wreak revenge on a grown-up Goldilocks.

As children move through the primary years, they need to develop their reading stamina and engage with an increasing variety and length of texts, being willing to tolerate the uncertainty when starting the process of getting into a new book. This is not to say that children should not also be allowed to revel in the pleasures of rereading favourite texts and authors, but this may be where the 'class reader' and your personal knowledge of literature comes again into its own. By sharing texts with the whole class and making individual recommendations, you can introduce your pupils to imaginative and exciting writing that they may not otherwise choose to encounter. It is worth persevering here and also seeking out the recommendations of others, including the children, which may mean that you have to challenge your own prejudices. Short stories, shorter fiction and longer novels should all be part of this varied reading diet.

A rewarding and challenging read for upper Key Stage 2 children, both as part of a unit of work or as a class novel, is David Almond's *Skellig* (1998), which has been a Carnegie Medal and Whitbread Children's Book Award winner. The plot focuses on something that the protagonist, Michael, finds at the back of the dilapidated garage that is part of the house he has just moved into. The other plot line concerns the health of Michael's baby sister, who has a heart problem. Throughout the book, Almond portrays Michael's uncertainties and worries about his sister in an authentic and touching way. Towards the end of the book, his sister recovers from an operation, and Mum and the baby return home:

> 'Welcome home, Mum,' I whispered, using the words I'd practised.
>
> She smiled at how nervous I was. She took my hand and led me back into the house, into the kitchen. She sat me on a chair and put the baby in my arms.
>
> 'Look how beautiful your sister is,' she said. 'Look how strong she is.'
>
> I lifted the baby higher. She arched her back as if she was about to dance or fly. She reached out, and scratched with her tiny nails at the skin on my face. She tugged at my lips and touched my tongue. She tasted of milk and salt and of something mysterious, sweet and sour all at once. She whimpered and gurgled. I held her closer and her dark eyes looked right into me, right into the place where all my dreams were, and she smiled.
>
> 'She'll have to keep going for check-ups,' Mum said. 'But they're sure the danger's gone, Michael. Your sister is really going to be all right.'
>
> We laid the baby on the table and sat around her. We didn't know what to say. Mum drank her tea. Dad let me have swigs of his beer. We just sat

there looking at each other and touching each other and we laughed and laughed and we cried and cried.

(p. 168)

The use of language in this passage is exquisite. The human senses of sight, touch, smell and taste overwhelm us. The image of the view through the eyes to 'the place where all my dreams were' is striking. The profound relief about the baby's recovery is made more poignant by the seemingly mundane language of 'swigs' of beer. An initial way to encourage discussion might be to ask children if anything in the passage resonates with their own lives: for example, being worried about a brother or sister.

By demonstrating that you do not have all the answers, you will be showing that literature can have different meanings depending on personal experience. You might start further questions with some straightforward literal ones to check understanding and then proceed to much more searching questions that require inference ☞:

- What happens in the passage?
- Which sentences describe the use of the senses?
- What does 'swigs' mean? Why does Almond use this word here?
- Why did they laugh *and* cry?
- Where is the place, 'where all my dreams were'? How do you think this works as a metaphor?

*Skellig* is a gripping story. Like all the best children's literature, encourages depth of thought in its readers through its recurring themes; celebration of life is contrasted with exploration about what happens to things when they die. Another theme concerns learning and education. Michael meets a new friend called Mina. One of their discussions is about owls and Almond uses the opportunity to present factual information about owls in the guise of the children's curiosity. During another of their conversations, we hear Mina's description of her learning:

'My mother educates me,' she said. 'We believe that schools inhibit the natural curiosity, creativity and intelligence of children. The mind needs to be opened out into the world, not shuttered down inside a gloomy classroom'.

(p. 47)

Since *Skellig's* publication in 1998, Almond has revisited Mina in a thought-provoking prequel, *My Name Is Mina* (Almond, 2011).

Reader-response perspectives have demonstrated convincingly that children bring their own experiences to make sense of texts, but this of course does not mean that teachers do not have a role to play. The texts you bring to the classroom and the activities you plan will enable readers to reflect on their reading and enhance meanings already made. Some suggestions for general strategies that you might use are:

- reading aloud;
- asking open-ended questions;
- encouraging oral response;
- role-play, improvisation and drama;
- focusing on the author/author studies;
- analysing story features: genres, plots, structures, characters, settings, styles and themes;
- comparing versions and adaptations of texts;
- cross-curricular activities, including art and DT;
- written response/writing in role, scripts, book-making;
- reading journals and reviews.

## Non-fiction

As Mallet (2010) points out, non-fiction has often been marginalised as a form of 'literature', as have more 'popular' types of texts such as comics and magazines. Maynard *et al.* (2007), investigating the reading habits of children aged 4–16 years from a total of 46 schools, 22 primary and 24 secondary, asked participant children about what they like to read, which factors affected their choice of books and who they would recommend books to. Conclusions drawn demonstrated the clear importance of magazines for reading for pleasure, and that non-fiction was relatively important to boys and girls, both for support with schoolwork and also for pleasure. Such texts are all part of the currency of children's reading and offer rich opportunities for developing purpose for and pleasure in reading in the classroom as well as opportunities to develop critical reading skills.

Non-fiction texts written for children have shown dramatic improvements – notably in terms of organisation and layout – over the last 40 years and now differ widely in mode and media. Some of the major publishers of non-fiction for children – Dorling Kindersley, Usborne, Kingfisher, Heinemann, Franklin Watts, Oxford University Press, Evans Brothers, A&C Black and Collins – have made available a wide range of non-fiction content in print and (increasingly) in multimedia ☞ formats. Many editions of non-fiction texts have CD-ROMs and links to websites which can add an extra dimension to children's engagement. Multimedia texts such as these involve a different approach to reading solely print text and can offer a motivating experience for young readers into accessing information.

It is important, therefore, not to have a narrow view of particular text types and technical details about their structures and language feature; 'real' non-fiction texts shift boundaries and often contain a mixture of types, forms and structures. A leaflet for a local attraction, for example, will contain elements of persuasion, report and instruction. If you consider the writing styles of Jamie Oliver, Delia Smith and Nigel Slater, it is clear that recipes can be written in a variety of instructional formats! Walker's *Read and Wonder* was a breakthrough series that provides text with a poetic narrative voice accompanied by informative captions. 'Stories' such as *Think of an Eel* (Wallace and Bostock, 2008) are told in lyrical style and use features of characterisation, pace, tension and drama which are more commonly associated with fiction texts and make the books very readable. The text in the book is accompanied with beautiful illustrations and there is an accompanying CD-ROM. As with picture book fiction, exploring the contribution of illustrators and their different styles to these texts is a very worthwhile activity with children.

For older children, the books by the award-winning Manning and Granström partnership are examples of non-fiction texts organised narratively. In the *Fly on the Wall* series published by Frances Lincoln, information about Greek, Egyptian, Roman and Viking life is provided through a cast of characters, animated with direct speech in a sketchbook format. The authors' *What Mr Darwin Saw* (Manning and Granström, 2009) combines diary entries, text boxes and illustrations to provide information about Darwin's life and the reception of his ideas, offering lots of opportunities for cross-curricular work. Marcia Williams is another author who incorporates an extensive range of text types, notably in *Archie's War* (2007) which, in scrapbook format, recounts a child's experience of the First World War. In *Lizzy Bennett's Diary* (2013), Williams recreates the story of *Pride and Prejudice* with Lizzy's drawings, pressed flowers, ribbons from her bonnet, hand-written notes to fold out and read, dance cards, invitations, and even a letter from Mr Darcy. Another favourite is Gravett's *Little Mouse's Big Book of Fears* (2007). This is an ingenious book, impossible to categorise because it combines so many different layers and text types and, in true postmodern fashion, invites readers to contribute to the book through drawing, writing or collage.

Other non-fiction text types to include in your classroom repertoire are comics, magazines and newspapers. *The Phoenix* anthology comic, published weekly contains serialised adventure stories, original comics, fascinating non-fiction and puzzles and is suitable for readers aged 6–12. *First News* is an award-winning weekly newspaper for young people that is published in both print and digital format. There are many non-fiction texts that exploit paper technology to enhance children's reading experiences. *Above and Below* (Hegarty and Clulow, 2016) is a stunning split-page book that enables readers to peek below the surface of the ocean to uncover a world of hidden wildlife. *The Pop-up Book of Ships* (Kentley and Hawcock, 2009) is a dramatic pop-up journey into the world's maritime past, presenting various maritime cultural innovations which emerge from the page in varying sizes.

As a final recommendation here, an old favourite, *The Way Things Work Now* (Macaulay and Ardley, 2016), has been revised and updated to embrace the latest developments in the world of machines, from touchscreens to 3D printer.

Each scientific principle is brilliantly explained – with the help of a charming, if rather slow-witted, woolly mammoth. For further support in selecting non-fiction texts, Part II of *Mallet's Choosing and Using Fiction and Non-fiction, 3–11* is an invaluable and comprehensive guide.

Many of the teaching strategies already outlined apply to non-fiction texts. Reading aloud, sharing your own and children's enthusiasm and discussing responses to a variety of texts beyond those that have been created as teaching tools will help to engage and motivate children with a sense of purpose and pleasure. Wray and Lewis (1997) advocate a model for teaching non-fiction which moves from teacher modelling through joint and scaffolded activity into independent activity. They recommend the use of oral discussion, writing frames and writing located in meaningful experiences. Mallet (2010) sum-marises her response to this model, adding greater emphasis to the active, social and oral experiences which need to accompany these stages, and also stresses that the process may not be linear. She asserts that:

> children need to be at the controls when it comes to using reference and indeed all information texts. We encourage this by making critical reading important from the earliest stages so that children can develop their own viewpoints and their own 'voice' when they speak and when they write.
> (Mallet, 2010: 374)

Finally, she states that the teacher's role is central. These are statements which apply to all the forms of texts that have been addressed in this chapter.

## Multimodal, digital and film texts

Traditional constructions of what is meant by reading and writing have focused on linear printed texts. Yet, to be literate in the twenty-first century requires experience of a wide range of written and visual modes of text. Bearne (2003: xvii) argues for broader constructions of reading to include digital and multi-literacies. She notes: 'children now have available to them many forms of text which include sound, voices, intonation, stance, gesture, movement, as well as print and image. These have changed the way young readers expect to read'. Examples of multimodal ☞ texts include picture books, graphic novels, films, comics, slide presentations, computer games and podcasts. In multimodal texts, meaning is constructed through the combination of these four modes:

1 **Written mode:** meaning and design of the words – font, size, layout, spacing, colour.
2 **Visual mode:** composition, movement, colour, light, space.
3 **Sound mode:** tone, pitch, pauses, choice of instrument.
4 **Gestural mode:** facial expression, holding of posture or expression, spac-ing of people.

Many of the texts that children experience and enjoy at home are multimodal, and research such as the Essex Writing Project (2003) has shown that they respond enthusiastically to texts and activities which involve the visual.

Multimodal texts can be examined with regard to their design and layout and how different combinations of representations convey meaning. Activities involving reading complex picture books in which word and image interact are very helpful in developing children's reading comprehension. Here, a visualiser is an extremely helpful tool to enable the whole class to engage with a text. Children can then apply the observed relationship between words, image, gesture and sound to their own constructions on paper and on screen. Hardware and software such as digital cameras, PowerPoint and Photo Story can be used by children to then create multimodal texts.

With the advent of digital books and new reading devices, texts have diversified. Kucirkova and Littleton (2016) define digital books as an umbrella term encompassing the various kinds of digital texts available for young readers, including e-books, ibooks, story apps, book apps and LeapReader books that offer some kind of interactivity. There are several new reading platforms that children can use to engage with these texts. Children can become immersed in story worlds on smartphones, tablets, Wiis, Leapsters, e-readers, X-boxes or DSEs. These devices support different formats and provide different affordances for story engagement. These texts offer ways into playful digital spaces that contribute to affective and interactive engagement and also improve digital literacy skills. They also provide opportunities for shared, collaborative and sustained engagement, which are key facets of reading for pleasure.

Nosy Crow is a multi-award-winning independent children's book and app publisher. They aim to create books and apps that encourage children to read for pleasure. All of their fiction appears simultaneously in print and e-book form and their paperback picture books come with a free digital audio reading. In 2015 Nosy Crow was awarded the first ever UKLA Digital Book award for Axel Scheffler's *Flip Flap Safari*, in which children are invited to create their own animals and play guessing games whilst using the 'read to me' feature if they choose. Nosy Crow again won the award the following year for their *Goldilocks and Little Bear* app, which extends the classic story in innovative ways, providing, for example, the possibility of switching between two stories by simply rotating the iPad. This feature, together with the combination of multimedia enhancements in the book (such as music, sound effects and animation), creates a reading experience that would not be possible with a printed book. There are also special interactive possibilities embedded in the app, such as dressing up the baby bear.

Just as we teach children to read written words, so we need to teach them to 'read' moving images. Film texts convey meaning through multiple modes: spoken language, music, performance, costume, lighting, editing, sound effects, etc. As Maine (2015) has shown, short narrative films are an important resource in helping to develop comprehension strategies and to engage meaningfully with texts, especially if this is done within the context of small group work where talk prompts around key comprehension strategies are provided. These strategies, as for print texts, are:

- summarising and determining importance;
- questioning and possibility thinking;
- clarifying and self-monitoring;
- predicting, hypothesising and reasoning;
- making connections and using prior knowledge;
- empathising and visualising.

A useful resource for short films is https://vimeo.com. Another very helpful resource is the Film Education website: www.filmeducation.org. It has a wide range of curriculum-focused resources to use with ages 4 to 19, most of which are free. There is a library of films and clips, stills, trailers, interviews and suggestions for activities related to current releases. The British Film Institute (BFI) produced some suggestions for structured teaching. The *Story Shorts* resource (British Film Institute, 2001) consisted of a series of films about five minutes long. *Story Shorts* suggested a set of opening questions that can be used to stimulate discussion: the '3 Cs', which are Camera, Colour and Character; and the '3 Ss', which are Story, Sound and Setting. Following whole group discussion supported by the Cs and Ss questions, suggestions were made for how the films could be used to stimulate other, more extended activities which take speaking and listening into writing and drawing.

## Practice points

- Start a reading journal to document texts you have read across a range of narrative, non-narrative, poetry and digital examples and include notes about possible ways to use these in the classroom.
- Provide frequent opportunities for literature of all types to be shared, and also moments of quiet reflective pleasure.
- Consider the interests and attitudes of the children in your class to build reading stamina for children at different stages of development.

## Glossary

**Cinquain** – a verse of five lines.

**Classics** – books that remain of interest to significant numbers of people long after their initial publication date. They are also regarded as of special significance.

**Concrete poem** – a poem which forms a shape that complements the meaning of the poem.

**Endpapers**– pages at the beginning and end of a picture book on which there may be illustrations or images relating to the theme of the book.

**Genre** – a kind or type of text bound by rules and conventions.

**Graphic novel** – a contentious term but generally understood as a full-length story published as a book in comic-strip format.

**Haiku** – a Japanese poem of 17 syllables, in three lines of five, seven, and five syllables. Traditionally evokes images of the natural world.

**Inference** – the understanding of textual meanings that goes beyond the literal to what is implied rather than made fully explicit.

**Intertextuality** – where allusions to other texts are made within a text.

**Metafictive** – a term applied to a text which, in creating the story, comments on the process of its creation.

**Multimedia** – a communication combining different sorts of media.

**Multimodal** – a combination of any of the four modes of communication (print, image [moving and still], sound, gesture).

**Postmodern picture books** – picture books which expand the conventional boundaries of picture book format, often containing non-linear structures, multiple perspectives and elements of playfulness, ambiguity and irony.

**Reader-response theory** – a theory which explores how readers respond to, and make sense of, texts.

### References

Alexander, R. J. (ed.). (2010). *Children, Their World, Their Education: Final Report and Recommendations of the Cambridge Primary Review*. London: Routledge.

Arizpe, E. and Styles, M. (2016). *Children Reading Picturebooks: Interpreting Visual Texts*. London: RoutledgeFalmer.

Barrs, M. and Cork, V. (2001). *The Reader in the Writer: The Links Between the Study of Literature and Writing Development at Key Stage 2*. London. CLPE.

Bearne, E. and Styles, M. (2010). 'Children's literature', in D. Wyse, R. Andrew and J. Hoffman (eds.), *The Routledge International Handbook of English, Language and Literacy Teaching*. London: Routledge.

Bearne, E., Dombey, H., Grainger, T. (2003). *Classroom Interaction in Literacy*. Maidenhead: Open University Press.

Benton, M. (1978). 'Poetry for children: A neglected art', *Children's Literature in Education*, 9(3): 111–126.

British Film Institute (2001) *Story Shorts*. BFI.

Commeyras, M., Shockley Bisplinghoff, B. and Olson, J. (eds). (2003). *Teachers as Readers: Perspectives on the Importance of Reading in Teachers' Classrooms and Lives*. Newark, DE: International Reading Association.

Coulthard, K. (2016). '"The words to say it". Young bilingual learners responding to visual texts', in E. Arizpe and M. Styles (eds.), *Children Reading Picturebooks: Interpreting Visual Texts*. London: RoutledgeFalmer.

Cremin, T., Mottram, M., Collins, F. and Powell, S. (2008). *Building Communities of Readers*. Leicester: UKLA.

Cremin, T., Mottram, M., Powell, S., Collins, R. and Safford, K. (2014). *Building Communities of Engaged Readers: Reading for Pleasure*. London: Routledge.

Fox, C. (1993) *At the Very Edge of the Forest: The Influence of Literature on Storytelling by Children*. London: Cassell.

Hall, K. (2003). *Listening to Stephen Read: Multiple Perspectives on Literacy*. Buckingham: Open University Press.

Hardy, B. (1977). 'Towards a poetics of fiction: An approach through narrative', in M. Meek, A. Warlow and G. Barton (eds.), *The Cool Web: The Pattern of Children's Reading*. London: The Bodley Head.

Hempel-Jorgensen, A., Cremin, T., Harris, D. and Chamberlain, L. (2017) *Understanding Boys' (Dis)engagement with Reading for Pleasure*. Milton Keynes, UK: Open University Press.

Hughes, T. (1967). *Poetry in the Making*. London: Faber and Faber.

Hunt, P. (1991) *Criticism, Theory and Children's Literature*. Oxford: Wiley Blackwell.

Izzo, D. (2011). *W. H. Auden Encyclopedia*. Jefferson, NC: McFarland & Company.

Kucirkova, C. and Littleton, K. (2016). 'Young children's reading for pleasure with digital books: Six key facets of engagement', *Cambridge Journal of Education*, 47(1): 67–84.

Lewis, D. (2001). *Reading Contemporary Picturebooks: Picturing the Text*. London: Routledge.

Maine, F. (2015) *Teaching Comprehension Through Reading and Responding to Film*. London: The United Kingdom Literacy Association.

McGeown, S. (2015). 'Understanding children's reading activities: Reading motivation, skill and child characteristics as predictors', *Journal of Research in Reading*, 39(1): 109–125.

Meek, M. (1988). *How Texts Teach What Readers Learn*. Stroud: Thimble Press.

Moss, G. and Washbrook, L. (2016). *Understanding the Gender Gap in Literacy and Language Development*. Bristol: Bristol working papers in education.

OECD (2010). *PISA 2009 Results: Learning to Learn – Student Engagement, Strategies and Practices* (Volume III). http://dx.doi.org/10.1787/9789264083943-en

Rosenblatt, L. (1985). 'Viewpoints: Transaction versus interaction: A terminological rescue operation', *Research in the Teaching of English*, 19(1): 96–107.

Sims Bishop, R. (1990). 'Mirrors, windows and sliding glass doors', *Perspectives: Choosing and Using Books for the Classroom* (6)3: n.p.

Sipe, L. and Pantaleo, S. (2008). *Postmodern Picture Books: Play, Parody, and Self-Referentiality*. London: Routledge.

Stevenson, R. L. (1885). *A Child's Garden of Verses. USA: Platt & Munk Publishers.*

Styles, Morag and Watson, V. (eds.). (1996). *Talking Pictures: Pictorial Texts and Young Readers.* London: Hodder & Stoughton.

Wray, D. and Lewis, M. (1997). *Extending Literacy.* London: Routledge.

## Books for children

Ahlberg, A. and Ahlberg, J. (1986). *The Jolly Postman or Other People's Letters.* London: Heinemann Educational Books Ltd.

Almond, D. (1998). *Skellig.* London: Hodder Children's Books.

Almond, D. (2011). *My Name Is Mina.* London: Hodder Children's Books.

Baker, J. (2002). *Window.* London: Walker Books.

Baker, J. (2004). *Belonging.* London: Walker Books.

Browne, A. (1983). *Gorilla.* London: Random Century.

Browne, A. (1985). *Willy the Champ.* London: Little Mammoth.

Browne, A. (1989). *Piggybook.* London: Reed Consumer Books.

Browne, A. (1992a). *The Tunnel.* London: Walker Books.

Browne, A. (1992b). *Zoo.* London: Random House.

Browne, A. (2004). *Into the Forest.* London: Walker Books.

Child, L. (2002). *Who's Afraid of the Big Bad Book?* London: Hodder Children's Books.

Cooke, T. (1994). *So Much!* London: Walker Books.

Duffy, C. A. (2010). *New and Collected Poems for Children.* London: Faber and Faber.

Extracts from *So Much!* reproduced by permission of the publisher, Walker Books Ltd, London. Text © 1994 Trish Cooke, illustrated by Helen Oxenbury.

Gravett, E. (2007). *Little Mouse's Big Book of Fears.* London: Macmillan.

Gray, K. and Parsons, G. (2003). *Billy's Bucket.* London: Red Fox.

Grey, M. (2004). *The Pea and the Princess.* London: Red Fox.

Grey, M. (2005). *Traction Man Is Here.* London: Jonathan Cape.

Hegarty, P. and Clulow, H. (2016). *Above and Below.* London: Caterpillar Books.

Hodgkinson, L. (2012). *Goldilocks and Just the One Bear.* London: Nosy Crow.

Janeczko, P. and Raschka, C. (illus). (2005). *A Poke in the I.* London: Walker Books.

Kentley, E. and Hawcock, D. (2009). *The Pop-Up Book of Ships.* London: Universe Publishing.

Macaulay, D. and Ardley, N. (2016). *The Way Things Work Now.* Boston: Houghton Mifflin.

Mallet, M. (2010). *Choosing and Using Fiction and Non-Fiction 3–11.* London: Routledge.

Manning, M. and Granström, B. (illus). (2009). *What Mr Darwin Saw*. London: Frances Lincoln.

Maynard, S., Mackay, S., Smyth, F. and Reynolds, K. (2007). *Young People's Reading in 2005: The Second Study of Young People's Reading Habits*. London: National Centre for Research in Children's Literature.

McGough, R. and Monks, L. (illus). (2004). *All the Best: The Selected Poems of Roger McGough*. London: Puffin.

Mitchell, A. and Ross, T. (2004). *Daft as a Doughnut*. London: Orchard.

Mitton, T. (2010). *Plum*. London: Barn Owl Books.

Nichols, G. (2005). *Everybody Got a Gift: New and Selected Poems*. London: A & C Black Publishers.

Patten, B. (ed.). (1999). *The Puffin Book of Utterly Brilliant Poetry*. London: Puffin.

Roberts, L. and Roberts, D. (illus). (2005). *Little Red: A Fizzingly Good Yarn*. New York: Abrams.

Rosen, M. and Howard, P. (illus). (2001). *The Walker Book of Classic Poetry and Poets*. London: Walker Books.

Styles, M. (1986). *You'll Love This Stuff*. Cambridge: Cambridge University Press.

Tan, S. (2007). *The Arrival*. London: Hodder Children's Books.

Wallace, K. and Bostock, M. (illus). (2008). *Think of an Eel*. London: Walker Books.

Williams, M. (2007). *Archie's War*. London: Walker Books.

Williams, M. (2013). *Lizzy Bennett's Diary*. London: Walker Books.

Woollvin, B. (2016). *Little Red*. London: Two Hoots.

Zephaniah, B. (2000). *Wicked World*. London: Puffin.

## Annotated bibliography

Arizpe, E. and Styles, M. (2016). *Children Reading Picturebooks: Interpreting Visual Texts*. London: RoutledgeFalmer.
A wonderful exploration of children's responses to contemporary picture books, highlighting the significance of visual literacy to children's engagement with literature.
L2 ★★

Gamble, N. (2013). *Exploring Children's Literature: Reading with Pleasure and Purpose*, 3rd edn. London: Sage.
An extremely comprehensive and readable exploration of the range of children's fiction with excellent suggestions for further reading.
L2 ★★

Mallet, M. (2010). *Choosing and Using Fiction and Non-fiction 3–11: A Comprehensive Guide for Teachers and Student Teachers*. London: Routledge.
A wonderful guide, packed full of useful advice and recommendations illustrated with insightful case studies.
L2 ★★

Meek, M. (1988). *How Texts Teach What Readers Learn*. Stroud: The Thimble Press.
A seminal text in the field of children's literature. Meek argues strongly for the importance of specific high-quality texts as one of the main things that will help children learn to read.
**L2** ★★

## Useful websites

www.booksforkeeps.co.uk/
www.booktrust.org.uk/books/children/
www.carnegiegreenaway.org.uk
http://childrenspoetryarchive.org
www.justimaginestorycentre.co.uk
www.lantanapublishing.com
www.literacytrust.org.uk
www.readingzone.com/home.php
https://researchrichpedagogies.org/research/reading-for-pleasure

# Chapter 10

# Listening to children read

Reading with children plays a key role in teachers' understanding of children's reading development. The chapter addresses this in the context of one-to-one reading. It includes thoughts about the kinds of conversations that take place with readers, and what this might reveal about a child's reading preferences in addition to their level of reading. Finally, the chapter considers the positive benefits of one-to-one reading at home with parents/carers in the context of home–school links.

The teacher's ability to interact with a small group or a whole class can be greatly enhanced if they understand the subtleties of reading with an individual child. In the early years in particular, the supportive, personal context of one-to-one reading is very important. The strategy can provide a natural link with the kinds of reading that many children experience at home with parents and other family members. In addition, if teachers are to advise parents/carers on effective ways of supporting their child's reading, they need to be knowledgeable and effective at one-to-one reading themselves. One-to-one reading is also important because of its significant value in both identifying and enabling extra support for those pupils who find reading challenging for a variety of reasons.

## Listening to children read on a one-to-one basis

Historically, teachers have always listened to children read. Campbell (1995: 132) originally called this practice *shared reading*: 'shared reading involves a child and a teacher (or other adult) reading together, in a one-to-one interaction, from a book'. Since then, shared reading as a term has been used to mean a whole class sharing an enlarged version of a book (→ Chapter 13, 'Classroom practices for reading'). The following example from Campbell (1990) is indicative of the original definition. The extract is from a one-to-one reading session that took place over a period of approximately five minutes and which ended with the teacher and the child discussing the story for a few minutes more and the child telling of her own pet's adventures.

Five-year-old Kirsty was sharing a new book with her teacher. First, the teacher read the story while Kirsty looked at the pictures. Then they read through the story again with the teacher asking Kirsty questions that drew her into conversation about different incidents in the story. Then, when the teacher felt Kirsty was ready, she asked her to read aloud:

*Teacher:*   Your turn to read it. All right, let's see.
*Kirsty:*    The dog sees a box.
*Teacher:*   Mmmh.
*Kirsty:*    He sniffs in the box.
*Teacher:*   He sniffs it, doesn't he?
*Kirsty:*    He kicks the box. He climbs in the box.
*Teacher:*   Oh, now what happens?
*Kirsty:*    He falls down the stairs. The dog falls out the box. The (hesitates)
*Teacher:*   The (pauses)
*Kirsty:*    The dog falls over.
*Teacher:*   He does, doesn't he? Then what does he try to do?

Original text:

> The dog sees the box.
> The dog sniffs the box.
> The dog kicks the box.
> The dog gets in the box. The dog gets out of the box.
> The dog falls over.
>
> (Campbell, 1990: 29–30)

Although Kirsty was not reading every word accurately, the meaning of the text had largely been retained, so the teacher did not correct the child, although she did make several interjections to support and encourage Kirsty's reading. When the child hesitated, the teacher simply restarted the sentence, then paused, prompting Kirsty to respond appropriately. Such one-to-one interactions can effectively take place with all levels and ages of children learning to read in the primary school. The level of text may change, but the principles of encouragement, support, discussion, instruction and enjoyment will not. Similar types of discussions should take place at the end of undertaking a miscue analysis or running record with an individual child, established reading procedures designed to similarly help clarify a child's level of reading ability and comprehension of the text (→ Chapter 21, 'Language and literacy difficulties'). For EAL learners, who may have well developed understanding of sound symbol correspondences, the opportunity to ask appropriate questions to draw out their comprehension of what they have just read is additionally key to understanding their level of reading.

Key pointers for reading with children on a one-to-one basis include not always choosing the latest reading scheme text, but perhaps selecting a book that you think will engage the child's interest or being guided by the child's own choice. If the child is already reading texts with minimal support, try to select those that the child is able to read without support, but which you consider to be of high quality (→ Chapter 9, 'Children's literature'). Read alongside the child, reading to them if necessary but allowing them to take the lead whenever possible. Discuss the book with the child, noting, for example, interest, understanding, awareness of bibliographical features and genre. If appropriate, focus on semantic, syntactic and graphophonic cueing strategies ☞ (and consider where and how these strategies could be developed, explicitly or implicitly, throughout the child's day (→ Chapter 13, 'Classroom practices for reading')).

Another strategy when working with children on a one-to-one basis is to conduct a reading 'interview' or 'conference'. The purpose of this strategy is to find out about the child's reading interests and attitudes. Questions might include:

*   Which books do you like to read/look at?
*   What is your favourite book? Why?
*   What do you like to read at home?
*   Do you go to the library?
*   What do you like to read? Why?
*   Do you enjoy reading? Why? Why not?
*   Who reads to you at home?
*   Where do you do most of your reading?

You will of course need to frame your reading interview questions to reflect the age and ability of the child, but it is important to remember that children should not equate reading solely with books, but also a range of different texts (non-fiction texts, poetry, magazines, comics, film, TV and digital media, to name but a few). You could conduct interviews with the whole class to help develop a positive reading environment in your classroom, something that an engaging book corner, full of books to encourage reading for pleasure, can support (DfE, 2013: 10). Choice across a range of texts should additionally support children's motivation to read. Tapping into children's interests is important, as research has shown for example how background knowledge of a particular topic impacts children's vocabulary learning as a result of their reading (Keafer, 2017).

As children's reading becomes more fluent, the priorities for one-to-one reading change. For the majority of children, efficient use of teachers' time can be made by gradually moving towards more emphasis on group reading and literature circles (→ Chapter 13, 'Classroom practices for reading'). Overall, there is less of a need for one-to-one sessions for fluent readers, but this does not mean that they should be completely abandoned. One of the most important things that can be discovered by talking to a child about their reading in the one-to-one session is their motivation for reading. Do they read for pleasure? If so, what kinds of texts are they interested in? A motivated reader is surely one who is likely to progress further in their understanding of English than one who is not motivated. The one-to-one session also allows for a more in-depth exploration of children's understanding of and response to particular texts, including their ability to decode the words, and any problems in these areas.

## Links between reading at home and reading at school

Early years settings and primary schools have encouraged parents/carers to read with their children by sending home a range of texts that the child has chosen to share. Parents/carers are also encouraged to listen to their children read their 'reading books' and to make records in reading diaries. In this way, parents/carers can support their children on a one-to-one basis. Teachers can also scaffold this process by sending home comments and key questions to prompt response and discussion. Many parents and carers engage in bedtime reading sessions with their children, one good example of an informal literacy-based activity at home. Holdaway's (1979) developmental approach allowed for a more formal framing of this type of experience:

1   The parent introduces a new story. The child is curious, so they ask questions and predict the things that are likely to happen next.
2   If the child likes the story, they ask for it to be reread immediately and/or later on. They participate more each time they hear the story. The number of questions they ask increases.
3   If the book is stored somewhere accessible, they use it later to play at reading and to re-enact the story independently, which gives them additional satisfaction.
4   Further rereadings result in the child becoming more familiar with the book.
5   Play-reading and re-enactment become closer to the language of the text.
6   The concepts, language and attributes of the story are extended into play.
7   The book may become an 'old favourite' and/or attention turns to the beginning of the cycle with a new book.

Parents should always be made aware of the fact that children will want to return to favourite books again and again, and that it is not problematic to be reading the same texts on a daily basis if they are requested. Favourite books do change eventually! Encourage children to share their weekend reading at the beginning of each week, and this may give you an insight into their reading preferences. Reading workshops for parents on how to help their child at home may encourage greater confidence in supporting their child's reading. Make sure they keep the bedtime story going; Helen read with her son every evening until he finished primary school at the age of 11 years. Let parents know their role in fuelling children's imagination with books and explain how this extends their vocabulary for writing. Encourage parents to introduce a range of text types when reading with their child.

### Paired reading

An even more structured method that research has shown to be effective is paired reading, first developed by Keith Topping (1987). Although the approach is structured, it is not overly technical – a pragmatic consideration that has to be taken into account with any method for parents. Paired reading involves the following stages:

1   Ideally, the child chooses the book or text to be read.
2   The book should be briefly discussed before reading aloud commences.
3   The adult and the child begin by reading the text aloud together.
4   The child may follow the text with a finger.
5   When the child wants to read alone, they indicate this by tapping the table or arm of the adult.
6   The adult ceases reading immediately and praises the child for signalling.
7   The child continues alone.
8   When a child makes a miscue, the adult supplies the word.
9   The child then repeats the word.
10  Praise is given for the correct reading of difficult words and for self-corrections ☞.
11  If a child is unable to read a word or correct an error in about five seconds, the adult and child return to reading in unison.
12  The child makes the signal when they feel confident enough to resume reading alone.
13  Further praise and encouragement are given at the end of the session.

One important feature of paired reading is that the adult provides a model of appropriate reading behaviour for the child alongside the child's own attempts at reading satisfactorily.

Topping has since developed the strategy of paired reading for the school setting, as he colourfully describes in the following extract:

> What is that noise? It is like a buzz. Let us go investigate. Down the corridors of the school we go. Getting nearer. It does indeed sound like a

beehive. But louder! By the classroom door now – and what do we find? Instead of the regular similar-age children, there is a mix of older and younger pupils, a couple of years between them. They are all matched up in pairs, older with younger. Each pair has their own book, each quite different to any other pair in the class. And each pair is reading – sometimes one out loud, sometimes both. That is where the buzz is coming from! How is the teacher coping with this chaos? Actually, where is the teacher? Oh, there she is – down low with a pair. Coaching, I guess – then she moves on to another pair. Seems unflustered. But there is more buzz than just one classroom – sure enough, the next classroom is at it as well. And the next! So many children reading at the same time, in an interactive way, without the expense of additional resources.

(Topping *et al.*, 2011: 3)

In paired reading, pairs of children (either same age or cross-age groups) choose their own books (or other reading materials) which are of high interest to themselves. The materials must be above the independent readability level of the tutee but not of the tutor. Both members of the pair should be able to see the book equally easily in a quiet, comfortable place. They are encouraged to talk about the book to develop shared enthusiasm and ensure the tutee really understands (comprehends) the content. Tutors support tutees through difficult text by reading together and by giving praise.

### Practice points

- Make time for one-to-one reading with all children in your class.
- Allow time for children to respond to the text they are reading, for example by asking questions to elicit their comprehension.
- Note the strategies children use to decode texts when they are reading aloud to you to help develop next steps as well as to find suitable texts to move on to.
- Provide opportunities for parents/carers to be involved with their children's reading to develop home–school links. This may include organising reading workshops for parents.

### Glossary

**Graphophonic cueing strategies** – the reader understands the relationship between letters and phonemes. They 'sound out' words using their knowledge of the alphabetic code.

**Self-corrections** – a child's oral correction of a word they mis-read.

**Semantic cueing strategies** – the reader understands the centrality of meaning to reading. They check the meaning of a word, showing awareness that the word makes sense within the sentence as a whole.

**Syntactic cueing strategies** – the reader understands the relationship between words and word order. They use their grammatical knowledge to predict the words in a sentence.

## References

Campbell, R. (1990). *Reading Together*. Buckingham: Open University Press.

Campbell, R. (1995). *Reading in the Early Years Handbook*. Buckingham: Open University Press.

DfE. (2013). *The National Curriculum*. London: DfE Publications.

Holdaway, D. (1979). *The Foundations of Literacy*. Portsmouth, NH: Heinemann Educational Books Ltd.

Keafer, T. (2017). 'The role of topic-related background knowledge in visual attention to illustration and children's word learning during shared book reading', *Journal of Research in Reading*. doi: 10.1111/1467-9817.12127

Topping, K. (1987). 'Paired reading: A powerful technique for parent use', *The Reading Teacher*, 40(7): 608–614.

Topping, K., Miller, D., Thurston, A., McGavock, K. and Conlin, N. (2011). 'Peer tutoring in reading in Scotland: Thinking big', *Literacy*, 45(1): 3–9.

## Annotated bibliography

Hall, K. (2003). *Listening to Stephen Read: Multiple Perspectives on Literacy*. Buckingham: Open University Press.

A wonderful, seminal example of different perspectives on a child's reading. **L2 ★★**

Lepola, J., Lynch., J., Kiuru, N., Laakkonen, E. and Niei, P. (2016). 'Early oral language comprehension, task orientation, and foundational reading skills as predictors of Grade 3 reading comprehension', *Reading Research Quarterly*, 51(4): 373–390.

A recent longitudinal 5-year study exploring links between children's oral language comprehension skills, reading fluency and reading comprehension. **L3 ★★★**

www.lovemybooks.co.uk/

*Lovemybooks* brings together brilliant books with creative reading activities specially designed to engage and excite young children's interest. Parent and practitioners can click on a topic or age group to find books. **L2 ★**

# Chapter 11

# Reading comprehension

In this chapter, we discuss the nature of reading comprehension and argue that it is an essential component of teaching reading from the earliest stages of children's learning. Reading comprehension is part of the holistic understanding of reading, which includes motivation for reading and the development of children's understanding of letters and phonemes.

Comprehension is the essence of reading because it entails readers' understanding the written word expressed in texts. Without comprehension, reading does not take place. However, understanding how reading comprehension develops is complex and, as Tennent (2015) suggests, there is no conclusive definition of this process because readers make sense of text in individual and unique ways. In this chapter, we provide an overview of research into this challenging and exciting area of teaching and provide examples of strategies to support your teaching of reading comprehension.

The thinking about reading comprehension in this chapter is informed by the position that comprehension should not be artificially and unduly separated from learning about the links between letters and phonemes. The importance of not neglecting reading comprehension at any stage is underlined by the finding that children who have poor reading comprehension skills are at risk of not only poor reading development but also generally poor educational attainment (Cain and Oakhill, 2006).

Children are naturally predisposed to make sense of life, and this includes making sense of texts. This meaning-making has been explained from a socio-cultural perspective as a *transaction* between text and reader: 'an event involving a particular individual and a particular text, happening at a particular time, under particular circumstances, in a particular social and cultural setting, and as part of the ongoing life of the individual and the group' (Rosenblatt, 1985: 100).

Comprehension, then, is an active process that children engage with from the earliest stages and throughout the primary age phase. It involves the following elements:

- engaging with the text;
- understanding the text;
- making connections with existing knowledge;
- critically evaluating the text;
- reflecting upon children's responses;
- self-regulating the comprehension process through making decisions about which strategies will help to clarify understanding.

The RAND Reading Study Group (RRSG, 2002) define reading comprehension as follows:

> the process of simultaneously extracting and constructing meaning through interaction and involvement with written language. We use the words *extracting* and *constructing* to emphasise both the importance and insufficiency of the text as a determinant of reading comprehension. Comprehension entails three elements:
>
> - The *reader* who is doing the comprehending
> - The *text* that is to be comprehended
> - The *activity* in which comprehension is a part.
>
> (RRSG, 2002: 11)

Key strategies to develop when teaching reading comprehension include the following (Keene and Zimmerman, 2007; Maine, 2015):

- Determining importance
- Summarising and synthesising
- Asking questions and hypothesising

Teachers model to children how to prioritise aspects of information the text provides. They also help children to identify key points in the text and explain their thinking in a variety of ways, for example through talk/diagrams/writing/illustrations. Questions operate at different levels and can take children deeper into texts. Different types of questions demand different levels of thinking, from lower-order to higher-order, making increasing cognitive demands on children. Another way of thinking about questions is in terms of developing children's understanding of the text to promote thinking at three levels:

1   Literal questions involve simple recall questions designed to help children recall/revise material already covered as well as simple comprehension

questions to elicit their understanding of main points and as a description of what they know.

2    Deductive or inferential questions ask children to become text detectives, deducing from the lines from the text and reading between the lines of the text (inference).

3    Evaluative or response questions ask children to evaluate a text, asking children to think about whether the text achieves its purpose, for example, or to make connections with other texts.

Questions can be initiated either by the teacher or the children to support their literal and higher-order thinking skills. Questions initiated by the children evolve from the content of the text before, during and after reading. This type of questioning is self-initiated. The reader might ask questions in relation to the following: to clarify meaning; to think about the text yet to be read; to show doubt about an idea in the text; to determine the author's intent, style and content of format; to locate a specific answer in a text; or to take their understanding deeper into the text. Younger children or less experienced readers could be asked what questions they would like to find the answers to in the context of a guided reading session and supported to develop strategies to find the answers leading to being able to generate questions independently.

*    Making predictions

Prediction involves encouraging children to think about how a text, or part of a text, might develop. Particularly with younger children, this might involve scrutinising the title, the front cover, the endpapers or the blurb and thinking about what these might suggest to gain clues about the content. The teacher's role here is to offer open-ended questions to prompt responses. Pausing during the reading of a text, for example, to review events and discuss character, plot, themes and setting can also allow children to consider what might happen next. Here are some examples of the kind of open-ended questions that can be used to support prediction and hypothesis:

*    'What do you think this book is about? Why/what makes you say that?'
*    'I wonder why . . .?'
*    'What do you think might happen next?'
*    'Why do you think the author/illustrator . . .?'
*    'What does this page/image tell you?'

It is important to model how to respond to these types of questions, and also to encourage personal responses so that children understand there is usually not one right answer.

*    Monitoring for meaning (clarifying)

This is another crucial element in comprehending texts. Have children understood vocabulary/key points in the text? This will require you to pre-read the text and identify vocabulary/contexts that your class may find challenging.

- Making connections to what is known

Through activating prior knowledge, children's understanding can be developed, thus helping them to make links between what they already know and new information they encounter within a text. A starting point could be a key word from the title or text, or an artefact associated with a book.

- Evoking visual and emotional images

Visualisation involves constructing mental images suggested by the text that will deepen children's understanding by encouraging them to return to interrogate the text more thoroughly, and to check or look for more details. Visualisation can be elicited through asking children to create a picture in their mind, using drawing or through employing drama techniques (→ Chapter 7).

- Engaging and responding

This is vital. Consider how you are going to 'hook' the children in to the texts you share and how you provide opportunities for active engagement.

## Developing inference

Inference is such an important area to develop in children's reading comprehension that it deserves further mention in its own right. Inference involves filling in the 'gaps' that are not explicitly stated by the author. Several types can be found in the literature:

- *Bridging inference,* where the reader semantically or conceptually relates the sentence being read to previous content.
- *Explanation-based inference,* where the reader infers that the cause of the event being read can be related to previous content found within the text.
- *Process inference,* where understanding increases as the reader reads further into the text – unearthing the steps of an event, for example.
- *Predictive inference,* where the reader predicts or forecasts possible upcoming events in the text.
- *Goal inference,* where the reader infers the motives for an action of a character.
- *Elaborative inference,* where the reader infers the facts of an event or the characteristics of a character not explicitly explained in the text.
- *Anaphoric inference* (see below).

Anaphoric inference is reference to a previously mentioned concept in consecutive sentences or within the same sentence. Anaphors can be explicit – for example, involving the repetition of key words – or implicit – for example, through the use of pronouns, synonyms and repeated nouns. An example of an explicit anaphor would be the repeated chorus which begins 'We're going on a bear hunt' in Michael Rosen's classic children's picture book of the same name. The chorus serves the purpose of regularly drawing the audience back to the main theme of the story. In summary, making inferences is the ability to look between and beyond the lines.

Chapter 13 outlines general classroom practices for reading that are commonly used as ways to develop children's comprehension of texts. These practices support children to think carefully and in greater depth about the texts they are exposed to and to develop their reading comprehension.

### Practice points

- Develop children's reading comprehension skills alongside their developing phonic knowledge, from the earliest stages.
- Practise devising open-ended comprehension questions across a range of genres.
- Listen to children's responses and consider implications for developing reading comprehension.

### References

Cain, K. and Oakhill, J. (2006). 'Profiles of children with specific reading comprehension difficulties', *British Journal of Educational Psychology*, 76: 683–696.

Keene, E. and Zimmerman, S. (2007). *Mosaic of Thought: The Power of Comprehension Strategy Instruction*, 2nd edn. Portsmouth, NH: Heinemann Educational Books Ltd.

Maine, F. (2015). *Teaching Comprehension Through Reading and Responding to Film*. Leicester: UKLA.

Rosenblatt, L. (1985). 'Viewpoints: Transaction versus interaction: A terminological rescue operation', *Research in the Teaching of English*, 19(1): 96–107.

RRSG. (2002). *Reading for Understanding: Toward an R & D Program in Reading Comprehension*. Santa Monica, CA: RAND.

Tennent, W. (2015). *Understanding Reading Comprehension*. London: Sage.

### Annotated bibliography

For additional ideas and teaching approaches to support the development of children's reading comprehension skills, useful sources are CLPE's Bookpower publications. See www.clpe.co.uk/publications for more information.
**L1** ★

National Institute of Child Health and Human Development. (2000). *Report of the National Reading Panel: Teaching Children to Read: An Evidence-Based Assessment of the Scientific Research Literature on Reading and Its Implications for Reading Instruction.* Washington, DC: *US Government Printing Office.*
One of the reports from this major initiative in the US was on reading comprehension.
**L3** ★★★

Tennent, W. (2015). *Understanding Reading Comprehension.* London: Sage.
A comprehensive guide to understanding the complexities of reading comprehension including practical case studies of teaching.
**L2** ★★

# Chapter 12

# Phonics

Phonics teaching is an integral part of the process of supporting early and developing readers. This chapter outlines the importance of developing phonological awareness ☞ and examines statutory guidance that you need to be aware of as you start teaching phonics. We end by sharing examples of phonics teaching schemes in the context of a systematic phonics programme.

The English language has 26 letters ☞ which are used in various combinations (graphemes) to represent approximately 44 phonemes (sounds) ☞ The exact number continues to be an area of debate, but the essential point is that all words in the English language are spoken using combinations of phonemes and written using combinations of graphemes. One of the challenges of teaching children to read – and also to write – is the lack of consistency between grapheme to phoneme correspondence in English (as opposed to languages such as Finnish or Italian which feature a 1:1 letter mapping to sound). An important concept to understand, and therefore to teach, is that in English, the same phoneme can be represented by different letters or groups of letters. For example, if you look at the vowel phoneme /ie/, which is the 'long I' sound, it can be represented by different graphemes, as in the following words: tried, light, my, shine and mind. It is therefore quite accurate to say that the /ie/ sound in the spoken word can be represented in the written word by the letters 'ie', 'igh' 'y', 'i–e' or 'i'. Consider too the way that the /sh/ phoneme is represented in the following words: appreciate, ocean, machine, moustache, fuchsia, conscious, extension, pressure, admission, sure, initiate, attention, luxury. It is also worth reflecting on the name of the university department, 'The Centre for Reading in Reading'!

Because of this complexity, it is essential that teachers understand how phonological awareness develops and how best to then support early and developing readers in making sense of written texts. These levels of awareness can be described at three levels: syllables, rhyme (onset–rime) and phoneme. In all languages so far studied, children develop syllable and rhyme/onset–rime awareness prior to schooling, whilst phoneme awareness develops as literacy is taught in schools (Goswami, 1995).

One of the ideas that has emerged from the research on phonics is the significance of onsets ☞ and rimes ☞, the beginnings and ends of syllables. An understandable mistake is to confuse 'rime' and 'rhyme': the following poem helps to illustrate this:

### Spellbound

I have a spelling chequer
It came with my PC
It plainly marks four my revue
Miss takes I cannot sea.
I've run this poem threw it
I'm shore your pleased too no;
It's letter perfect in it's weigh
My chequer tolled me sew.
                    (Vandal, 1996: 14)

The *rhymes* in lines 2 and 4, and 6 and 8, are present because the 'rime' of the words 'C' (letter names are also words, in this case spelled 'cee') and 'sea', and 'no' and 'sew' are the same. This poem nicely illustrates the problems that we can have when representing phonemes with letters.

Using the concept of onset and rime, Goswami (1995: 139) emphasised the importance of reading by analogy: 'Analogies in reading involve using the spelling-sound pattern of one word, such as *beak*, as a basis for working out the spelling-sound correspondence of a new word, such as *peak*'. Children's development of phonological understanding tends to proceed from the ability to identify syllables, then onsets and rimes, and finally the ability to segment phonemes. The use of analogies draws on children's early recognition of onsets and rimes.

One of the important aspects of onset and rime is that when young children learn nursery rhymes and simple songs, their awareness of sounds is raised and it is often their attention to the rime of the words that is strong. Because this is the case, it has been argued that teaching which emphasises onset and rime can be beneficial, particularly if it is linked to the different ways that onsets and rimes can be written down. Children's understanding of rime seems to be part of a normal developmental process, whereas the ability to segment phonemes does not come so naturally and therefore needs to be taught specifically.

## Teaching phonics

As you saw in Chapter 8, the teaching of reading – and particularly the teaching of phonics – continues to arouse intense debate. Most countries for whom English is the main language include some phonics teaching as part of their reading curriculum. However, the trend in recent years has been for governments to require a particular form of phonics teaching called *synthetic phonics*. Simply put, synthetic phonics programmes focus first and foremost on the smallest parts or units of sound, the phonemes, and the ways that letters represent phonemes. Synthetic phonics consists of:

- identifying sounds (phonemes) in spoken words;
- recognising the common spellings of each phoneme;
- blending phonemes into words for reading;
- segmenting words into phonemes for spelling.

The 'synthetic' part of the term 'synthetic phonics' comes from the part played by synthesising (blending) in reading. Children are taught grapheme–phoneme (letter–sound) correspondences and how to use their phonological knowledge to work words out.

Synthetic phonics teaching emphasises that when children are reading, they should look at the graphemes (letters) of a word from left to right, convert them into phonemes and blend the phonemes to work out the spoken forms of the words. For example, if children see the word *hat*, they need to know what phoneme to say for each grapheme (/h/ – /a/ – /t/) and then to be able to blend those phonemes together into a recognisable word. However, knowing the phoneme represented by a particular letter, e.g. the letter 'a' in 'hat, requires knowledge of how the letter 'a' works in combination with other letters. In this case, its connection with the letter 't' results in the 'short a' phoneme. As a result, the teaching of phonemes and the links with letters cannot be used in an exclusive, decontextualised way. Children need to learn to understand the links between combinations of letters and word structures. Once words have been learned through phonics teaching and through practice in real reading, the words become known at sight and can then be read without sounding out and blending. For spelling, children are taught to segment spoken words into phonemes and write down graphemes for those phonemes. For example, if children want to write *hat*, then they need to be able to split it into the phonemes /h/ – /a/ – /t/ and use the appropriate letters.

*Analytic phonics*, on the other hand, is taken to refer to larger-unit phonics programmes which introduce children to whole words as a context for analysing their component parts, with an emphasis on the larger subparts of words (i.e. onsets and rimes), which draw attention to analogies and spelling patterns as well as phonemes.

In England, systematic synthetic phonics is viewed by the government as the prime strategy for teaching children to read and is enshrined in the National Curriculum (DfE, 2013). Official guidance has laid out core criteria and recommended programmes. These programmes emphasise teaching pupils to learn about:

- grapheme/phoneme (letter/sound) correspondences (the alphabetic principle) in a clearly defined, incremental sequence;
- to apply the highly important skill of blending (synthesising) phonemes, in order, all through a word to read it;
- to apply the skills of segmenting words into their constituent phonemes to spell; and that
- blending and segmenting are reversible processes.

(DfE, 2010: 3)

Schools have available a number of systematic synthetic programmes to use. A programme originally approved by government is Letters and Sounds (DfES, 2007a), a six-phase programme which builds first and foremost on children's early speaking and listening skills and then develops word reading and spelling skills:

Letters and Sounds is designed as a time-limited programme of phonic work aimed at securing fluent word recognition skills for reading by the end of Key Stage 1, although the teaching and learning of spelling, which children generally find harder than reading, will continue . . . it enables children to see the relationship between reading and spelling from an early age.

(DfE, 2007b: 3)

Shapiro and Solity (2015) found that whole word learning combined with synthetic phonics was beneficial for some children when the Early Reading Research (ERR) method was used compared with the Letters and Sounds method. Tables 12.1 to 12.4 outline progression through the phases recommended in Letter and Sounds, showing how a systematic approach goes from the simple to the more complex. Starting at the simplest level means that learning to read and spell some words in English can be similar to learning words in other languages that have more regular phoneme–grapheme correspondences. Children can thus grasp the basic workings of alphabetic writing before they have to start dealing with the complexities which are unavoidable in English.

Other popular schemes are commercially produced. Examples that you may see used in schools include: *Floppy's Phonics*, *Jolly Phonics*, *Phonics Bug*, *Phonics International* and *Read Write Inc.* All of these fulfil the government criteria outlined above to varying degrees. However, there is a wealth of evidence

*Table 12.1* Phonemes to graphemes (consonants)

| Phoneme | Grapheme(s) | Correspondences found in many different words | High-frequency words containing rare or unique correspondences (graphemes are underlined) |
|---|---|---|---|
| | | Sample words | |
| /b/ | b, bb | bat, rabbit | |
| /k/ | c, k, ck | cat, kit, duck | s<u>ch</u>ool, mos<u>qu</u>ito |
| /d/ | d, dd, -ed | dog, muddy, pulled | |
| /f/ | f, ff, ph | fan, puff, photo | rou<u>gh</u> |
| /g/ | g, gg | go, bigger | |
| /h/ | h | hen | <u>wh</u>o |
| /j/ | j, g, dg | jet, giant, badge | |
| /l/ | l, ll | leg, bell | |
| /m/ | m, mm | map, hammer | la<u>mb</u>, autu<u>mn</u> |
| /n/ | n, nn | net, funny | <u>gn</u>at, <u>kn</u>ock |
| /p/ | p, pp | pen, happy | |
| /r/ | r, rr | rat, carrot | <u>wr</u>ite, <u>rh</u>yme |
| /s/ | s, ss, c | sun, miss, cell | <u>sc</u>ent, li<u>st</u>en |
| /t/ | t, tt, -ed | tap, butter, jumped | <u>Th</u>omas, dou<u>bt</u> |
| /v/ | v | van | o<u>f</u> |
| /w/ | w | wig | pen<u>gu</u>in, <u>o</u>ne |
| /y/ | y | yes | on<u>i</u>on |
| /z/ | z, zz, s, se, ze | zip, buzz, is, please, breeze | sci<u>ss</u>ors, <u>x</u>ylophone |
| /sh/ | sh, s, ss, t (before -ion and -ial) | shop, sure, mission, mention, partial | spe<u>ci</u>al, <u>ch</u>ef, o<u>ce</u>an |
| /ch/ | ch, tch | chip, catch | |
| /th/ | th | thin | |
| **/th/** | th | then | brea<u>the</u> |
| /ng/ | ng, n (before k) | ring, pink | to<u>ngue</u> |
| /zh/ | s (before -ion and -ure) | vision, measure | u<u>s</u>ual, bei<u>ge</u> |

pointing to the fact that pre-school children acquire a range of sophisticated understandings and that, whilst debate continues over the efficacy of a purely synthetic phonics approach, research shows that a systematic approach (not necessarily synthetic) to the teaching of phonics will support the development of children's reading. Understanding how the English language works, and its

*Table 12.2* Phonemes to graphemes (vowels)

| Phoneme | Grapheme(s) | Sample words | High-frequency words containing rare or unique correspondences (graphemes are underlined) |
| --- | --- | --- | --- |
| /a/ | a | ant | |
| /e/ | e, ea | egg, head | s<u>ai</u>d, s<u>ay</u>s, fri<u>e</u>nd, l<u>eo</u>pard, <u>a</u>ny |
| /i/ | i, y | in, gym | w<u>o</u>men, b<u>u</u>sy, b<u>ui</u>ld, pr<u>e</u>tty, engi<u>ne</u> |
| /o/ | o, a | on, was | |
| /u/ | u, o, o-e | up, son, come | y<u>ou</u>ng, d<u>oe</u>s, bl<u>oo</u>d |
| /ai/ | ai, ay, a-e | rain, day, make | th<u>ey</u>, v<u>ei</u>l, w<u>eigh</u>, str<u>aigh</u>t |
| /ee/ | ee, ea, e, ie | feet, sea, he, chief | th<u>e</u>se, p<u>eo</u>ple |
| /igh/ | igh, ie, y, i-e, i | night, tie, my, like, find | h<u>eigh</u>t, <u>eye</u>, I, goodb<u>ye</u>, t<u>y</u>pe |
| /oa/ | oa, ow, o, oe, o-e | boat, grow, toe, go, home | <u>oh</u>, th<u>ou</u>gh, f<u>ol</u>k |
| **/oo/** | oo, ew, ue, u-e | boot, grew, blue, rule | t<u>o</u>, s<u>ou</u>p, thr<u>ough</u>, tw<u>o</u>, l<u>o</u>se |
| /oo/ | oo, u | look, put | c<u>ou</u>ld |
| /ar/ | ar, a | farm, father | c<u>al</u>m, <u>are</u>, <u>au</u>nt, h<u>ear</u>t |
| /or/ | or, aw, au, ore, al | for, saw, Paul, more, talk | c<u>augh</u>t, th<u>ough</u>t, f<u>our</u>, d<u>oor</u>, br<u>oa</u>d |
| /ur/ | ur, er, ir, or (after 'w') | hurt, her, girl, work | l<u>ear</u>n, j<u>our</u>ney, w<u>ere</u> |
| /ow/ | ow, ou | cow, out | dr<u>ough</u>t |
| /oi/ | oi, oy | coin, boy | |
| /air/ | air, are, ear | fair, care, bear | th<u>ere</u> |
| /ear/ | ear, eer, ere | dear, deer, here | p<u>ier</u> |
| /ure/ | | | s<u>ure</u>, p<u>oor</u>, t<u>our</u> |
| /ə/ | many different graphemes | corn<u>er</u>, pill<u>ar</u>, mot<u>or</u>, fam<u>ou</u>s, fav<u>our</u>, murm<u>ur</u>, <u>a</u>bout, cott<u>on</u>, mount<u>ai</u>n, poss<u>i</u>ble, happ<u>en</u>, cen<u>tre</u>, thor<u>ough</u>, pict<u>ure</u>, cupb<u>oa</u>rd . . . and others |

complexity at the syllabic, rhyme and phonemic levels, is key to teaching phonics systematically and effectively. Any phonics programme should be contextualised in language as a whole for real purposes. As discussed in Chapters 8 and 9, this context should be rich in literature, should encourage reading for pleasure (for powerful evidence of long term effects see Sulivan and Brown, 2015) and provide opportunities for talking about texts.

*Table 12.3* Graphemes to phonemes (consonants)

| | Correspondences found in many different words | | Correspondences found in some high-frequency words but not in many/ any other words |
|---|---|---|---|
| **Grapheme** | **Phoneme(s)** | **Sample words** | |
| b, bb | /b/ | bat, rabbit | lamb, debt |
| c | /k/, /s/ | cat, cell | special |
| cc | /k/, /ks/ | account, success | |
| ch | /ch/ | chip | school, chef |
| ck | /k/ | duck | |
| d, dd | /d/ | dog, muddy | |
| dg | /j/ | badge | |
| f, ff | /f/ | fan, puff | *of* |
| g | /g/, /j/ | go, gem | |
| gg | /g/, /j/ | bigger, suggest | |
| gh | /g/, /-/ | ghost, high | rough |
| gn | /n/ | gnat, sign | |
| gu | /g/ | | guard |
| h | /h/ | hen | honest |
| j | /j/ | jet | |
| k | /k/ | kit | |
| kn | /n/ | knot | |
| l | /l/ | leg | half |
| ll | /l/ | bell | |
| le | /l/ or /l/ | paddle | |
| m, mm | /m/ | map, hammer | |
| mb | /m/ | | lamb |
| mn | /m/ | | autumn |
| n | /n/, /ng/ | net, pink | |
| nn | /n/ | funny | |
| ng | /ng/, /ng+g/, /n+j/ | ring, finger, danger | |
| p, pp | /p/ | pen, happy | |
| ph | /f/ | photo | |
| qu | /kw/ | quiz | mosquito |
| r, rr | /r/ | rat, carrot | |
| rh | /r/ | | rhyme |
| s | /s/, /z/ | sun, is | sure, measure |
| ss | /s/, /sh/ | miss, mission | |
| sc | /s/ | scent | |

| Grapheme | Phoneme(s) | Correspondences found in many different words — Sample words | Correspondences found in some high-frequency words but not in many/any other words |
|---|---|---|---|
| se | /s/, /z/ | mouse, please | |
| sh | /sh/ | shop | |
| t, tt | /t/ | tap, butter | listen |
| tch | /ch/ | catch | |
| th | /th/, /**th**/ | thin, then | Thomas |
| v | /v/ | van | |
| w | /w/ | wig | answer |
| wh | /w/ or /hw/ | when | who |
| wr | /r/ | write | |
| x | /ks/ /gz/ | box, exam | xylophone |
| y | /y/, /i/ (/ee/), /igh/ | yes, gym, very, fly | |
| ye, y-e | | | goodbye, type |
| z, zz | /z/ | zip, buzz | |

*Table 12.4* Graphemes to phonemes (vowels)

| Grapheme | Phoneme(s) | Correspondences found in many different words — Sample words | Correspondences found in some high-frequency words but not in many/any other words |
|---|---|---|---|
| a | /a/, /o/, /ar/ | ant, was, father | water, any |
| a-e | /ai/ | make | |
| ai | /ai/ | rain | *said* |
| air | /air/ | hair | |
| al, all | /al/, /orl/, /or/ | Val, shall, always, all, talk | half |
| ar | /ar/ | farm | war |
| are | /air/ | care | *are* |
| au | /or/ | Paul | aunt |
| augh | | | caught, laugh |
| aw | /or/ | saw | |
| ay | /ai/ | say | says |
| e | /e/, /ee/ | egg, he | |

(Continued)

*Table 12.4* (Continued)

| | Correspondences found in many different words | | Correspondences found in some high-frequency words but not in many/ any other words |
|---|---|---|---|
| **Grapheme** | **Phoneme(s)** | **Sample words** | |
| ea | /ee/, /e/ | bead, head | great |
| ear | /ear/ | hear | learn, heart |
| ed | /d/, /t/, /∂d/ | turned, jumped, landed | |
| ee | /ee/ | bee | |
| e-e | /ee/ | these | |
| eer | /ear/ | deer | |
| ei | /ee/ | receive | veil, leisure |
| eigh | /ai/ | eight | height |
| er | /ur/ | her | |
| ere | /ear/ | here | *were, there* |
| eu | **/yoo/** | Euston | |
| ew | **/yoo/**, **/oo/** | few, flew | sew |
| ey | /i/ (/ee/) | donkey | *they* |
| i | /i/, /igh/ | in, mind | |
| ie | /igh/, /ee/, /i/ | tie, chief, babies | friend |
| i-e | /igh/, /i/, /ee/ | like, engine, machine | |
| igh | /igh/ | night | |
| ir | /ur/ | girl | |
| o | /o/, /oa/, /u/ | on, go, won | *do*, wolf |
| oa | /oa/ | boat | broad |
| oe | /oa/ | toe | shoe |
| o-e | /oa/, /u/ | home, come | |
| oi | /oi/ | coin | |
| oo | **/oo/**, **/oo/** | boot, look | blood |
| or | /or/, /ur/ | for | work |
| ou | /ow/, **/oo/** | out, you | *could*, young, shoulder |
| our | /ow ∂/, /or/ | our, your | journey, tour |
| ow | /ow/, /oa/ | cow, slow | |
| oy | /oi/ | boy | |
| u | /u/, /oo/ | up, put | |
| ue | **/oo/**, **/yoo/** | clue, cue | |
| u-e | **/oo/**, **/yoo/** | rude, cute | |
| ui | | | build, fruit |
| ur | /ur/ | fur | |
| uy | | | buy |

## Practice points

- Make a clear distinction between sounds and letter names.
- Help children to understand that various letter combinations (graphemes) can produce the same phoneme.
- Contextualise phonics teaching in real texts, sentences and words.

## Glossary

**Grapheme** – written representation of a sound.

**Onset** – any consonant sounds that come before the vowel in a syllable.

**Phoneme** – the smallest unit of sound in a word.

**Phonological awareness** – understanding of the links between sounds and symbols.

**Rime** – the vowel and any consonants that follow the onset in a syllable.

## References

Department for Education. (2010). *Phonics Teaching Materials: Core Criteria and the Self-Assessment Process*. Crown copyright.

Department for Education (DfE). (2013). *The National Curriculum in England: Framework Document. September 2014*. London: Department for Education.

Department for Education and Skills (DfES). (2007a). *Letters and Sounds*. London: DfES Publications.

Department for Education and Skills (DfES). (2007b). *Letters and Sounds: Notes of Guidance for Practitioners and Teachers*. London: DfES Publications.

Goswami, U. (1995). 'Phonological development and reading: What is analogy, and what is not?', *Journal of Research in Reading*, 18(2): 139–145.

Shapiro, L. and Solity, J. (2015). 'Differing effects of two synthetic phonics programmes on early reading development', *British Journal of Educational Psychology*. doi: 10.1111/bjep.12097

Sulivan, A. and Brown, M. (2015). 'Reading for pleasure and progress in vocabulary and mathematics', *British Educational Research Journal*, 41(6): 971–991.

Vandal, N. (1996). 'Spellbound', in J. Foster (compiler), *Crack Another Yolk and Other Word Play Poems*. Oxford: Oxford University Press.

## Annotated bibliography

Goouch, K. and Lambirth, A. (eds.). (2011). *Teaching Early Reading and Phonics*. London: Sage.

A theory-based practical approach to teaching early reading and phonics with many examples of suggested activities.

L2 **

Davis, A. (2017). *A Critique of Pure Teaching Methods and the Case of Synthetic Phonics*. London: Bloomsbury.

Raises some important philosophical questions about the problems with synthetic phonics implemented as the sole method of reading.

**L3** ★★★

Torgerson, C., Brooks, G., Gascoine, G., and Higgins, S. (2018). 'Phonics: reading policy and the evidence of effectiveness from a systematic "tertiary" review', *Research Papers in Education*. doi:10.1080/02671522.2017.1420816

This review found that balanced reading teaching is likely to be effective, and that there is insufficient evidence to support one phonics approach over others.

**L3** ★★★

# Chapter 13

# Classroom practices for reading

This chapter illustrates some of the practical techniques that teachers need to adopt to support children in developing as enthusiastic and independent readers. We start by outlining two significant strategies: shared and guided reading. We discuss the importance of independent reading time, including reading journals and literature circles, and emphasise the importance of reading aloud. We end with thoughts on organising a classroom environment for reading to enable these practices to take place.

A teacher's knowledge of children's literature and other texts for children influences many of the decisions to be made about the teaching of literacy. This knowledge enables teachers to select appropriate texts to use during the structured routines of shared and guided reading discussed in this chapter. It will enable teachers, too, in selecting a wide range of class readers ☞ and texts for the classroom library or book corner which will motivate and interest the children. Effective teaching of reading requires work with the whole class through shared reading, with groups of children in guided reading sessions as well as with individual children.

## Shared reading

*Shared reading* has become a term used to describe a whole class reading experience designed to form a bridge between the teacher reading to pupils and supporting independent reading by children. Its main feature is that the teacher is the 'expert', modelling what it means to be an experienced reader. Shared reading provides a wide range of opportunities for teachers to model effective reading strategies, explicitly teaching a skill or approach and then drawing children into the process through questioning, paired talk, drama and other response activities.

The key features of shared reading are as follows:

- the modelling of reading processes, including phonics and word reading strategies and comprehension;
- teaching which is informed by lesson objectives;
- high-quality teacher–pupil interaction about the text;
- discussions about the text focusing on a range of comprehension strategies (see → Chapter 11, 'Reading comprehension');
- preparation for group and independent activities of the lesson;
- a preparation for writing (see → Chapter 22, 'Planning').

Texts chosen for shared reading should be able to be shared by the whole class, either through displaying on a screen or with multiple copies for individuals or pairs, a big book ☞ or other enlarged text. They should be a little above the attainment level of some of the class because the teacher is reading the text for them; this is an opportunity to extend children's thinking and knowledge. For younger children especially, if it is the first time that you have shared the book, it is appropriate for you to read it without comment from the children. This gives them the opportunity to hear the text as a whole. If the text is a story or poem, it gives you the opportunity to offer a dramatic reading to highlight the memorable features and to motivate the class. On subsequent readings, you will engage the children in discussion about a range of things that are of interest in the text and highlight specific learning objectives.

An important feature of the shared read is that it should be used to introduce a main lesson activity. It is our view that this works best if the activity involves further active engagement with the text, exploration of the issues that the text raises, and active practical work to extend understanding (refer to the Follow-up Activities below for *Where's My Teddy?*)

Our advice for successful shared reading is to keep sessions relatively brief. Fifteen minutes should be ample time to engage the children in a text. It is also important that you remember to elicit the children's responses to texts, not just require them to answer the questions that you have planned in relation to your lesson objectives.

### Exploring shared reading in KS1

An example of a shared reading approach to use in KS1 is *Where's My Teddy?*, written and illustrated by Jez Alborough.

### Where's My Teddy? *(written and illustrated by Jez Alborough)*

- Discuss cover features. Discuss pictures, author/illustrator and title, predicting what the book is about. Read the blurb.
- Read the book and ask questions to encourage prediction throughout: e.g. a page that says 'WHAT'S THAT?' waits for page turn for answer.

- Ask specific questions related to the children's lives/experiences: e.g. do you ever feel scared in the dark?
- Post-it notes to cover key words, e.g. rhyming words. Ask for suggestions.
- Re-enact Teddy's actions from the text in sequence.
- Dialogue pairs: try and remember what Eddy and the giant teddy say to each other when they first meet. Think of different versions.
- Think about story structure.
- Write a sentence on a card and rearrange the words. Do a similar activity using the IWB (Interactive White Board).
- Find the rhyming words in the text. Make a list.
- This is a great text for looking at some basic punctuation.
- Do read the book for enjoyment and general discussion on one or more occasions.

Some suggestions for follow-up activities:

- Take a trip to a wood and get the children to bring their teddies. Record some of the children's thoughts about the wood (organisation for writing). Take photographs.
- Write a letter from Eddy to the Big Bear asking him or her to become friends.
- Take children into the ICT suite to write their own stories, experimenting with different fonts, lower-case and upper-case letters, etc. for effect.
- Design a 'Lost Teddy' poster to be stuck around the neighbourhood.

### Exploring shared reading in KS2

By the time children enter their later primary years, the focus of a shared read moves to focus more exclusively on comprehension strategies, although there will be some children who need further support in word reading skills. These need to be addressed in guided and intervention work. The learning objectives and reading strategies to address during shared reads include predicting/hypothesising, visualising, making connections, questioning, clarifying, inferring, synthesising and exploring vocabulary in relation to meaning. Shared reading in KS2 provides teaching opportunities to model what it means to be more fluent and experienced readers. The Centre for Literacy in Primary Education (CLPE) reading scales provide detailed description and research evidence to outline children's development in reading. See: www.clpe.org.uk/sites/default/files/CLPE%20READING%20SCALE%20REBRAND.pdf

## Guided reading

The term 'guided reading' was first introduced in England through the National Literacy Strategy (DfEE, 1998) and is in common use in English classrooms. The practice is used to describe the teaching of reading in small groups and, from its inception, the purpose has been to enable young readers to gain meaning through text. Practices across the primary age range will necessarily differ, but here we outline principles to enable the 'deliberate and explicit teaching of reading strategies that support the comprehension of text' (Tennent *et al.*, 2016: 6).

The goal of guided reading, following on from the shared whole class reading experiences outlined above, is to enable children to then become independent readers who are able to read, understand and appreciate texts on their own without the teacher's help. We argue that guided reading is a key teaching strategy for supporting young readers because it enables teachers to adapt teaching to the strengths and needs of all pupils through close listening, discussion and assessment in a collaborative environment. The small group aspect is essential because all children can be actively involved.

Guided reading sessions are dialogic experiences which enable children to talk about their reading and share responses and approaches to reading a wide range of texts. In these sessions, the teacher manages the group and monitors progress, interacting with the group and individual children as appropriate. The teacher will have identified a clear learning objective and outcome for the session which will extend the range of pupils' reading strategies. Whilst for younger readers, sessions may include time for children to read independently, with the teacher responding to readers individually, as children become more proficient and fluent, guided reading sessions will shift towards a greater emphasis on response and discussion. One of the many positive features of small group work such as guided reading is that children often find it easier to contribute to discussions in that context as opposed to the whole class.

The key features of guided reading are as follows:

- a group of children (maximum of six) work together on the same text;
- the text is matched to the reading needs of the group in order to support progress;
- multiple copies of the same text are used (one each if possible);
- new texts or reflections on a known text, or section of text, read previously are discussed, based on the learning objective;
- the teacher guides children to discuss the text, focusing on different reading strategies (word reading/comprehension).

### Generic teaching sequence and questions for guided reading

1   Starting points:
    - How will you introduce the text if it is new?
    - Which objectives/comprehension strategies will you introduce?

    When selecting objectives for the group, consider both curriculum *and* learning objectives. In addition, consider:

- which reading strategies are established;
- which reading strategies need to be practised and consolidated;
- which reading strategies need demonstration and development;
- what the next steps are for the group to develop their understanding of text.

2 Reading and responding:
   - What parts of the text will you use to highlight and practise these strategies?
   - What key language will you use to elicit responses?

   The language of your interaction is the heart of any guided reading session. Allow time for the children to respond to the text, develop and justify opinions and explore personal preferences.

3 Reflecting and evaluating:
   - How will you find out what the children have learnt?
   - How will you clarify the strategies taught?

   Assessment of the children's learning during guided reading will inform the next steps for your planning. This will involve the selection of appropriate objective(s) for the next session in the context of the reading targets. Future planning for the group should also ensure opportunities for children to follow up their guided reading sessions with further reading of the text and activities to consolidate learning and assessment

4. Selecting texts:
   Teachers need to make informed choices about the texts they select for guided reading. Texts should be of appropriate quality, should interest and excite children and should offer opportunities to work towards specific objectives. A guided reading text should not present too many difficulties in terms of contextual or vocabulary challenge, because otherwise both meaning and motivation may be lost. It is worth remembering, too, that texts for guided reading can be cross-curricular; for example, guided reading could be taught as part of a science topic.

## Organisation for guided reading

Many schools operate guided reading as a separate lesson rather than as part of an English/literacy lesson. The class is frequently organised into five guided reading groups. Whilst the guided group are working with the teacher, the other groups could be:

- preparing for the next guided session;
- following up from the previous guided session;
- having a supported reading session with a teaching assistant (any group);
- reading from a variety of other texts/selected texts related to classroom topics;
- listening to stories;
- reading for pleasure.

## Independent reading time, reading journals, literature circles and reading aloud

The structured classroom practices for teaching reading outlined above are all key to developing children's word reading and reading comprehension. However, other opportunities and routines should all form part of a school's holistic provision in enabling young readers to develop as enthusiastic and independent readers.

A whole class session in which each child in the class reads a text of their choice used to be called *quiet reading*. This has attracted a number of acronyms over the years, such as ERIC (Everyone Reads in Class); USSR (Uninterrupted Silent Sustained Reading); and DEAR (Drop Everything and Read), and was sometimes characterised by the teacher modelling the process of individual reading by reading their own book alongside the class. Recent research (Cremin *et al.*, 2014) has indicated the significance that authentic independent reading time plays in children's thoughtful engagement with text.

Reading journals provide children with the opportunity to document their reading experiences and responses and can be used in a variety of ways. They can take the form of a book or folder but might also be kept as an audio diary or on a computer. Journals can provide space for reflection and evaluation as well as speculation and exploration of ideas. They can provide teachers with information about pupils' thinking and comprehension skills as they engage with text. There are many formats for reading journals. Much will depend on the age and ability of the children, personal preference and how reading is organised. It is important to be clear about how the journal is to be used and its purpose. The teacher could model how to use a journal during shared and guided reading. With younger children or less confident writers, the journal could be a whole class book where the teacher takes responsibility for the writing process and children can concentrate on articulating ideas and responses. In some classes, each guided group could have its own collaborative journal. This is useful when children are being introduced to journals or have not yet developed confidence to maintain individual journals.

Literature circles are another way to explore children's understanding of text, improve their reading skills and encourage the development of positive attitudes towards reading. Harste *et al.* (1988) first used the term 'literature circle', which is seen as a way of encouraging enjoyment of reading and talking about books. In a literature circle, a group of children all read the same text and, sometimes with the support of reading journals and/or their teacher, come together to talk about their reading and to read more of the text together. For older children, literature circles are similar to adult reading groups in that the text is read independently and most of the circle time is spent discussing the group members' responses. Younger children (aged 5 to 7) are encouraged to read the text aloud around the group and then talk about it. Literature circles are therefore an active and creative way to give children time to read and talk about the texts with each other and with adults.

The importance of reading aloud as part of daily classroom practice in early years and primary classrooms cannot be over-emphasised. Reading aloud to, with and between children creates a sense of community, building the class repertoire of 'books in common' and a shared reading history (Cremin *et al.*, 2014). We argue that opportunities for this type of informal text sharing, through 'story time' and read aloud sessions of other types of texts, must be a part of daily classroom practice.

## Organising the classroom environment for reading

An important consideration in any teacher's classroom is how spaces for reading and reading resources are utilised to the best possible effect. There are many advantages to a comfortable and attractive carpeted area which allows the class to sit together and discuss ideas. It has to be acknowledged, however, that the size and physical shape of many classrooms makes this difficult to achieve. Whether there is a carpeted area or not, classroom organisation needs to accommodate a range of teaching approaches, including shared, guided and independent reading opportunities. In early years settings, designated classroom areas that support language and literacy work are common. These often include a reading area or book corner with comfortable seating such as cushions, listening points and message boards as well as role-play areas and outdoor spaces for reading. These are also very valuable for older children.

Whether you have a carpeted reading area or not, the display and storage of books are important considerations. Teachers need to know about the books in their reading area and to display them in a way that will entice children. Practical issues, such as how often books will need to be changed, should also be considered. There should be a wide range of fiction, non-fiction and poetry, including books made by the class. Multiple copies of reference books such as atlases, dictionaries and thesauri should also be clearly labelled and accessible. Magazines, comics and newspapers should also be featured. Many classrooms now have iPads, too, where children can access a range of texts and reference materials. Children can be encouraged to bring any reading materials from home to school on a regular basis. This can serve a number of purposes: it shows that you value children's own reading choices; it motivates many children; as the teacher you get an insight into their reading interests; and the children share their interests with their peers, who may read more as a consequence. An open mind about the quality of reading materials coupled with a critical appreciation of the qualities of different texts is a good foundation for teaching reading. For further ideas about how teachers can create effective environments for reading, see: https://researchrichpedagogies.org/research/theme/social-reading-environments.

## Practice points

- Select texts of the highest quality as a vital first step in your teaching of reading.
- Develop your knowledge of the reading materials in your classroom and how these can be used to model reading objectives in shared and guided reading.
- Provide a range of structured and less formal reading experiences and routines with the children in your class.

## Glossary

**Big book** – an enlarged text which the teacher shares with the class, usually during shared reads.

**Class reader** – a text, often fiction, which the teacher reads aloud to the class. This is normally carried out as a regular session outside of the main English teaching.

## References

Cremin, T., Mottram, M., Collins, F., Powell, S. and Safford, K. (2014). *Building Communities of Engaged Readers: Reading for Pleasure*. London and New York: Routledge.

Department for Education and Employment (DfEE). (1998). *The National Literacy Strategy Framework for Teaching*. Sudbury: DfEE Publications.

Harste, J., Short, K. and Burke, C. (1988). *Creating Classrooms for Authors: The Reading–Writing Connection*. Portsmouth: Heinemann.

Tennent, W., Reedy, D., Hobsbaum, A. and Gamble, N. (2016). *Guiding Readers–Layers of Meaning: A Handbook for Teaching Reading Comprehension to 7–11 Year Olds*. London: IOE Press.

## Annotated bibliography

Graham, J. and Kelly, A. (2007). *Reading Under Control: Teaching Reading in the Primary School*, 3rd edn. London: David Fulton.
A very useful account with a particularly strong section on 'Reading Routines' which develops a number of the points about reading that we touch on in this book.
**L1 ★★**

Tennent, W., Reedy, D., Hobsbaum, A. and Gamble, N. (2016). *Guiding Readers–Layers of Meaning: A Handbook for Teaching Reading Comprehension to 7–11 Year Olds*. London: IOE Press.

An excellent resource which includes an overview of the research into guided reading and case studies of examples for different year groups.
**L2** ★

Skidmore, D., Perez-Parent, M. and Arnfield, D. (2003). 'Teacher–pupil dialogue in the guided reading session', *Reading Literacy and Language*, 37(2): 47–53.
Important research evidence about how teacher–pupil dialogue can help or hinder during guided reading.
**L2** ★★★

# Part IV

# Writing

# Chapter 14

# The development of writing

Historically, the teaching of writing has been much less of a focus than the teaching of reading. However, just as we illustrated for reading, in order to teach writing effectively it is necessary to be aware of how children learn. We return to the evidence from case studies of children in order to look at writing development. This picture of development is followed by a large section on the teaching of writing, including some of the different views that have been expressed in relation to the importance of creativity, expression and choice.

It is important to understand the typical stages of development that children pass through in their writing. This knowledge helps you to pitch your planning and interaction at an appropriate level for the children you are teaching. People who have already experienced such development as teachers and parents are in an advantageous position. However, teachers who are inexperienced need to grasp the fundamental aspects of such development. One of the reasons for this is that it heightens your awareness of what to look for when you have the opportunity to interact with young writers.

As we showed in Chapter 8, there are a number of in-depth case studies of individual children that can help in acquiring knowledge about children's development. Studies of individual children do not act as a blueprint for all children: one of the important things that such case studies show us is that children's experiences vary greatly. However, if we focus on certain key concepts and significant milestones, these can be applied to larger groups of children. These milestones are likely to happen at roughly the same age for many children, but there will be significant numbers of children whose development is different. Once again, the stages of development are based on our analysis of case studies of children's writing development, which more frequently feature young children's development than older children. Tables 14.1 to 14.3 illustrate the development of children's writing through primary school.

*Table 14.1* Expectations for a child's writing at age 4

| What you can expect | What you can do to help |
| --- | --- |
| Understands distinction between print and pictures | Talk about the differences between pictures and print. Show what you do when you write and tell children that you are writing. |
| Plays at writing | Provide a range of accessible resources. Encourage the use of writing as part of role-play. |
| Assigns meaning to own mark-making | Ask children about their writing and discuss its meaning with them. Set them challenges to write things for you, such as little notes. |
| Often chooses to write names and lists | Help children to write their name properly. Encourage them to sign their name on greetings cards. |
| Uses invented spelling | Encourage children to have a go at writing and spelling in their own way. Once they have this confidence, help them move towards conventional spellings. |
| Has knowledge of letter shapes, particularly those in child's name | Teach children how to form the letters properly. Teach them how to write their name. |
| Recognises some punctuation marks | Help them to recognise the difference between letters and punctuation marks. |
| Knows about direction and orientation of print | Talk to children about left and right, top and bottom. Use your finger to point as you read from time to time. Ask questions to encourage children to show you their knowledge about orientation of print. |

*Table 14.2* Expectations for a child's writing at age 7

| What you can expect | What you can do to help |
| --- | --- |
| Occasional interest in copying known texts | Encourage this provided it does not become the main form of writing over time. Use the opportunity to help with letter formation and whole word memory. |
| Range of genres of chosen writing more limited, reflecting specific interests and motivation | Encourage children to explore the things that they are interested in and to write about those topics. |
| Able to write longer texts such as stories | Children's stamina for writing improves as the conventions like handwriting and spelling get a little easier. They will still need help with structuring their texts as they try to control these longer forms. |
| Understands the need to make changes to writing | Help children to see how redrafting writing can help them to get better outcomes. |
| Understands that writing is constructed in sentences | Explain that a sentence is something that makes complete sense on its own. |
| Word segmentation secure and all phonemes represented in invented spellings | Help children by engaging them with the visual aspects of words. Word games, word chunks, etc. should be the focus to help them understand English spelling. |
| Use of punctuation for meaning. Full stops used conventionally | Help children to organise their writing in sentences and to remember to check for capital letters and full stops. |
| Handwritten print of lower- and upper-case letter shapes secure | Keep an eye on letter formation and remind children from time to time if they are not forming letters conventionally. |

*Table 14.3* Expectations for a child's writing at age 11

| What you can expect | What you can do to help |
| --- | --- |
| Using information sources and writing to learn | Support the skills of note-taking and/or tabulating information, etc. |
| Will redraft composition as well as transcription elements | Help children to see the value of redrafting to improve the final product. Support their proof reading skills. |
| Able to successfully control a range of text forms and have developed expertise in favourites | Encourage experimentation to find types of writing that they enjoy. |
| Length of writing increasing | Help children to control the larger structural elements such as headings and paragraphs. |
| Growing understanding of levels of formality in writing | Discuss differences between things like emails to friends and family as opposed to formal letters. |
| Standard spelling most of the time. Efficient use of dictionaries and spell checking | Help children to enjoy the wealth of information contained in dictionaries. Show them how to use standard adult dictionaries. |
| Basic punctuation secure. Aware of a range of other marks | Encourage use of full range of punctuation. Enjoy spotting things like the 'grocer's apostrophe', e.g. apple's and pears. |
| Presentation and fluency of handwriting differentiated for purpose | Support handwriting with good-quality pens and other implements. Encourage proper typing when using computer keyboard. |

We explained in Chapter 8 that linear models of development are necessary but not sufficient models for understanding reading development; the same is true for writing development, as the next section reveals.

## Cognitive models of writing

Universal patterns of behaviour reflecting a common set of cognitive processing decisions on the part of children have been identified, as is evident from children making marks which reflect the written language of their culture when asked to write, and composing 'signs' when they first start to make associations between the making of marks and the representation of personal meanings (Borzone de Manrique and Signorini, 1998; Levin and Bus, 2003; Scheuer *et al.*, 2006a; Yamagata, 2007; Yang and Noel, 2006). Children's early writing also displays the features of form common to writing in almost any language such as linearity, directionality and presence of distinguishable units.

Working memory is increasingly seen as an important function of the brain which has contributed to our understanding of language processing. It has been proposed that working memory consists of three parts: a phonological loop for storing verbal information; a visuospatial sketchpad for visual information;

and a central executive which regulates the other two parts (Hayes, 2006). There is disagreement as to whether some parts of working memory are not involved in the basic decisions of writing processing such as planning, reading and editing, or whether all three parts of working memory are involved at all times (Hayes, 2006). An attractive feature of Hayes' theory is the representation of the individual's cognitive resources as part of 'the task environment' (see Flower and Hayes, 1981).

Neuroscience has enabled us to think further about the potential role of working memory. Berninger and Winn's (2006) model, informed by neuroscientific research, portrays text generation as the two elements of 'transcription' and 'executive functions', both controlled by a third element: working memory. Berninger and Winn (op. cit.) portray working memory as activating long-term memory during planning, composing, reviewing and revising, and activating short-term memory during reviewing and revising output. However, they appropriately caution that most neuroscientific research has addressed transcription processes, with a few studies tackling word generation with and without the constraints of sentence context, but studies have not been undertaken at discourse-level.

Research has also examined the extent to which the effort required by transcription can compromise other aspects of writing (McCutchen, 2006). Evidence suggests that for beginner writers in particular, the heavy demands on working memory lead to limitations of writing output. For example, when trying to compose sentences, handwriting may not be fluent enough for children to record everything they want to say before they start to forget some of their original thoughts (Graham *et al.*, 2008). Being able to write letters automatically and legibly therefore has implications for children's writing proficiency (Ritchey, 2008). Only when automaticity with basic writing skills and handwriting is achieved will children be able to fully focus cognitive resources on aspects such as spelling and compositional demands (Kellogg, 1996). This does not, however, imply that the teaching of basic writing skills should necessarily happen before, or to the exclusion of, the teaching of other aspects of writing.

## The writing environment

Young children show natural curiosity about the nature of written language through living, observing and participating in an environment in which others use print for various purposes (Purcell-Gates, 1996; Dyson, 2001). Factors such as a print-rich environment, informal instruction and reading ability have some effect on the emergence of writing (Dyson, 2001, 2008; Neuman and Dickinson, 2001; Nunes and Bryant, 2004). Children's earliest discoveries about written language are therefore learned through active engagement with both their social and cultural worlds (Rowe, 1994; Gee, 2001; Pellegrini, 2001; Makin and Jones-Diaz, 2002; Barrat-Pugh, 2002; Compton-Lilly, 2006). Neuman and Roskos (1997) argued that participation in writing and reading practices represents an important phase of literacy learning because children come to understand that print is meaningful, and participation enables them to practise what written language is for and how it works. Embedding writing experiences in meaningful activity models several distinctive features about it for young children such as: written text conveys a message; writing is made up of separate words that correspond to spoken utterances; words are made up of individual letters; and, in English, texts are read from left to right. Several research studies have suggested that as a result of such individual and shared exploration, children are able to test their hypotheses about the forms and functions of written language in situational contexts from a very early age (Ferreiro and Teberosky, 1982; Barrat-Pugh, 2002; Rowe, 2008).

Young children arrive at school with different understandings of what writing is because of their varying exposure to writing experiences (Barrat-Pugh, 2002). Whilst there is a regularity which characterises literacy development, children reach developmental milestones through a variety of different routes (Pellegrini, 2001); each child's pathway into literacy is a distinctive journey shaped by personal, social and cultural factors (Martello, 2002; Rowe, 2008; Scheuer et al., 2006b). Aspects of literacy that children see as most relevant will differ according to the circumstances of their upbringing: 'learning, especially learning an expressive system like written language, is not divorced from one's identity and history but, of necessity, embedded within it' (Dyson, 2001: 139). For writing to become part of a child's communicative repertoire, children need to be in an environment that allows them opportunities to write. It is the quality and frequency of the literacy-related interactions and activities that children experience at home that makes a difference to children's short- and long-term outcomes (as shown in Wells, 1986; Dunsmuir and Blatchford, 2004).

Cognitive theory and socio-cultural theory seem to converge on the idea that young children's writing features variable mental processing during writing, dependent on the form of writing the child is attempting to use and the point in time during the writing process. This in turn is influenced by the task and the writing experiences that are part of young children's lives outside formal educational settings.

## Towards conventional writing

Drawing is an important precursor to and part of emergent writing (e.g. Ferreiro, 1986; Levin and Bus, 2003; Yamagata, 2007; Yang and Noel, 2006; Lancaster, 2003. Even if children have not learned how to write conventionally, they are able to distinguish between the two systems of drawing and writing and are therefore able to produce marks based on and associated with the features of each. Levin and Bus (2003: 891) showed that in the early stages the child writes by *drawing* the two-dimensional object (known as 'print') until they begin to understand that writing represents meaning primarily by phonological units of language. As children explore the features of writing, the discovery that some features are distinctive helps children to organise their written materials; for example, moving from a discontinuous linear pattern to a small number of distinguishable elements. It is in this respect that the graphic patterns of writing are being reproduced. Later writing is distinguishable from early drawings in terms of properties such as linearity and segmentation into units.

Yamagata (2007) investigated the process by which representational activity and knowledge about drawing and letter and number writing emerge in Japanese children aged between 21 and 46 months. An example of representational activity in relation to writing is the child's name written in the top left-hand corner of the paper: at this stage, the meaning of what is written is perhaps determined by the place where it appears, or by the child's intention as a writer, rather than by its linguistic features. The main findings of the study were: a) the recognition of representational systems increases with age; b) representational activities correspondingly develop with age through several phases; and c) that while children over 3 can recognise each system correctly, this is not necessarily related to representational activity. By the age of 4 years, writing has been grasped internally by the child as a particular activity that produces a specific formal output distinct from drawing, in that it is linear and discrete (Yang and Noel, 2006).

A child's knowledge of their name plays a significant role in their early writing development prior to phonological understanding (Haney, 2002). Names provide a way for children to make sense of the print world as they learn to recognise their own name; names also become a natural focus for them as they begin to explore written language (Bloodgood, 1999; Haney, 2002; Blair and Savage, 2006; Yang and Noel, 2006). Bloodgood (1999) found that the name has the potential to enable children to connect literacy strands in a meaningful way. When faced with a writing task, the child problem-solves, typically by applying and using what they know in order to make meaning of the situation. If all or some of the letters in their name are the only letters they know how to reproduce, children will often reorchestrate that same set of letters intentionally to produce an infinite number of words. Significantly, whilst each message might look similar, children consider them to be different (Ferreiro and Teberosky, 1982).

As young children make the important transition from emergent to conventional writing (building on their knowledge of drawing and writing, and the writing of their name), they are usually receiving formal writing instruction at school. We have established that children have considerable understanding of how to communicate meaningfully via intentional marks they make on the page, and that the transcriptional demands of conventional writing are a particular challenge. One of the fundamental aspects that children must understand in order to progress to conventional writing is knowledge of the alphabetic code (Blair and Savage, 2006), which Goswami (2008) suggests develops as a result of direct teaching rather than through experience prior to schooling, a point which could beneficially be further explored through research. Once the rules of the alphabetic code are internalised, children are able to move from a preoccupation with code acquisition according to conventional grapheme–phoneme correspondence rules to a concern with orthography (Scheuer *et al.*, 2006b). As far as English orthography is concerned, this requires children to begin learning about the complexity of English spelling; English is a language that is uniquely difficult in this regard, especially in comparison with other languages such as Spanish (→ Chapter 8).

Tolchinsky (2006) argues that children's writing develops at many levels simultaneously and that rather than following a continuum or linear line of development, 'what children come to know about texts guides and constrains their knowledge of letters and words, and what they grasp about letter-sound correspondences guides and constrains their way of writing texts' (op. cit.: 87). These findings provide a challenge to over-prescriptive linear models of development that are commonly adopted in the context of teaching policy documents and school assessment systems.

## The teaching of writing

In Chapter 8, 'The development of reading', we described how the pedagogy ☞ of reading teaching had been dominated by the 'reading wars'. As far as writing is concerned, it is much more difficult to identify a central theme to the discussions about teaching. In part, this reflects the fact that writing has traditionally attracted less attention than reading: less research is devoted to writing and there are fewer publications on the subject. Writing also seems to attract less attention in the media, although standards of grammar and spelling (→ Chapters 16 and 18) recurrently hit the news. However, overall the disagreements in relation to the teaching of writing have tended to centre on the amount of creativity and self-expression that is desirable and how these should be balanced with acquiring the necessary writing skills. As we work through a number of key moments in the history of writing pedagogy, you will see that this central point about creativity and skills will recur.

Frank Smith (1982: 20) made the distinction shown in Table 14.4 between the composition of writing and transcription. As you will see, the extent to

*Table 14.4* The composition and transcription of writing

| Composition (author) | Transcription (secretary) |
| --- | --- |
| Getting ideas | Physical effort of writing |
| Selecting words | Spelling |
| Grammar | Capitalisation |
| | Punctuation |
| | Paragraphs |
| | Legibility |

*Source*: Smith (1982: 20).

which pupils are allowed to experience the processes of 'getting ideas' is one of the battlegrounds in the teaching of writing.

One aspect of composition is the extent to which children should be required to plan their writing. The process that professional writers go through sheds light on the issue of planning. Carter (1999) collected together the thoughts of a number of fiction writers and included reflections on the routines that they used for writing. Helen Cresswell, a prolific and talented author for both children and adults, describes her way of composing:

> With most of my books I simply write a title and a sentence, and I set off and the road leads to where it finishes. All my books are like journeys or explorations. Behind my desk I used to have this saying by Leo Rosten pinned up on the wall that went 'When you don't know where a road leads, it sure as hell will take you there.' When I first read that, I thought, that's exactly it! That's what happens when I start on my books – I really don't know what's going to happen; it's quite dangerous, in a way. I often put off starting because it seems a bit scary. Yet at the end of the day, I feel that a story has gone where it's meant to have gone.
>
> (Carter, 1999: 118)

There are other writers who carry out written plans in detail before they write a word:

> Unlike novelists like Brian Moore, who write to discover what happens to their characters, Iris Murdoch writes nothing until she knows how the story will develop:
>
> > 'I plan in enormous detail down to the last conversation before I write the first sentences. So it takes a long time to invent it.'
> >
> > (Harthill, 1989: 87)

In a systematic analysis of interviews with some of the world's greatest writers, Wyse (2017) shows the ways in which planning is a complex process, beginning with significant overarching goals along with multiple influences that drive creativity. Professional writers have different approaches to planning, so it is logical to assume that children need a range of ways into writing. It is not a good idea to insist that every piece of writing that a child does should be preceded by a written plan. Many will benefit from frequent opportunities to start writing with the minimum of delay.

## Creative writing

As a reaction against rather formal approaches, 'creative writing' flourished in the 1960s. One of the most famous texts from this time is Alex Clegg's book *The Excitement of Writing*. Clegg recognised the extensive use – and potentially damaging effect – of published English schemes. As an alternative, he showed examples of children's writing

> taken from schools which are deliberately encouraging each child to draw sensitively on his own store of words and to delight in setting down his own ideas in a way which is personal to him and stimulating to those who read what he has written.
>
> (1964: 4)

The use of artefacts and first-hand experiences is also a well-established means of stimulating writing through a creative writing approach which the following examples illustrate.

A maths lesson was postponed for an hour when it began to hail and the children's collective attention was focused on the sudden change in light, the noise, the heightened sense of unease and danger. Standing outside, underneath a canopy, they began to write what they saw. Rebecca (aged 8) wrote:

### Hail

Suddenly, the light changed
Crisp, bright, yellow
I rushed outside and stood
Waiting impatiently
Just as the hail fell
Heavy ricochets
The air smelt, strangely
And the breath was sucked
From my lips
The wind changed the weather vane
And made the bushes dance
In a moment it was gone
The air hung grey empty
The clanging flagpole signalled
All clear

Many children are not able to sustain observational writing in this way, but they should still feel their work has worth and potential. A class of children on a field trip to the Northumbrian coast sat beside the harbour watching the day developing. One girl wrote a series of unrelated observations which neither inspired nor interested her. She was asked to select two or three elements of her writing which might 'feel right' together, and to express them in the smallest space which achieved what she wanted to say. Charlotte wrote:

Soon the tide
And the birds will follow

It took Charlotte some considerable time to arrive at these two lines, and it brings to mind an anecdote from Oscar Wilde. He once said that he had spent the morning working on a piece of writing and by lunch had added a comma. In the afternoon he took it out again.

Protheroe (1978) provided a very useful summary of the impact of creative writing, and his paper also signalled some of the criticisms that were emerging. Overall, he felt that the creative writing movement was an important one and that 'the emphasis on personal, imaginative writing [needed] to be maintained and extended' (1978: 18). But he felt the model had some weaknesses. One of these weaknesses was the restriction on the forms of writing that were used. The teacher provided a stimulus (such as a piece of music or visual art) which was followed by an immediate response, and this implied brief personal forms of writing such as a short descriptive sketch or a brief poem. The model did not encourage the writing of other forms such as argument, plays, or even short stories. Protheroe recommended that:

> the stimulated writing is to be seen *not* as the end-product, but as a stage in a process. Pupils need to be helped to develop their work, and to learn from each other as well as from the teacher.

> (ibid.: 18)

By the end of the 1970s, concerns were growing about the emphasis on 'feeling' in writing teaching and the fact that much of the creative stimuli required an immediate response which did not allow for suitable reworking or redrafting. Allen (1980) pointed out that too much focus on expressive writing could lead to a lack of emphasis on more 'abstract modes'. At this time, it was suggested that the teaching of writing required tighter structures that were deemed to be missing from the creative writing ideas.

One of the influential thinkers of the period, James Britton, proposed that writing could be categorised into several key forms (Figure 14.1). Britton offers a scientific report as one example of transactional writing ☞. He argues that this kind of writing 'may elicit the statement of other views, of counterarguments or corroborations or modifications, and is thus part of a chain of interactions between people' (1970: 175). He contrasted this with poetic writing, in which the reader is invited to share a particular verbal construct ☞. The sharing of the writer's thoughts in poetic writing does not 'elicit interaction' in the same way that transactional writing does.

Transactional/ ◄────────/Expressive/ ────────►/Poetic/
1            2          3          4          5

*Figure 14.1* Britton's categorisation of forms of writing.

*Source:* Reproduced from *Language and Learning.* Harmondsworth: Allen Lane/The Penguin Press, 1970, second edition 1972. Copyright © James Britton, 1970.

Britton suggested that most of children's writing produced in the primary school is expressive writing. But it develops, through Britton's transitional categories (2 and 4), towards transactional and poetic forms as they gain greater experience and control over their writing. Britton argued that children's expressive writing needs to adapt to the more public writing of transactional and poetic forms. Transactional writing needs to be more explicit, for the unknown reader. Poetic writing, on the other hand, emphasises implicit meanings in order to create 'sounds, words, images, ideas, events, feelings' (1970: 177). At this time there was a feeling that expressive writing could and should be a foundation for other more abstract forms. However, overall, Allen (1980) maintains that the mid- to late 1970s were characterised by uncertainty and lack of consensus on approaches to the teaching of writing.

## Developmental writing

The creative writing movement can be seen as linked with philosophies such as those of Rousseau, who advocated that children's free expression was vital. But there was a lack of research evidence to support claims about children's 'natural' development. One of the reasons that in-depth case studies of individual children became important was that they documented children's natural development as language users. This kind of data was also collected from larger groups of children. Harste *et al.* (1984) were able to extend our knowledge of children's writing by looking at 3- and 4-year-olds. Their conclusions signalled concern about the lack of 'uninterrupted' writing in most early years settings. One of the striking features of their work was the researchers' ability to focus on the positive features of early writing rather than the deficits: an extract from 'Lessons from Latrice' – a chapter from their book – is shown in Figure 14.2.

The researchers initially confessed to being more unsure about Latrice's writing than any of the other children they studied: she was developmentally the least experienced child that they encountered. The researchers asked Latrice to write her name and anything else that she could write; she was then asked to draw a picture of herself. By positively and actively searching for evidence of Latrice's achievements, they were able to understand her writing in great depth. The following is a list of some of the knowledge that Latrice had already acquired:

- Latrice was aware of how to use writing implements and paper.
- She understood and demonstrated the difference between writing and pictures.
- She switched between writing and drawing as a strategy to maintain the flow of her writing.
- Each new mark represented a new or different concept.

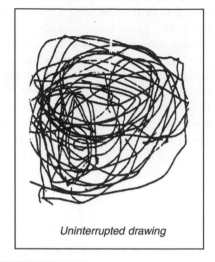

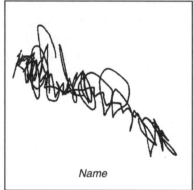

*Figure 14.2* Samples of Latrice's writing.

Source: Reproduced from Harste *et al.* (1984) *Language Stories and Literary Lessons*, Oxford: Heinemann. Used with permission.

- She had developed some knowledge of the importance of space in relation to text.
- She was aware of the permanence of meaning in relation to written language.

Another important point that Harste *et al.* made is that judgements about children's writing based on the final product do not give us enough information about their writing achievements. It is only by analysing the process of writing, in addition to the product, that valid information can be gathered.

The research evidence on children's natural literacy development led to new theories on writing pedagogy. It was argued that as children seemed to develop to a large extent by using their own natural curiosity and ability, perhaps formal teaching should take account of this reality. The theories of 'emergent literacy' developed alongside approaches such as 'developmental writing'. The use of the term 'emergent literacy' in education was popularised by Hall (1987) in his book *The Emergence of Literacy*. The basis of the philosophy is the notion of the child as an active and motivated learner who experiments with a wide range of written forms out of a sense of curiosity and a desire to learn. Hall described emergent literacy as follows:

> It implies that development takes place from within the child . . . 'emergence' is a gradual process. For something to emerge there has to be something there in the first place. Where emergent literacy is concerned this means the fundamental abilities children have, and use, to make sense of the world . . . things usually only emerge if the conditions are right. Where emergent literacy is concerned that means in contexts which support, facilitate enquiry, respect performance and provide opportunities for engagement in real literacy acts.
>
> (1987: 9)

The theory of emergent literacy was very closely linked with the practice of developmental writing. The following list identifies some of the key features of developmental writing and was influenced by Browne's (1996: 21) points that characterise such writing:

- Builds on children's literacy experience prior to coming to school.
- Encourages independent writing from day one of the nursery.
- Modelling is provided by physical resources and the actions of the teacher.
- Transcription errors are dealt with after the meaning has been established. A smaller number of errors are corrected but each one in more detail.
- Learning to write developmentally can be slow but results in motivation for future writing.
- Writing tasks emphasise purpose and real reasons.
- Children have time to develop pieces of writing in depth.
- The confidence to take risks is encouraged.

Developmental writing differs from the creative writing of the 1960s and 1970s in two main ways. Both approaches share the recognition that children must be given opportunities to carry out uninterrupted writing which uses their previous knowledge and experience. However, with developmental writing there is a stronger expectation that the teacher will interact, particularly with individual children, in order to take learning forward. The second difference relates to the first in that the teacher's interaction during developmental writing is based on a high level of knowledge about common developmental patterns in the children's writing, and this informs the focus of their inter- action. With these clearer pictures of development came different and more realistic expectations of children's learning.

## The process approach to writing

The uncertainty of the 1970s was finally transformed by the process writing of the 1980s. The work of the New Zealander Donald Graves became very influ- ential, culminating in international recognition for his work and great demand for him as a keynote speaker. Czerniewska (1992: 85) described Graves as 'one of the most seductive writers in the history of writing pedagogy'. Graves's approach to writing became known as the 'process approach' and had a sig- nificant influence on the teaching of writing in the UK. It is difficult to assess exactly how many schools and teachers took up the approach in the UK but, for example, the National Writing Project and the Language in the National Curriculum Project both involved many schools in the UK, and it is clear from their reports of practice that the process approach was influential. Frank Smith was also very popular at the time, and although his theories on read- ing have attracted some severe criticism, his theories on writing, particularly the separation between composition and transcription, have remained largely unchallenged.

> It has been argued that writing is learned by writing, by reading, and by perceiving oneself as a writer. The practice of writing develops interest and with the help of a more able collaborator provides opportunity for discovering conventions relevant to what is being written . . . None of this can be taught. But also none of this implies that there is no role for a teacher. Teachers must play a central part if children are to become writ- ers, ensuring that they are exposed to informative and stimulating demon- strations and helping and encouraging them to read and to write. Teachers are influential, as models as well as guides, as children explore and discover the worlds of writing – or decide that writing is something they will never voluntarily do inside school or out.
>
> (Smith, 1982: 201)

Smith expresses some of the key ideas of the process approach and particu- larly the notion of children being regarded as writers from the start. However, the idea of the teacher as primarily a demonstrator, as role model and as an 'encourager' has received repeated criticism because of the perception that this does not involve direct instruction. Graves's work (which fitted with Smith's ideas) developed classroom routines which turned such theories into a practical reality for many teachers.

One of the fundamental principles of Graves's process approach was downplayed in the UK. He was quite clear that children needed to be offered choices in their writing.

> Children who are fed topics, story starters, lead sentences, even opening paragraphs as a steady diet for three or four years, rightfully panic when topics have to come from them . . . Writers who do not learn to choose topics wisely lose out on the strong link between voice and subject . . . The data show that writers who learn to choose topics well make the most significant growth in both information and skills at the point of best topic. With best topic the child exercises strongest control, establishes ownership, and with ownership, pride in the piece.
>
> (Graves, 1983: 21)

This choice was not the restricted kind offered when a teacher has decided the form of writing. Graves advocated that children should select the topic and form of the writing. Wyse carried out the first research and book-length publication in the UK about the process approach at a time when few researchers were focusing on the teaching of writing. The book that resulted from the research, *Primary Writing* (Wyse, 1998), included evidence of the subtly different ways that teachers in England used the process approach compared to those characterised by Graves. Many years later multidisciplinary evidence, including from randomised controlled trials (RCTs), clearly shows the benefits and relevance of the process approach for writers across the life course (see Wyse, 2017).

## Genre theory

In the late 1980s, the popularity and optimism of the process approach began to be attacked mainly by a group of Australian academics called the 'genre theorists'. The tide began to turn away from the importance of self-expression towards greater emphasis on skills and direct instruction. The three authors who perhaps have been referred to most in relation to genre theory are J. R. Martin, Frances Christie and Joan Rothery. One of the key texts from 1987 was *The Place of Genre in Learning*, in which these three authors put forward some of their ideas as a response to other authors in the book. They also offered some criticisms of the process approach. A very recent update to the work suggests that over the intervening 30 years these categories have only had minor modifications. The categories are now seen as: stories; factual texts; and unusually, a separate category for 'arguments'. Story genres are subdivided into recount (purpose: recounting events) and narrative (resolving a complication). Factual texts are divided into description (describing specific things); report (classifying and describing general things); explanation (explaining sequences of events); and procedure (how to do an activity).

In a section of Martin *et al.*'s chapter, they examined the notion of 'freedom' during the process approach. They asked a series of important questions:

> What is freedom? Is a progressive process writing classroom really free? Does allowing children to choose their own topics, biting one's tongue in conferences and encouraging ownership, actually encourage the development of children's writing abilities?
>
> (Martin *et al.*, 1987: 77)

To answer these questions, the authors reported on a school in the Australian Northern Territory with a large population of Aboriginal children. They claimed that over the course of the year the children had only written about one of four topics: '(a) visiting friends and relatives; (b) going hunting for bush tucker; (c) sporting events; (d) movies or TV shows they have seen' (ibid.: 77). This example is used to cast doubts on the effectiveness of the process approach, claiming that the range of forms that children choose is limited. However, as Wyse's (1998) research showed, the process approach can have the opposite effect. The following is a snapshot of children's writing carried out during a writing workshop. It also gives a contextual background pointing to the origin of the idea and indication of the nature of teacher support given during a writing conference:

*Computer games and how to cheat*
The two pupils came up with the idea. The teacher suggested a survey of other children in the school who might be able to offer ways of getting through the levels on computer games. The teacher also suggested a format which would serve as a framework for the writing about each game.

*A book of patterns*
Self-generated idea with the teacher offering guidance on the amount of text that would be required and the nature of that text.

*Tools mania*
A flair for practical design technology projects resulted in one of the pair of pupils choosing this topic which involved writing a manual for the use of tools. Both pupils found the necessary expository writing a challenge.

*The new girl*
The girl herself was new to the school and this title may have provided her with a means of exploring some of her own feeling when she first arrived.

*Manchester United Fanzine*
This was a particularly welcome project as it involved three girls working on an interest they had in football. It was an opportunity to challenge the

stereotypes connected with football. The teacher set a strict deadline as the project seemed to be growing too big and also suggested the girls send the finished magazine to the football club to see what they thought.

### Football story
The pupil worked unaided, only requesting the teacher's support to check transcription.

### A book of children's games
Using a book from home, the pupil chose her favourite games and transcribed them in her own words.

### Secret messages
Various secret messages were included in the book which the reader had to work out. This was aimed at the younger children and involved a series of descriptions of unknown objects which the reader had to find around the school.

### Kitten for Nicole
This was an advanced piece of narrative; the teacher made minor suggestions for improving the ending. Unfortunately the child decided she didn't like the text and started on a new one without publishing this.

### Book for young children
The two boys used pop-art style cartoons for the illustrations as a means of appealing to the younger children. The teacher gave some input on the kinds of material that were likely to appeal to the younger children. One of the pair tended to let the other do most of the work and the teacher encouraged the sharing out of tasks.

### Football magazine
There had been an epidemic of football magazines and the teacher made a decision that this was to be the last one for a time in order to ensure a balance of forms. The two boys used ideas from various professional magazines combining photographs with their own text.

### Information about trains
Great interest in one of the school's information books which included impressive pull-out sections was the stimulus for this text. At the time the work in progress consisted of a large drawing of a train. The teacher had concerns that concentration on the drawing could become a strategy for avoiding writing.

### The magic coat
An expertly presented dual-language story which had been written with help from the child's mother for the Urdu script. The home computer

had also been used to create borders and titles. The teacher's role simply involved taking an interest in the progress.

*Catchphrase*

A pupil's doodling had given the teacher an idea for an activity which involved devising catchphrases based on the television programme. This pupil decided to compile a book of her own catchphrases.

*Chinwag*

Originally two pupils had been encouraged to devise and sell a school magazine. This included market research around the school, design, word processing, editing other children's contributions, selling, accounting, etc. This was a large-scale project and the original editors felt they would like to delegate the responsibility for the second issue to someone else, so two new editors took over.

*Newspaper*

The idea came from the two pupils but coincided fortuitously with a competition organised by the local paper encouraging students to design their own paper. The children asked various people around the school to offer stories. Layout became an important issue. The children brought in their own camera and took pictures to illustrate their text. BBC and Acorn computers were both used, necessitating understanding of two different word processors.

*Modern fairy tale*

The two pupils were struggling for an idea so the teacher suggested they contact another school to find out the kinds of books they liked with a view to writing one for them. The school was in a deprived area and had many more bilingual children than the two pupils were used to. They realised that their initial questionnaire would need modification if it was to be used again. The children at the other school expressed a preference for traditional stories so the two pupils decided to write a modern fairy tale. They were encouraged by the teacher to ask the opinion of bilingual peers on suitable subject matter and some information about India.

*Joke book*

The two pupils surveyed the children in the school for good jokes. This was a popular title and had been done before in the course of the year.

*Knightrider*

A book based on the favourite television programme of the pupil.

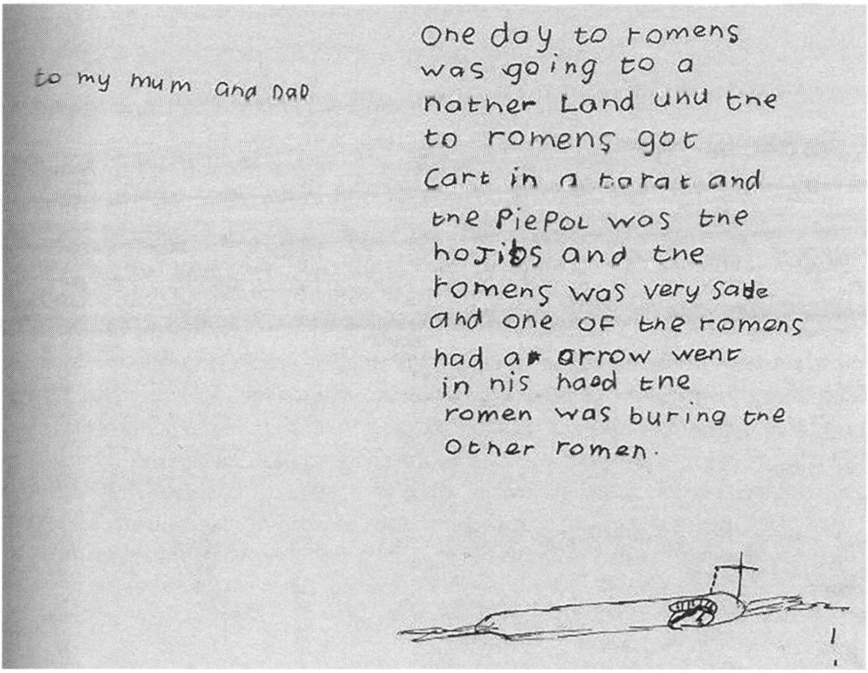

to my mum and Dad

One day to romens
was going to a
nather Land und the
to romens got
Cart in a torat and
the PiePoL was the
hoJibs and the
romens was very Sade
and one of the romens
had a arrow went
in his haod the
romen was buring the
Other romen.

*Figure 14.3* An example of Neil's writing.

In an inner-city school in London, during the weekly writing workshop, one of the children developed a short book. Figure 14.3 shows the opening page.

Neil's writing shows a number of notable features: it is epic in scale – Romans going to another land; Neil has invented a new race – the Hojibs; as an opening it grabs the reader's attention through conflict, death and burial (although rather sudden for an opening); the illustration is touching, and appropriately connected to the writing; it is written with a clear audience in mind: 'mum and dad'. There are also a range of transcription issues which we address later in the chapters on spelling and on punctuation.

It can be seen from these examples that the children were involved in a large range of ideas and formats. Many of the ideas are firmly rooted in the children's interests and culture. A significant proportion of the texts involved children collaborating in twos or threes as well as those children who wrote individually. The flexibility of the workshop allowed for a range of groupings that were influenced by the piece of writing concerned and the children's social needs. This organisation also reflected the nature of language and literacy as a social phenomenon.

Writing workshops offer the potential for a much greater range of texts created using the children's intrinsic motivation. Another major benefit is the opportunity for study in depth over a long period of time. Set written tasks often have a deadline; too often this can be to start and finish on the same day. With writing workshops, the session is timetabled and the children decide on the task. This means that the children are thinking about their writing prior to the day itself. Often they will be working on texts at home (an important test of their interest in school activities), which they bring in to continue. Having the time to continue with a text for as long as it takes is an important principle. The result can be texts which are longer and written with more thought.

In spite of a number of significant criticisms (Barrs, 1991; Cairney, 1992), the views of the genre theorists proved to be influential. Consequently, between 1998 and 2010 genre theories were a dominant feature of the National Literary Strategy in England. There was an equal emphasis on fiction and non-fiction that had been informed by the view that there was too much story writing happening in primary schools. The goals for written composition no longer emphasised personal choice, writing to interest and excite readers, finding a vehicle for expression, writing to explore cross-curricular themes, writing as art, but were much more about the analysis of genre structures. The importance of writing for real purposes and reasons in order to communicate meaning was replaced by an emphasis on textual analysis as the main stimulus for composition.

## Developing written argument

A significant review of studies of the writing of argument in non-fiction forms found that the following were important:

- a writing process model in which students are encouraged to plan, draft, edit and revise their writing;
- self-motivation (personal target-setting as part of self-regulated strategy development);
- some degree of cognitive reasoning training in addition to the natural cognitive development that takes place with maturation;
- peer collaboration, thus modelling a dialogue that (it is hoped) will become internal and constitute 'thought'.

(Andrews *et al.*, 2006: 32)

Andrews *et al.* also suggested some specific interventions that were successful, including support to use the structures and devices that aid the composition of argumentative writing; the use of oral argument to inform the written argument; identification of explicit goals, including the audience for the writing; teacher modelling; and the teacher coaching writing during the process. Andrews *et al.* also point out that the recommendations were not universally shared by the studies that they looked at in their systematic review of research studies.

One of the limitations of these outcomes is that the recommendations for practice cannot be related to the writing of fiction or poetry. At the heart of these and other forms is the use of imagination, and the extent of the originality and quality of ideas are paramount concerns. But these are measurable only if children are actually given choices over the topic and form of their writing. The links between genre theories, structured teaching and individuality were explicitly addressed by Donovan and Smolkin (2002). Their study examined the use of scaffolding in a range of writing tasks, including story writing and non-fiction writing. One of their key findings, based on evidence that writers' personal interests could result in improved writing, was about the importance of 'author aim', which was explained as a keen sense of the audience for the writing linked with personal intentions and motives:

> Author aim reintroduces individuality to the writing landscape, a point with which certain Systemic Functional linguists [the theoretical tradition to which the genre theorists were linked] were not particularly comfortable . . . we are not distressed by the idea of instructing children in form. We are, however, concerned that individuals, authors, and their aims receive so little focus in considerations of structure-based instruction.
> (ibid.: 462)

This chapter has explored the teaching of writing from a multidisciplinary, cognitive and socio–cultural perspective consistent with the book's theoretical framework (→ Chapter 2). Its theoretical ideas can also be seen reflected in recent experimental trial research. Research such as that by Graham (2010) has provided experimental trial evidence that the combination of a focus on writing processes (such as the process approach) with instruction for writing strategies is the most effective way to teach writing. As we argue in this book, pupil ownership (which is related to motivation) is the other vital aspect of such writing teaching. Is it important for children to have opportunities to make decisions? Research has addressed that question. However, this matter is also a question of values. You may feel that offering genuine choices frequently during a child's early years and primary schooling is ethically necessary and that this could result in children being more motivated to write.

## Combining approaches and strategies for writing

There are a growing number of RCTs carried out to examine effectiveness of teaching in schools, including in the teaching of writing. Steve Graham has pioneered this kind of research in relation to writing teaching, including not only carrying out RCTs but also systematic reviews and meta analyses that accumulate the findings from multiple experimental trials that address the same research questions in order to see across more than one study how effective a particular approach is. In his most recent work, he has compiled *tertiary reviews* (multiple systematic reviews) which now give very clear indications of how writing can best be taught in schools. Graham and Harris arrive at a series of main recommendations that summarise much complexity across multiple experimental studies (Graham and Harris, 2017).

The first recommendation is the simple but powerful idea that young people must write regularly, and thus, increasing the amount of time actually writing improves outcomes. This appears obvious, but sometimes simply doing more of an activity doesn't necessarily result in optimal gains in achievement. As part of pupils writing more, studies have shown that the process approach in particular has important elements that support writing development, with a moderate overall *effect size* of 0.34, and a high effect size for primary/elementary pupils of 0.48 (as opposed to 0.25 for secondary pupils).

*Effect sizes* go beyond simply establishing whether an approach has worked or not, to an indication of how well it worked, through their measure of the *extent of difference* between comparison groups in experimental studies (Higgins *et al.* 2012). An effect size from 0.01 to 0.1 is considered low effect and equivalent to between zero and two months of education progress, as measured by standardised tests appropriate to the nature of learning measured. An effect size from 0.26 to 0.44, equivalent to a range of three to six months' progress, is considered moderate. An effect size of 0.5 to 1.0 is described as high effect to very high effect, within the range of six months' progress to one year's progress.

The second Graham recommendation is that teachers need to create an appropriately supportive writing environment in the classroom. A supportive writing environment is one that includes teachers communicating enthusiasm about writing and providing specific positive feedback about students' efforts. Support from teachers should be just enough to help students succeed but not so much that students' independent thinking is compromised. Teachers set clear goals for writing and have high expectations, but also adopt teaching to meet the different needs of individual students.

Evidence-based practice in the teaching of writing also requires pupils' skills, strategies and knowledge to be developed. The knowledge required includes important high-level aspects such as creativity. Consistent with one of the features of the process approach, it has been shown that teaching about planning, revising and editing text is important but that so are the lower level details, such as how paragraphs work.

Not only do students need to learn about strategies, but they also need to 'self-regulate' by being able to reflect on their own writing and their success or otherwise at using the strategies that they have been taught. In addition to compositional knowledge and skills, transcription aspects also need teaching, such as handwriting and typing, spelling and vocabulary linked with particular types of writing. Attention to sentence construction, for example through sentence combining exercises where simple sentences are combined to produce complex sentences, has been shown to be effective. Traditional grammar teaching has *not* been shown to be effective in enhancing writing development. But teachers work constantly with children, including with the use of feedback and marking, to help children and young people use sentences to communicate effectively.

One important evidence-based practice that applies not just to writing but to highly quality teaching more generally is effective assessment of pupils' learning by teachers to inform teaching. This has become known as *assessment for learning*. The reason for the specification of 'for learning' is that assessment is used for a range of purposes in education; for example, it is sometimes used to hold teachers to account on the basis of correlations between pupil test scores and teacher performance. Research on the use of assessment information to hold teachers to account, known as high stakes assessment, has shown a number of problems for education, whereas assessment for learning has shown clear benefits. Consistent with assessment for learning, there is evidence from writing teaching research that teacher feedback to their pupils about their writing is a vital part of learning to write. This feedback needs to include positive, specific feedback, not just bland general praise. Writing is also improved when students are taught to evaluate their own writing, and when peers are taught to give each other appropriate feedback about their writing. There is also some evidence that in limited ways computers can give feedback about writing and that this can be beneficial.

Computer software such as word processing software has been shown to enhance writing. Yet in spite of this evidence, and in spite of the proliferation of computer devices in society, the bulk of writing in schools is still pencil or pen and paper. And particularly rare is the practice of starting a first draft of writing on a computer and then continuing this through to completion.

Steve Graham's final main recommendation is that pupils' writing benefits not only from pupils experiencing writing, supported by teachers, and by pupils' own reflections on their writing, but also by using writing to learn as part of all the subjects of the school curriculum. Part of the benefit in this case is in writing for a variety of purposes, and hence the need to experience different forms of writing associated with different subject areas, such as the science experiment report or the history essay.

### Practice points

- Base your practice on research evidence of what work as much as possible.
- Make decisions on how and when you will offer choices.
- Use your observations to adjust your planning for writing so that children's needs are met.

## Glossary

**Construct** – in this context the word is a noun – as opposed to a verb – and means a specific way of thinking about something.
**Pedagogy** – approaches to teaching.
**Transactional writing** – concerned with getting things done, e.g. information, instructions, persuasion, etc.

## References

Allen, D. (1980). *English Teaching Since 1965: How Much Growth?* London: Heinemann Educational Books Ltd.

Andrews, R., Torgerson, C. J., Low, G., McGuinn, N. and Robinson, A. (2006). 'Teaching argumentative non-fiction writing to 7–14 year olds: A systematic review of the evidence of successful practice', Technical report. *Research Evidence in Education Library*. Retrieved 29 January 2007, from http://eppi.ioe.ac.uk/cms/

Barrat-Pugh, C. (2002). 'Children as writers', in L. Makin and C. Jones-Diaz (eds.), *Literacies in Early Childhood: Challenging Views, Challenging Practice*. New South Wales: MacLennan & Petty.

Barrs, M. (1991). 'Genre theory: What's it all about?', *Language Matters*, 92(1): 9–16.

Berninger, V. W. and Winn, W. D. (2006). 'Implications of advancements in brain research and technology for writing development, writing instruction, and educational evolution', in S. Graham, C. A. MacArthur and J. Fitzgerald (eds.), *Handbook of Writing Research*. New York: The Guilford Press.

Blair, R. and Savage, R. (2006). 'Name writing but not environmental print recognition is related to letter-sound knowledge and phonological awareness in pre-readers', *Reading and Writing*, 19: 991–1016.

Bloodgood, J. (1999). 'What's in a name? Children's name writing and literacy acquisition', *Reading Research Quarterly*, 34(3): 342–367.

Borzone de Manrique, A. M. and Signorini, A. (1998). 'Emergent writing forms in Spanish', *Reading and Writing*, 10: 499–517.

Britton, J. (1970). *Language and Learning*. Harmondsworth: Penguin.

Browne, A. (1996). *Developing Language and Literacy 3–8*. London: Paul Chapman.

Cairney, T. (1992). 'Mountain or mole hill: The genre debate viewed from "Down Under"', *Reading*, 26(1): 23–29.

Carter, J. (1999). *Talking Books: Children's Authors Talk about the Craft, Creativity and Process of Writing*. London: Routledge.

Clegg, A. B. (1964). *The Excitement of Writing*. London: Chatto & Windus.

Compton-Lilly, C. (2006). 'Identity, childhood culture, and literacy learning: A case study', *Journal of Early Childhood Literacy*, 6(1): 57–76.

Czerniewska, P. (1992). *Learning about Writing*. Oxford: Blackwell.

Donovan, C. and Smolkin, L. (2002). 'Children's genre knowledge: An examination of K–5 students' performance on multiple tasks providing differing levels of scaffolding', *Reading Research Quarterly*, 37(4): 428–465.

Dunsmuir, S. and Blatchford, P. (2004). 'Predictors of writing competence in 4- to 7-year old children', *British Journal of Educational Psychology*, 74: 461–483.

Dyson, A. (2001). 'Writing and children's symbolic repertoires: Development unhinged', in S. B. Neuman and D. K. Dickinson (eds.), *Handbook of Early Literacy Research*. New York: The Guilford Press.

Dyson, A. (2008). 'Staying in the (curricular) lines: Practice constraints and possibilities in childhood writing', *Written Communication*, 25(1): 119–157.

Ferreiro, E. (1986). 'The interplay between information and assimilation in beginning literacy', in W. Teale and E. Sulzby (eds.), *Emergent Literacy: Writing and Reading*. Norwood, NJ: Ablex.

Ferreiro, E. and Teberosky, A. (1982). *Literacy Before Schooling*. Portsmouth, NH: Heinemann Educational Books Ltd.

Flower, L. and Hayes, J. R. (1981). 'A cognitive process theory of writing', *College English*, 46(2): 99–117.

Gee, J. P. (2001). 'A sociocultural perspective in early literacy development', in S. B. Neuman and D. K. Dickinson (eds.), *Handbook of Early Literacy Research*. New York: The Guilford Press.

Goswami, U. (2008). *Cognitive Development: The Learning Brain*. Hove: Psychology Press.

Graham, S. (2010). 'Facilitating writing development', in D. Wyse, R. Andrews and J. Hoffman (eds.), *The Routledge International Handbook of English, Language and Literacy Teaching*. London: Routledge.

Graham, S., and Harris, K. (2017). 'Evidence-based writing practices: A meta-analysis of existing meta-analyses', in R. Redondo, K. Harris, and M. Braeksma (eds.), *Design Principles for Teaching Effective Writing*. Leiden, the Netherlands: BRILL.

Graham, S., Harris, K. R., Mason, L., Fink-Chorzempa, B., Moran, S. and Saddler, B. (2008). 'How do primary grade teachers teach handwriting?', *Reading and Writing*, 21: 49–69.

Graham, S., Morphy, P., Harris, K., Fink-Chorzempa, B., Saddler, S. M., and Mason, L. (2008). 'Teaching spelling in the primary grades: A national survey of instructional practices and adaptations', *American Educational research Journal*, 45: 796–825.

Graves, D. H. (1983). *Writing: Teachers and Children at Work*. Portsmouth, NH: Heinemann Educational Books Ltd.

Hall, N. (1987). *The Emergence of Literacy*. Sevenoaks: Hodder & Stoughton.

Haney, M. R. (2002). 'Name writing: A window into the emergent literacy skills of young children', *Early Childhood Education Journal*, 30(2): 101–105.

Harste, J. C., Woodward, V. A. and Burke, C. L. (1984). *Language Stories and Literacy Lessons*. Portsmouth, NH: Heinemann Educational Books Ltd.

Harthill, R. (1989). *Writers Revealed: Eight Contemporary Novelists Talk about Faith, Religion and God*. New York: Peter Bedrick Books.

Hayes, J. R. (2006). 'New directions in writing theory', in C. MacArthur, S. Graham and J. Fitzgerald (eds.), *Handbook of Writing Research*. New York: The Guilford Press.

Higgins, S., Kokotsaki, D. and Coe, R. (2012). *The Teaching and Learning Toolkit: Technical Appendices*. London: Education Endowment Foundation and The Sutton Trust.

Kellogg, R. T. (1996). 'A model of working memory in writing', in C. M. Levy and S. Ransdall (eds.), *The Science of Writing: Theories, Methods, Individual Differences and Applications*. Hillsdale, NJ: Lawrence Erlbaum Associates.

Lancaster, L. (2003). 'Moving into literacy: How it all begins', in N. Hall, J. Larson and J. Marsh (eds.), *Handbook of Early Childhood Literacy*. London: Sage.

Levin, I. and Bus, A. G. (2003). 'How is emergent writing based on drawing? Analyses of children's products and their sorting by children and mothers', *Child Psychology*, 39(5): 891–905.

Makin, L. and Jones-Diaz, C. (2002). *Literacies in Early Childhood: Challenging Views, Challenging Practice*. New South Wales: MacLennan & Petty.

Martello, J. (2002). 'Many roads through many modes: Becoming literate in early childhood', in L. Makin and C. Jones-Diaz (eds.), *Literacies in Early Childhood: Challenging Views, Challenging Practice*. New South Wales: MacLennan & Petty.

Martin, J. R., Christie, F. and Rothery, J. (1987). 'Social processes in education: A reply to Sawyer and Watson (and others)', in I. Reid (ed.), *The Place of Genre in Learning*. Victoria: Deakin University.

McCutchen, D. (2006). 'Cognitive factors in the development of children's writing', in C. A. MacArthur, S. Graham and J. Fitzgerald (eds.), *Handbook of Writing Research*. New York: The Guilford Press.

Neuman, S. B. and Dickinson, D. K. (eds.). (2001). *Handbook of Early Literacy Research*. New York: The Guilford Press.

Neuman, S. B. and Roskos, K. (1997). 'Literacy knowledge in practice: Contexts of participation in young writers and readers', *Reading Research Quarterly*, 32(1): 10–32.

Nunes, T. and Bryant, P. (2004). *Handbook of Children's Literacy*. Dordrecht: Kluwer Academic Publishers.

Pellegrini, A. D. (2001). 'Some theoretical and methodological considerations in studying literacy in social context', in S. B. Neuman and D. K. Dickinson (eds.), *Handbook of Early Literacy Research*. New York: The Guilford Press.

Protheroe, R. (1978). 'When in doubt, write a poem', *English in Education*, 12(1): 9–21.

Purcell-Gates, V. (1996). 'Stories, coupons, and the TV guide: Relationships between home literacy experiences and emergent literacy knowledge', *Reading Research Quarterly*, 31(4): 406–428.

Ritchey, K. D. (2008). 'The building blocks of writing: Learning to write letters and spell words', *Reading and Writing*, 21: 27–47.

Rowe, D. W. (1994). *Preschoolers as Authors: Literacy Learning in the Social World of the Classroom*. New York: Hampton Press.

Rowe, D. W. (2008). 'The social construction of intentionality: Two-year-olds' and adults' participation at a preschool writing center', *Research in Teaching*, 42(4): 387–434.

Scheuer, N., De la Cruz, M., Pozo, J. I., Huarte, M. F. and Sola, G. (2006a). 'The mind is not a black box: Children's ideas about the writing process', *Learning and Instruction*, 16: 72–85.

Scheuer, N., De la Cruz, M., Pozo, J. I. and Neira, S. (2006b). 'Children's autobiographies of learning to write', *British Journal of Educational Psychology*, 76: 709–725.

Smith, F. (1982). *Writing and the Writer*. Portsmouth, NH: Heinemann Educational Books Ltd.

Tolchinsky, L. (2006). 'The emergence of writing', in C. A. MacArthur, S. Graham and J. Fitzgerald (eds.), *Handbook of Writing Research*. New York: The Guilford Press.

Wells, G. (1986). *The Meaning Makers: Children Learning Language and Using Language to Learn*. London: Hodder & Stoughton.

Wyse, D. (1998). *Primary Writing*. Buckingham: Open University Press.

Wyse, D. (2017). *How Writing Works: From the Invention of the Alphabet to the Rise of Social Media*. Cambridge: Cambridge University Press.

Yamagata, K. (2007). 'Differential emergence of representational systems: Drawings, letters, and numerals', *Cognitive Development*, 22: 244–257.

Yang, H. C. and Noel, A. M. (2006). 'The developmental characteristics of four-and-five-year-old pre-schoolers' drawings: An analysis of scribbles, placement patterns, emergent writing, and name writing in archived spontaneous drawing samples', *Journal of Early Childhood Literacy*, 6(2): 145–162.

## Annotated bibliography

Donovan, C. and Smolkin, L. (2002). 'Children's genre knowledge: An examination of K–5 students' performance on multiple tasks providing differing levels of scaffolding', *Reading Research Quarterly*, 37(4): 428–465.

This study provides evidence in relation to the importance of pupil choice in relation to their writing.

L3 ★★★

MacArthur, C., Graham, S. and Fitzgerald, J. (eds.). (2016). *Handbook of writing Research*, 2nd ed. New York: The Guilford Press.
An impressive overview of research on writing.
**L3** ★★★

Wyse, D. (2017). *How Writing Works: From the Invention of the Alphabet to the Rise of Social Media.* Cambridge: Cambridge University Press.
A multidisciplinary analysis of writing and its processes. Includes comparison with music composition as part of multidisciplinary focus.
**L3** ★★★

# Chapter 15

# Classroom practices for writing

The aim of teaching writing is to develop motivated and independent writers: those who write with understanding and purpose. This chapter illustrates some of the practical techniques that teachers need to adopt to support this aim. We explore ways that writing can be stimulated using a range of inclusive approaches, including the use of writing journals.

Working on the Northumbrian coastline, a group of children studied one village's fading relationship with the sea. One child looked at the slight film of oil on the surface of the water and wrote:

> Anchored kittiwakes bob calmly
> on the vinegar water
> A bitter scent lingers in the air.
> Sweet shards of crystal nuzzle
> into the knotted rocks.
> A lilted tongue tilts to its side
> whispering
> tish
> tish

The child's perception of the water's surface is something that would have been difficult to predict, and responses such as these become crucial starting points for creative poetry writing as they allow metaphors to be played with, expanded and explored linguistically. At another (stormier) part of the coast, other children variously described the sea as a cobra, a lion, a porpoise and a wolf, developing animal metaphors and similes which were often insightful and occasionally surprising. Waves were variously described as 'carelessly turquoise', 'hypnotising', 'pearl diamonds' and 'silk sheets'. The noise of the water became a lullaby, a quarrel, a whisper, a growl, a lisp and a roar.

Observations such as these offer powerful starting points for discussion and for further investigations into poetry and children's instinctive use of metaphor and simile.

As primary teachers typically spend longer periods of time with the same children, it is possible for them to develop methods of writing which build on shared previous experiences. After a period of working on the development of new images to describe observations, one of the authors (Russell) arrived at school one morning after a particularly heavy frost. He took his Year 3 class into some woodland adjacent to the playground. A girl wrote:

> Sour frost swirls through the air,
> mist killing the sun.
> A solid surface
> protecting the undergrowth.
> The ice crumbles on frozen puddles, spikes on branches
> frozen
> like fingers trying to crack the air.
> Sun beaming through a line
> of gleaming frost,
> lost
> in a crystal clear desert of ice.
> Cracked and empty.

Consider the following comment from a 10-year-old boy, following his experience of a school visit to a museum: 'I started off with no clue at all what to do, but then some thoughts came into my head and it just went down my arm to the pen and onto the page . . .' How do we move children from having limited ideas about what to write to being thoroughly engaged in the writing process, like this child? One of the key questions when planning the teaching of writing is, 'What kind of stimulus should I offer?' In other words, the teacher has to decide what kind of encouragement, activities and experiences children need in order to help them to write. These decisions should be affected by consideration of children's interests and motivation. Most teachers make the sensible assumption that when children are not motivated, they do not learn as well as they could. In the case of the child who progressed from 'no clue' to fluent ideas for writing, the class were writing in the genre of biography, inspired by the story of John the Baptist. The children had visited Ely Stained Glass Museum, which is located in Ely Cathedral. They viewed a stained-glass image of John the Baptist's beheading and heard the story being told before enacting it in the museum, complete with authentic clothing, banquet food, and even a papier mâché severed head of John the Baptist on a silver plate. All the children had a role, and as a result of the gripping story and its enactment they were inspired! Evidence of the positive impact on writing of a combination of educational visits and an emphasis on the process of writing was also

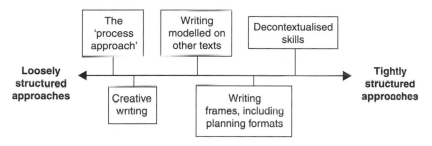

*Figure 15.1* Writing approaches continuum.

found in a larger scale study that included an experimental trial to evaluate impact on writing quality (Torgerson *et al.*, 2014).

It is, of course, impossible to undertake a school visit every time you want to stimulate children's writing, so when planning activities it is helpful to think of a continuum between loosely structured and tightly structured approaches to writing stimuli (see Figure 15.1).

Tightly structured approaches prescribe very clearly the kind of texts or extracts of texts that children are expected to produce. For example, the National Curriculum includes the statutory requirements for the teaching of spelling. Loosely structured approaches, however, give children much more opportunity to make choices about their writing. As an example of loosely structured approaches, the *process approach* has an enviable track record in motivating children to write and is based on establishing authorship in the classroom supported by a publishing process. The most important feature of the 'process approach' to teaching writing is that children are offered choice over what to write. This is not just choice within a genre ☞ prescribed by the teacher, but real choice over the topic and type of writing. The key elements of the process approach are regular *writing workshops*; a publishing cycle in the classroom; writing seen as a process that requires drafting and redrafting; extended opportunities for writing to completion; writing with real audiences in mind; mini-lessons to work on writing skills and understanding; and one-to-one interaction with the teacher to support writing. The National Curriculum requires children in Years 5 and 6 to identify audiences and purposes and to select appropriate forms of writing, requirements that could be addressed as part of a process approach to writing. Approaches such as the process approach can also be incorporated within a 'mantle of the expert' context where children write 'in role' as part of the teacher's stimulus (see → Chapter 7).

A number of features act as motivating factors for children in preparation for writing and during writing (Lambirth and Goouch, 2006). These include:

- authentic contexts for writing;
- choice of content, genre and audience;

- reference to 'more knowledgeable others';
- access to models for writing;
- freedom to use cultural connections;
- access to support for skills development;
- access to a range and choice of resources for writing.

This list highlights the importance of preparing children for writing. Stimuli you might use include: visual images, including pictures, photographs and film; music; children's literature; artefacts; first-hand experiences, including school trips; writing warm-ups; writing games; role-play and drama; talk before writing; or story boxes which include a selection of resources related to a particular text. For example, if you were to use children's literature as a stimulus, texts that are a core focus for reading can also be used to stimulate writing. Thinking about different versions, sequels and taking the characters into a new context are all examples of ways to use high-quality texts as a stimulus (→ Chapter 9, 'Children's literature'). Drama and role-play are also used to stimulate writing, sometimes linked to a core text or simply as a way of generating ideas (→ Chapter 7, 'Drama'). The use of artefacts and first-hand experiences can lead to writing that is either loosely structured or tightly structured. Artefacts and other physical resources are a well-established means of stimulating writing through a creative writing approach, as in the example of the Play Leaning and Narrative Skills (PLaNS) project (2013–15). Here, primary school teachers used LEGO sets to scaffold inspiration for children's narratives and preparation ahead of writing. Children worked together in groups to create stories and develop their writing in several ways, for example through creating 3D scene-representative storyboards (see annotated bibliography for website).

Children need to be taught strategies that will help their writing. For example, learning to summarise texts that have been read helps writing because it can be used as a way to illustrate that writing requires conciseness and accuracy. Other strategies children need to be taught include how to plan, write, revise and edit writing. Structural concepts – such as headings and paragraphs – also need to be taught.

The teaching of strategies for writing can be supported by modelling the writing using a flipchart or board (interactive or traditional) and interacting with children on a one-to-one basis to discuss their writing and help them improve it. For children in the earliest stages, teachers can also act as *scribes* for children's ideas.

## Writing warm-ups

In recent years, preparing young children to write has included taking into account physical aspects. It is necessary for children to strengthen relevant muscles to support holding a pencil comfortably and without discomfort. Learning to hold a pencil and use it to form recognisable letters is a demanding skill that all children will develop at their own rates and in their own ways; however, teachers are under increased pressure to raise outcomes in writing before children move into Key Stage 1. *Dough Gym*, designed by Alistair Bryce-Clegg, aims to strengthen and develop children's fine and gross motor dexterity, balance and hand-eye coordination. It is a physical daily intervention that involves the manipulation of a piece of playdough through a series of exercises often performed to a catchy disco beat. Although convincing research evidence of its success is yet to be acquired, the importance of the development of the physical skills of writing is recognised.

Writing warm-ups can also involve actual writing, such as those devised by Pie Corbett. These are quick-fire activities designed to warm up the brain and get children in a creative mood (e.g. see https://fileserver.booktrust.org.uk/usr/resources/527/piecorbettpriandsec.pdf). One such warm-up is called *Ink Waster*, where children are given a topic and asked to write as much as they can in say, one minute. They should be encouraged to write as rapidly as possible to limber up and free the mind. Possible guidance can include these suggestions that Helen has used with teachers:

- Use warm-ups to generate discussion and interest.
- Use warm-ups to generate what children know as opposed to what they don't know.
- Consider the writing implements they use for this exercise; children like 'beautiful writing pens'.
- Understand that the nature of the writing in these warm-ups is 'fit for purpose'.
- Don't worry about things like spelling for the warm up
- Have fun!

## Strategies for teaching writing

The current programme of study for English in the National Curriculum (DfE, 2013) distinguishes two main elements of writing: composition and transcription (see → Chapter 14 for the origins of this distinction). One way of understanding writing is to see it in its developmental context, perceiving it as a journey from early engagement with the writing process towards writing fluently using conventional text, and as a skill that has to be mastered over a long period of time. In a review of effective instructional practices for teaching writing to elementary grade pupils (5–6-year-olds), Graham et al. (2012) argued that it is important to identify writing pedagogies with *evidence* of effectiveness, as this provides teachers with instructional practices that can potentially improve the quality of their instruction and children's writing. For example, research has shown that writing composition strategies, including the components of the processes of writing, should be modelled by teachers and supported through pupils' opportunities to practice (Education Endowment Foundation, 2017).

### Shared writing

Shared writing is a term now used to describe a whole class writing experience designed to form a bridge between the teacher demonstrating the writing process to pupils and independent writing by children. There are three levels of shared writing which provide a range of opportunities for teachers to model effective strategies, explicitly teach a skill or approach and draw children into the process through questioning, talk, drama and other response activities. The three levels are:

1    Demonstration, where the teacher is the 'expert', modelling how to write a particular genre of text or a particular feature, giving a running commentary on what they are doing, and why.
2    Scribing, where the teacher involves the pupils in the composition, writing down their suggestions to create a shared piece. Examples of suggestions might include word choices, sentence construction, characterisation and editing.
3    Supported writing, where the teacher gives responsibility for word choices or a sentence to the class. Children work in pairs with whiteboards and hold their whiteboards up for the teacher to see.

### Guided writing

Guided writing is a small group approach for which children are grouped on the basis of ability and need (although the research evidence on the merits of ability grouping is not strong). Guided writing normally follows on from shared writing and addresses the specific identified writing development needs of each group, with a specific focus at whole text, sentence or word level. As with guided reading, sessions involve groups of six children working with the teacher or teaching assistant for a period of 20 minutes. The focus may be general, for example understanding the concept of sentences. Or it might be more specific to a particular genre, such as the use of descriptive language when writing a story or thinking about the structure of a report. Guided writing may be used to further address elements of the shared writing session for less confident writers and extend shared writing to challenge more able writers.

### Writing frames

Writing frames are a device for supporting some children, particularly in the context of non-fiction writing. Lewis and Wray explain the notion of writing frames:

> Writing frames are outline structures, enabling children to produce non-fiction writing in the different generic forms. Given these structures or

skeleton outlines of starters, connectives ☞ and sentence modifiers, children can concentrate on communicating what they want to say. As they practise building their writing around the frames, they become increasingly familiar with the generic forms.

(Lewis and Wray, 1995: 53)

One of the important ideas behind writing frames is that they are intended to support writing in meaningful contexts – the sort of contexts where appropriate audiences and purposes have been identified. We would also point out that children need to experience the extensive reading of any genre that they are trying to write themselves. Prior to using writing frames, teacher modelling ☞ and shared construction of texts are important.

Lewis and Wray are quite clear that 'using the frames for the direct teaching of generic structures in skills-centred lessons' is inappropriate. There are six writing frames: recount, instruction, report, procedure, explanation, discussion and persuasion, which arose from genre theory.

The following example shows how a child used the persuasive writing frame: the child's written response appears in italics within the frame text.

Although not everybody would agree, I want to argue that
*Children should not wear school uniform.*

I have several reasons for arguing for this point of view. My first reason is
*That they feel more comfortable in clothes which they choose to wear. They would feel more relaxed and be able to work better and concentrate more on their work.*

Another reason is
*There wouldn't be the problem of parents not wanting to buy uniforms because they think they are too expensive.*

Furthermore
*Sometimes you might wake up and find your two lots of uniform in the wash.*

Therefore, although some people argue that
*Children might take it past the limits.*

I think I have shown that
*Children should be able to choose their clothing just as adults do, as long as they wear sensible clothes.*

(ibid.: 85)

The writing frames are designed to be flexibly applied, and it is intended that children should move towards independence. This means that the form

I would like to persuade you that …
There are several points I want to make to support my point of view. Firstly …
These words and phrases might help you:
because
therefore
you can see
a supporting argument
this shows that
another piece of evidence is

*Figure 15.2* Part of writing frame with suggested connectives.
*Source:* Lewis and Wray (1995: 133).

of the frame can be modified to offer a different level of support. If we return to persuasive writing, Figure 15.2 gives an example that includes a list of connectives.

The teacher's goal is to foster children's independence in writing. For younger children, providing opportunities to write in role-play areas can support a positive approach to developing independence in writing through cultivating a 'can do' attitude. For older children, a way of fostering growing independence is through the use of writing journals. A writing journal often takes the form of an exercise book, given to a child for the sole purpose of encouraging their independent writing. Ideally time is allocated, as part of the school day, for the children to use them. Graham and Johnson (2003) suggest 20–30 minutes, two or three times a week. The children are given flexibility over where to work. Importantly, teachers may not provide a written response to the children's writing. They do not mark the children's writing in any way and indeed, in some cases, will read the writing only with the permission of the child. At the end of the session, children are invited to share their pieces of writing with others in the class. Lambirth and Goouch (2006) argue that writing journals form part of a pedagogy derived from an understanding of how children can be motivated to express themselves independently of teachers. Writing journals can also allow children to bring elements of their home cultures into the classroom through writing and can be particularly helpful for EAL learners who can be given an opportunity to write in their first language. Writing journals are an inclusive tool where all children can try out their developing written English skills without fear of 'failure'.

## Practice points

- Use a range of approaches to stimulate writing. Do not underestimate the value of warming up for writing, including the role of talk when preparing children for writing.
- Evaluate the extent to which children are motivated by writing and seek to improve their motivation. Children need to understand the reasons for writing.
- Incorporate times when children can write freely of their own accord, for example in their own writing journal and/or as part of a regular writing workshop.

## Glossary

**Connectives** – words and phrases whose main purpose is to connect phrases, sentences and other large units of text.

**Genre** – a form of writing such as newspaper, story, poem. In education, this can refer to spoken forms as well as written forms.

**Teacher modelling** – the teacher shows by demonstration and speaking their thoughts aloud how writing is composed.

## References

Department for Education (DfE). (2013). *The National Curriculum in England: Framework Document. September 2014*. London: Department for Education.

Education Endowment Foundation. (2017). *Improving Literacy in Key Stage Two: Guidance Report*. London: Education Endowment Foundation.

Graham, L. and Johnson, A. (2003). *Children's Writing Journals*. Royston: United Kingdom Literacy Association.

Graham, S., McKeown, D., Kiuhara, S. and Harris, K. R. (2012). 'A meta-analysis of writing instruction in the elementary grades', *Journal of Educational Psychology*, 104(4): 879–896.

Lambirth, A. and Goouch, C. (2006). 'Golden times of writing: The creative compliance of writing journals', *Literacy*, 40(3): 146–152.

Lewis, M. and Wray, D. (1995). *Developing Children's Non-Fiction Writing: Working with Writing Frames*. Leamington Spa: Scholastic.

Torgerson, D., Torgerson, C., Ainsworth, H., Buckley, H., Heaps, C., Hewitt, C. and Mitchell, N. (2014). *Improving Writing Quality: Evaluation Report and Executive Summary*. London: Education Endowment Foundation.

### Annotated bibliography

Bearne, E., Chamberlain, L., Cremin, T. and Mottram, M. (2011). *Teaching Writing Effectively: Reviewing Practice.* Leicester: UKLA.

Teaching for learning, planning for writing and how to offer choice and foster independence in writing.

**L2** ★★

Brownhill, S. (2014). *Getting Children Writing: Story Ideas for Children Aged 3–11.* London: Sage.

This book is a practical guide designed to stimulate story writing in the early years and primary classroom through offering a collection of novel and effective ideas which can be used by educators to energise, excite and motivate children to willingly write stories across the 3–11 age phase.

**L1** ★

Play, Learning, and Narrative Skills Project (2013–2015) www.educ.cam. ac.uk/research/projects/plans/

An exciting initiative to help children prepare their thoughts and ideas ahead of writing in a particular genre through the use of LEGO kits.

**L1** ★★

# Chapter 16

# Grammar

The value of many forms of grammar teaching continues to be in doubt. This chapter reminds us that questions about grammar teaching have been around for a long time. We outline the difference between 'descriptive' grammar and 'prescriptive' grammar. An examination of the important idea of 'knowledge about language' is concluded with the point that playing with language in ways that are meaningful and fun is the key to grammar teaching.

The word 'grammar' itself is used in two very distinct ways: prescriptively and descriptively. Prescriptively, the term is used to prescribe how language should be used; descriptively, the term is used to describe how the language actually is used. Prescriptive grammarians ☞ believe that English grammar is a fixed and unchanging series of rules which should be applied to the language. For prescriptive grammarians, expressions such as *I ain't done nothing wrong* or *We was going to the supermarket* are quite simply wrong. To understand this rather better, it is necessary to consider two other related questions that often get muddled up with grammar in the public discourse ☞: the question of style and the question of Standard English.

Many complaints about incorrect grammar are actually complaints about style. Split infinitives ☞ are a case in point. There is nothing *grammatically* wrong with a sentence like: *I am hoping to quickly finish writing this paragraph*. It makes perfect sense, but it might be thought stylistically preferable to write: *I am hoping to finish writing this paragraph quickly*, or even to write: *I am hoping quickly to finish writing this paragraph*. However, if I were to write: *Hoping writing I paragraph finish to quickly am*, there would be something grammatically wrong with that!

So far as Standard English is concerned, an accident of history meant that, when printing developed, it was the Anglian regional dialect that was written down (→ Chapter 1, 'The history of English, language and literacy'). Because it was written, it became the 'standard'. It would thus be more accurate to describe Standard English as the standard *dialect*. Other dialects are then described as 'non-standard'. Standard English is distinguished from non-standard dialects by features of vocabulary and features of grammar. In addition, middle-class speech tends to keep some of the grammatical features of the written form, particularly with regard to the use of negatives and the use of some verb forms. Thus, matters of class and matters of dialect have come to be linked.

From the point of view of *prescriptive* grammarians, the grammar of Standard English is 'good' or 'correct' grammar, and the grammar of non-standard dialects is 'bad' or 'incorrect' grammar. So, for example, children who say *I ain't done nothing* often have their language 'corrected' by their teachers, and even by their own parents, on the ground that it is 'bad' grammar, or indeed, more generally, that it is just 'bad' English.

*Descriptive* grammarians are interested in describing how the language actually is used rather than how it ought to be used. Thus, a descriptive grammarian will note that a middle-class speaker, using the standard dialect known as Standard English, may say: *We were pleased to see you*, and that a working-class speaker using a working-class Cockney dialect may say: *We was pleased to see you*. Both examples are grammatical within their own dialects, both examples make perfect sense, and in neither example is there any ambiguity. The idea that a plural subject takes a plural verb is true only of the middle-class standard dialect, not of the working-class Cockney dialect. To put it another way, the plural form of the verb in middle-class Standard English is *were*, while the plural form of the verb in working-class Cockney is *was*.

Let us now offer this simple working definition: *grammar is an account of the relationship between words in a sentence*. In the light of this definition, what the grammarian has to do is to look for regular patterns of word use in the language and give labels to them. However, some of the relationships are pretty complicated, and describing them is not easy. The definitions many of us half-remember from our own primary school days – 'a noun is a naming word', 'a verb is a doing word' – are at best unhelpful and at worst downright misleading. Though meaning has a part to play in determining the relationship between words, parts of speech are not defined in terms of word meaning; they are defined, rather, in terms of the function of the words within the sentence. To illustrate, think about the word *present*.

*Present* can be a verb:

I **present** you with this tennis racket as a reward for your services.

or a noun:

Thank you for my birthday **present**, I've always wanted socks!

or an adjective:

In the **present** circumstances I feel unable to proceed.

It will be clear, then, that teaching grammar has its problems. Confusion can occur at a number of levels: between prescriptive and descriptive approaches, between questions of grammar proper and questions of style, and around issues of variation between standard and non-standard dialects.

## Grammar. For writing?

One of the most long running debates about grammar has been whether the teaching of grammar has a positive impact on pupils' writing. Wyse (2001) concluded:

> The findings from international research clearly indicate that the teaching of grammar (using a range of models) has negligible positive effects on improving secondary pupils' writing (p. 422) . . . The one area where research has indicated that there may be some specific benefit for syntactic maturity is in sentence combining.
>
> (p. 423)

This finding was independently verified by Andrews *et al.* (2004) in their study, which used systematic review methodology developed by the EPPI Systematic Review Centre at the Institute for Education (IOE), University of London:

> The results of the present in-depth review point to one clear conclusion: that there is no high quality evidence to counter the prevailing belief that the teaching of the principles underlying and informing word order or 'syntax' has virtually no influence on the writing quality or accuracy of 5 to 16 year-olds. This conclusion remains the case whether the syntax teaching is based on the 'traditional' approach of emphasising word order and parts of speech, or on the 'transformational' approach, which is based on generative-transformational grammar.
>
> (p. 4)

The study also cited Wyse (2001) as one of only three extensive reviews of the subject in a 100-year period (and the first for about 30 years).

Although it has been established that grammar teaching does not impact positively on pupils' writing, it is clear that pupils do need to learn to control language as one part of learning to write. Building on his earlier work from 2001, Wyse's (2006) research, which used in-depth analysis of pupils' word choices during the process of writing, proposed a pedagogical theory of con-textualised teaching of language for writing:

> Although the notion that grammar teaching should be contextualized at text-level is common, such theories have not hitherto had a strong

empirical base. The research reported in this paper begins to offer empirical evidence to support such theory.

(p. 44)

One of the key findings was the idea that pupils' text-level thinking interacted with sentence-level and word-level thinking, sometimes resulting in unconventional language use (or what would be called 'errors' by some).

As part of a large scale study, Myhill *et al.* (2011) claimed they had 'robust evidence from the data in favour of the use of grammar in an embedded way within the teaching of writing' but also that 'the study certainly does not suggest that this would be of universal benefit' (p. 162). This caveat was included because the study found larger gains for pupils with higher attainment. The idea of use of grammar in an embedded way can clearly be linked to the idea of grammar teaching contextualised at text-level that Wyse (2006) had theorised some five years earlier. The principles of the Myhill *et al.* approach reveal a rich, and admirable, contextualised approach to writing:

- The grammatical metalanguage is used but it is always explained through examples and patterns.
- Links are always made between the feature introduced and how it might enhance the writing being tackled.
- The use of 'imitation': offering model patterns for students to play with and then use in their own writing.
- The inclusion of activities which encourage talking about language and effects.
- The use of authentic examples from authentic texts.
- The use of activities which support students in making choices and being designers of writing.
- The encouragement of language play, experimentation and games.

(Myhill *et al.*, 2011: 148)

A point not explained in Myhill *et al.* is that in spite of an explicit focus on grammar as part of the National Literacy Strategy (including the Grammar for Writing resources that seems in significant ways similar to the approach in the Myhill *et al.* study) implemented in England between 1998 and 2010, the gains for writing shown in statutory test scores were negligible, nor was there any evidence of significant gains compared to earlier periods that did not include explicit grammar teaching. It is also important to remember that the NLS grammar approach was mainly focused on primary schools. Contrary to their claims, Myhill *et al.* have not provided evidence to contradict the well-established view that grammar teaching (apart from sentence combining) has minimal positive impacts on pupils' writing. As the research was carried out in secondary schools, it is also important that it is not cited in support of more grammar teaching in primary and early years settings.

In a rekindling of the debate about grammar, a paper in 2017 demonstrated that the only approach to grammar in primary education that had a robust evidence based was *sentence-combining*. The paper also showed a complete mismatch between the traditional grammar teaching in England's National Curriculum, such as learning of grammatical terms, and the research evidence on what works in the teaching of writing (Wyse and Torgerson, 2017). It was concluded that:

> the current evidence from randomised controlled trials does not support the widespread use of grammar teaching for improving writing among native English speaking children. Based on the experimental trial and meta-analysis evidence about writing teaching more generally (e.g., in tables 1 and 2), our hypotheses are that supporting primary/elementary pupils' grammar is most likely to require teachers intervening during the writing process, and interacting to discuss the use of grammar in relation to the overall purpose of the writing task and the purpose of the writing. The necessity to use technical terms with pupils such as subordinate clause or subjunctive remains a question open to research, but it is doubtful that attention to such terms is beneficial. It is probable that adopting every-day language to discuss improvements in the use of grammar in writing will be more beneficial. Small group and whole class teaching that includes a focus on the actual use of grammar in real examples of writing, including professionally produced pieces, realistic examples produced by teachers including 'think aloud' live drafting of text, and drafts of pupils' writing, may also be more effective.
>
> (Wyse and Torgerson, 2017: 1043)

The findings of the paper resulted in more disagreements which erupted onto the tes online (see annotated bibliography below), which had picked up the findings of the research.

## Knowledge about language

The proposal to teach children explicitly about language raises some questions:

- What are the reasons for teaching primary children about language?
- What are the benefits to be gained?
- What should be taught, and at what ages?
- How should it be taught?
- What is the place of terminology?
- Where does grammar fit into all of this?

Cox (1991) suggested that there are two justifications for teaching children explicitly about language. The first is that it will be beneficial to their language use in general. The second is that it is essential to children's understanding of their social and cultural environment, given the role language plays in society. A third suggestion, related to the second, is that 'language should be studied in its own right as a rich and fascinating example of human behaviour' (LINC, 1992: 1).

Although the Language in the National Curriculum (LINC) project was carried out many years ago now, its recommendations about teaching knowledge about language are still valid. The starting point is that children should be encouraged to discuss language use in meaningful contexts that engage their interest. Here are some examples of work done in primary schools under the auspices of the LINC project, all of which were extremely productive in getting children to think and talk about language itself (all examples from Bain *et al.*, 1992):

- Making word lists.
- Compiling dictionaries including slang and dialect dictionaries.
- Discussing language variation and social context. For example, how does a mobile phone text message compare with a formal letter?
- Discussing accent and Standard English.
- Compiling personal language histories and language profiles.
- Capitalising on the language resource of the multilingual classroom, beginning with in-depth knowledge of languages spoken, read, written and their social contexts including religious ones.
- Role-play and drama, particularly to explore levels of formality and their links with language.
- The history and use of language in the local environment.
- Collecting and writing jokes.
- Collaborative writing which involves discussion about features of writing and language.
- Media work including the language of adverts.
- Book-making.

One of the aspects of language variation is the way that spoken language differs from written language, as Table 16.1 shows.

Many of the examples in the list above could be, and indeed were, done with the youngest children, and terminology was learned in context as and when the children needed it in their work.

Effective grammar teaching will involve pupils playing with and exploring language in ways that are meaningful and fun. Teachers will need both to understand the issues and to be confident in naming relevant terms themselves, so that they can then use them with confidence in everyday discussion with pupils. If teachers use the terms correctly with the children all the time, the children will learn what the terms refer to, even if they are not able to define them to the satisfaction of a linguistics expert until they are older. Dry as

*Table 16.1* Differences between speech and print

| Speech | Print |
| --- | --- |
| Requires other speakers to be present at time of speech (unless recorded) | Readers are not present at the time of writing |
| Speakers take turns | Main writer works alone |
| Instant and cannot be changed | Can be composed and reworked |
| Can be incomplete and make sense because of shared understanding of conversation | Writing usually doesn't make sense if it is not complete |
| Intonation, pitch and body language used to support meaning | Font effects such as italics used to support meaning |
| Organised in communicative units | Organised in sentences |
| Separated by pauses in flow of sounds | Separated by punctuation |
| Words integrated within streams of words | Words demarcated by spaces |
| Consists of phonemes (sounds) | Consists of graphemes (letters) |
| Accent and dialect recognisable features of speakers | Accent and dialect not features unless used as a deliberate device in fiction or poetry |
| Tends to be informal | Tends to be formal |

dust, decontextualised, old-fashioned grammar exercises of the *underline the noun* variety do not work and put more children off than they help. In addition, textbooks that do not discuss dialect variation, that confuse matters of style with matters of grammar and that take a prescriptive, 'correct English' approach throughout are to be avoided.

## Practice points

- Plan grammar teaching to enthuse children about the way that language works.
- Engage children's curiosity about language through work on, for example, accent and dialect.
- Use technical language about language (metalanguage) only if it actually helps learning, not simply to teach technical terms.

## Glossary

**Grammarian** – someone who studies grammar.
**Infinitive** – part of a verb that is used with 'to': e.g. to *go* boldly.

**Public discourse** – discussions and debates in the public domain, particularly seen through the media.

### References

Andrews, R., Torgerson, C., Beverton, S., Locke, T., Low, G., Robinson, A. and Zhu, D. (2004). 'The effect of grammar teaching (syntax) in English on 5 to 16 year olds' accuracy and quality in written composition', *Research Evidence in Education Library* Retrieved 5 February 2007, from http://eppi. ioe.ac.uk/cms/

Bain, R., Fitzgerald, B. and Taylor, M. (1992). *Looking into Language: Classroom Approaches to Knowledge about Language*. Sevenoaks: Hodder & Stoughton.

Cox, B. (1991). *Cox on Cox: An English Curriculum for the 1990s*. London: Hodder & Stoughton.

Language in the National Curriculum (LINC). (1992). 'Materials for professional development', unpublished.

Myhill, D., Jones, S., Lines, H. and Watson, A. (2011). 'Re-thinking grammar: The impact of embedded grammar teaching on students' writing and students' metalinguistic understanding', *Research Papers in Education*, 27(2): 139–166.

Wyse, D. (2001). 'Grammar. For writing?: A critical review of empirical evidence', *British Journal of Educational Studies*, 49(4): 411–427.

Wyse, D. (2006). 'Pupils' word choices and the teaching of grammar', *Cambridge Journal of Education*, 36(1): 31–47.

Wyse, D. and Torgerson, C. (2017). 'Experimental trials and "what works?" in education: The case of grammar for writing', *British Educational Research Journal*, 43(6): 1019–1047. doi: 10.1002/berj.3315(30)

### Annotated bibliography

Carter, R. and McCarthy, M. (2006). *Cambridge Grammar of English*. Cambridge: Cambridge University Press.
A comprehensive account of English grammar that uses examples of oral and printed language in modern use (corpus data) in order to describe and explain the way the language works.
L2 ★★

Cremin, T. and Oliver, L. (2017). 'Teachers as writers: A systematic review', *Research Papers in Education*, 32(3): 269–295. doi: 10.1080/02671522.2016.1187664
Reflects Cremin's passion for teachers being writers, although the evidence was still not conclusive at the time of this systematic review.
L3 ★★★

Wyse, D. and Torgerson, C. (2017). 'Experimental trials and 'what works?' in education: The case of grammar for writing', *British Educational Research Journal*, 43(6): 1019–1047. doi: 10.1002/berj.3315(30)

Examines the way in which governments rely on ideology rather than evidence when it comes to topics such as grammar.

**L3** ★★★

www.tes.com/news/school-news/breaking-views/grammar-teaching-national-curriculum-has-it-failed-test

Picks up points about grammar in England's national curriculum. Illustrates the way in which grammar causes debate.

**L1** ★★

# Chapter 17

# Punctuation

Two key ideas that underpin this chapter are linguistic and non-linguistic punctuation. The examples of punctuation that are given reveal that the communication of unambiguous meaning is central to punctuation, just as it is to writing more generally. Information about how punctuation works is linked with how best to help children learn to punctuate their writing.

Before reading this chapter, look again at Neil's writing about the Romans and the Hojibs (Figure 14.3 in Chapter 14), and this time, rather than focus on the composition elements, think how might you help him to improve his punctuation. You can revisit your ideas once you have finished reading this chapter.

Hall (1998) was one of the first education researchers to look in depth at the teaching and learning of punctuation in schools. His book includes an overview of the history of punctuation. For example, in 1700, Richard Browne said: 'What is the use of stops or points in reading and writing? To distinguish sense; by resting so long as the stop you meet with doth permit' (Hall, 1998: 2). In Roman times, it was the readers who inserted the punctuation into texts, not the writers: this was related to the need to declaim texts orally. Since that time, the function of punctuation has changed. However, the idea that punctuation is primarily designed to support oral reading – through pauses – still persists. Hall reiterates that punctuation is no different to writing in general in that the generation of *meaning is* the primary function of written language: one of the main points that he makes is that punctuation is learned most successfully in the context of 'rich and meaningful writing experiences' (1998: 9).

Hall usefully differentiates between 'non-linguistic punctuation' and 'linguistic punctuation'. An example of non-linguistic punctuation is where a child puts full stops at the end of every line of a piece of writing rather than at the end of the sentences. This illustrates the child's belief that punctuation is to do with position and space rather than to indicate meaning and structure. With regard to non-linguistic punctuation, the idea of 'resistance to punctuation' is discussed. One of the reasons that children can remain resistant to using punctuation appropriately is exacerbated by teachers whose comments are often directed to naming and procedures rather than explanation:

> As already indicated, teacher comments which are simply directed to the placing of punctuation rather than to explaining its function can leave the child with no sense of purpose. Yet research suggests that teacher practices are, probably quite unconsciously, dominated by procedure rather than explanation.
>
> (ibid.: 5)

Standard punctuation is linked with grammar and sentence structure. The necessary understanding of these complex concepts does not happen suddenly. Children gradually begin to realise that spatial concepts either do not work or do not match what they see in their reading material. Hall illustrates this with an example from his research:

> three children who were jointly composing a piece of text which was being scribed for them . . . After two lines, one child insisted on having a full stop at the end of each line. The rest of the piece was written with no more punctuation. Then there was a scramble for the pen and one child wanted to put a full stop after 'lot' on line three.

*Derek:*   You're not supposed to put full stops in the middle.
*Rachel:*   You are!
*Derek:*   No, they're supposed to be at the end, Ooh!
*Rachel:*   You are, Derek.
*Fatima:*   Yeah. So that's how you know that (meaning the 'and' at the end of line 3) goes with that (meaning line 4).

> (ibid.: 12)

An early piece of punctuation research was carried out by Hutchinson (1987), who worked closely with a child, Danny, on a piece of writing that was a retelling of *Come Away from the Water, Shirley* by John Burningham (1977). Danny wrote as he would speak:

> shirley and her mum and dad were going to the seaside and her dad told her to go and play with the other children and her dad didn no she went saling with a dog and a pirate ship was foiling and the pirates corght her.

An important aspect of teachers' knowledge is recognising the significant differences between speech and writing. Hutchinson (1987) made the point that

Danny's reliance on speech in this extract resulted in the use of conjunctions ☞ where full stops would be more appropriate in writing. However, he also pointed out that an analysis based on one text in one context is not sufficient for assessing understanding. The following day when they revisited the writing, Danny added some full stops to his work. Another point to make here is that the process of redrafting can enable children to reconsider their use of punctuation.

One of the main uses of punctuation is to avoid ambiguity. Children need to learn to understand the role that punctuation plays in clarifying intended meaning in written language. And they must learn to recognise that punctuation plays a crucial role in reading, whether someone is reading out loud or silently. Use of the wrong mark or putting a mark in the wrong place in a sentence can change the meaning of a sentence completely. The following example shows how completely different meanings can be created by different punctuation:

PRIVATE. NO SWIMMING ALLOWED.
PRIVATE? NO! SWIMMING ALLOWED.

Part of the intrigue of punctuation is spotting the frequent 'errors' that occur. The apostrophe is a bit of a classic in this regard. The following professionally printed banner appeared on a pub wall:

Qs monster meals won't scare you but the portion's might

One apostrophe is missing, one is correct, and one is incorrect. It is worth reminding ourselves of the common types of apostrophe:

1   contraction: didn't = did not;
2   possession singular: the cat's tail, the child's book;
3   possession plural: the cats' tails, the children's books (as 'children' is an irregular plural form, the apostrophe comes before the 's');
4   possession with name ending in 's': Donald Graves' book. Common errors include: this first happened in the 60's. – 60s is plural, not a contraction; was that it's name? – because of the confusion with it's (as in: It's (it is) my party and I'll cry if I want to). This possessive form is irregular and does *not* have an apostrophe (was that its name?).

The research evidence showing the importance of helping children with sentence construction (Education Endowment Foundation, 2017) suggests that punctuation should be part of this teaching. The most common punctuation marks in English are: capital letters and full stops, question marks, commas, colons and semi-colons, exclamation marks and quotation marks. Table 17.1 shows the punctuation that children will need to learn in each year group

Table 17.1 Punctuation in England's national curriculum

| Year | Punctuation | Technical terminology |
|------|-------------|----------------------|
| 1 | Separation of **words** with spaces<br>Introduction to capital letters, full stops, question marks to demarcate **sentences**<br>Capital letters for names and for the personal **pronoun** I | letter<br>capital letter<br>word<br>singular<br>plural<br>sentence<br>punctuation<br>full stop<br>question mark<br>exclamation mark |
| 2 | Use of capital letters, full stops, question marks and exclamation marks to demarcate sentences<br>Commas to separate items in a list<br>Apostrophes to mark where letters are missing in spelling and to mark singular possession in nouns [for example, the girl's name] | apostrophe<br>comma |
| 3 | Introduction to inverted commas to punctuate direct speech | inverted commas<br>(or 'speech marks') |
| 4 | Use of inverted commas and other **punctuatio**n to indicate [for example, a comma after the reporting clause; end punctuation with inverted commas: *The conductor shouted, 'Sit down!'*]<br>**Apostrophes** to mark **plural** possession [for example, *the girl's name, the girls' name*]<br>The use of commas after **fronted adverbials** | |
| 5 | Brackets, dashes or commas to indicate parenthesis<br>Use of commas to clarify meaning or avoid ambiguity | parenthesis<br>brackets<br>dash |
| 6 | Use of the semi-colon, colon and dash to mark the boundary between independent **clauses** [for example, *It's raining; I'm fed up*]<br>Use of a colon to introduce a list<br>**Punctuation** of bullet points to list information<br>How hyphens can be used to avoid ambiguity [for example *man eating shark* versus *man-eating* shark, or *recover* versus *re-cover*] | ellipsis<br>hyphen<br>colon<br>semi-colon<br>bullet points |

across Key Stage 1 and Key Stage 2 (drawn from Appendix 2 in the National Curriculum: DfE, 2013).

Punctuation can be taught through shared and guided reading and writing sessions, during the process of writing and as discrete mini-lessons. Children should be encouraged to self- and peer-assess their work. One of the most effective ways of improving punctuation is to work with someone who is proficient at proofreading. All professionally published materials pass through a proofreading stage: this book will be passed to a copy-editor and a proof-reader; newspaper articles go to sub-editors who work on style and presentation, etc. This is in part a recognition that proofreading is often more efficient if it is not carried out solely by the author, who often is primarily concerned with the composition. This also reflects the idea that it is sensible to separate composition and transcription in the various stages of the writing process.

### Practice points

- Use real texts (electronic and printed) to draw children's attention to the range of punctuation and the ways in which it is used.
- Help children to understand that one of the main reasons for punctuation is to improve clarity and avoid ambiguity.
- Encourage children to see checking punctuation as part of the proofreading process.
- Encourage children to read their writing out loud, taking into account their use of punctuation, or to ask a partner to read it out loud.

### Glossary

**Conjunction** – a type of word that is used mainly to link clauses in a sentence, e.g. She was very happy *because* John asked for help with his maths.

### References

Burningham, J. (1977). *Come Away from the Water, Shirley*. London: Cape.

Department for Education (DfE). (2013). *The National Curriculum in England: Framework Document. September 2014*. London: Department for Education.

Education Endowment Foundation. (2017). *Improving Literacy in Key Stage Two: Guidance Report*. London: Education Endowment Foundation.

Hall, N. (1998). *Punctuation in the Primary School*. Reading: University of Reading, Reading and Language Information Centre.

Hutchinson, D. (1987). 'Developing concepts of sentence structure and punctuation', *Curriculum*, 8(3): 13–16.

### Annotated bibliography

Education Endowment Foundation. (2017). *Improving Literacy in Key Stage Two: Guidance Report*. London: Education Endowment Foundation.
This report and a similar report for KS1 are a valuable contribution to evidence about what works in the teaching of literacy.
L2 ★★★

Hall, N. and Robinson, A. (1996). *Learning About Punctuation*. Clevedon: Multilingual Matters.

A range of contributors offer their thoughts on the teaching of punctuation. Includes a study that looked at the development of a group of 8- and 9-year-old children.

**L2** ★★

Medwell, J., Wray, D., Moore, G. and Griffiths, V. (2017). *Primary English: Knowledge and Understanding*, 8th edn. London: Sage.

A strong general introduction to knowledge about language. Designed for trainee teachers, it is also useful for qualified teachers who need support with teaching the technicalities of the English language.

**L1** ★

# Chapter 18

# Spelling

Spelling is a feature of most primary classrooms, but it can be one aspect of literacy instruction that relies on testing rather than teaching. Building the distinction between spelling being 'caught or taught', we show how children can be taught how to become competent and conventional spellers. The chapter considers the complexities of learning to spell in English alongside pedagogical approaches to teaching spelling. We conclude with reflections on the appropriate use of spelling tests and spelling homework.

Most people would agree that they read words more accurately than they spell them. Children's natural desire to make sense of the world can be seen in their attempts to master English spelling. The younger pupils are, the truer this is, as the following examples of children's spelling attempts illustrate:

    my klone
    ches and unen

The first is a famous footballer originally from Liverpool; the second is a flavour of crisps. It is the phonic irregularities of English that make it demanding for children to learn and for teachers to teach. You can also find some creative, yet at the same time plausible, attempts to spell in our example of Neil's writing in Chapters 5 and 14. Perhaps the most striking of these is the use of the phrase 'two romens'. The child will have known that to say one Roman is standard English, yet the plural of man is 'men'; hence to write 'two romens' is a plausible, logical deduction on their part. Your ability to understand children's sophisticated attempts at spelling (that may at first glance seem incomprehensible) is a skill you will acquire as part of your initial training. Not only will you learn to 'decode' their attempts, but you will also learn how to understand the child's thinking that lies behind their attempts to make meaning.

Spelling, particularly spelling in the English language, has been dominated by history and convention. In many cases, non-standard English spelling communicates meaning without ambiguity. However, there are strong societal pressures for correct or standard spelling, and it is true that until children have mastered standard spelling they cannot be said to be fully competent writers. This is emphasised nowhere more than in England's National Curriculum, which includes an entire 25-page appendix dedicated to spelling alone (Appendix 1 DfE, 2013). In the appendix, spelling is acknowledged as a skill that must be taught effectively, and includes word-lists for children in each year of Key Stage 2 that are statutory:

> The lists are a mixture of words pupils frequently use in their writing and those which they often misspell. Some of the listed words may be thought of as quite challenging, but the 100 words in each list can easily be taught within the four years of key stage 2 alongside other words that teachers consider appropriate.
>
> (DfE, 2013: 50)

Whilst an emphasis on *invented spelling* or the 'have a go' approach might be one useful strategy for learners at the early stages of development, we argue that this emphasis should not be to the exclusion of other more direct teaching approaches because different children will respond better to different approaches depending on their stages of development. Teaching spelling requires a clear focus on words and the letter patterns that words are built from. In the early stages, phonological knowledge is important, but quite soon children need to learn that English spelling has complex patterns of sound–symbol correspondences. The ability to use standard spelling requires understanding of elements such as word stems, word functions and grammatical meanings supported by visual memory. One of the key elements of the educator's approach to spelling should be to encourage children to see patterns in related groups of words (Schlagel, 2007). In summary, in order to write well, children need to be able to use standard spellings with fluency (Graham *et al.*, 2008).

In her seminal book *Spelling: Caught or Taught?*, Peters (1985) argued that in the past spelling had not been taught effectively. She suggested that spelling is a particular skill, or set of skills, that requires direct instruction for the majority of school pupils. Children who are not taught to spell properly often develop a poor self-image as far as spelling is concerned and lack self-confidence in their writing as a whole. Crucially, she emphasised the significance of children acquiring visual strategies ☞ rather than auditory strategies ☞ in learning to spell, though she also acknowledged the usefulness of kinaesthetic strategies ☞. She strongly recommended the 'Look–Cover–Write–Check' approach:

1   Look carefully at the word, noting particular features such as familiar letter strings, suffixes, etc. and memorise it by saying the word silently, thinking of the meaning of the word and trying to picture it in the mind's eye.

2    Cover the word.
3    Write the word from memory.
4    Check that the word written is correct by matching it with the original. If the spelling is incorrect, the whole process should be repeated.

Peters' research also led her to believe that there is a direct correlation between confident, clear and carefully formed handwriting and the development of competent spelling (→ Chapter 19, 'Handwriting'). Her approach has since been extended to include the importance of saying the word out loud before writing it down: 'Look – Cover – Say – Write – Check'. When saying the word, children can segment it into their constituent phonemes.

## Theories of spelling development

Classic theories of spelling development have traditionally focused on spelling as developing in stages. One example of this is Gentry (1982), who suggested that there were five stages of development for spelling. His work is helpful when considering that children's spelling ability develops over a long period of time; it should always be acknowledged that children will make attempts to spell based on what they know and understand about the English language system. While these may not always be conventional attempts, what is produced is a helpful indicator for teachers about children's spelling knowledge.

The first stage is the *pre-communicative* stage, when young children are making their first attempts at communicating through writing. The writing may contain a mixture of actual letters, numerals and invented symbols and, as such, it will be unreadable, although the writer might be able to explain what they intended to write.

When children are at the second stage – that is, the *semi-phonetic* stage – they are beginning to understand that letters represent sounds and show some knowledge of the alphabet and of letter formation. Some words will be abbreviated or the initial letter might be used to indicate the whole word.

At the *phonetic* stage, children concentrate on the sound–symbol correspondences, their words become more complete and they gain an understanding of word division. They can cope with simple letter strings such as *-nd, -ing* and *-ed* but have trouble with less regular strings such as *-er, -ll* and *-gh*.

During the *transitional* stage, children become less dependent on sound–symbol strategies. With the experience of reading and direct spelling instruction, they become more aware of the visual aspects of words. They indicate an awareness of the accepted letter strings and basic writing conventions of the English writing system and have an increasing number of correctly spelled words to draw upon.

Finally, the fully competent speller emerges at the *conventional* (or *correct*) stage. Conventional spellings are being produced competently and confidently almost all the time and there is evidence of the effective use of visual strategies and knowledge of word structure. Children at this stage have an understanding of basic rules and patterns of English and a wide spelling vocabulary. They can distinguish homographs ☞, such as 'tear' and 'tear', and homophones ☞, such as 'pear' and 'pair', and they are increasingly able to cope with uncommon and irregular spelling patterns. The idea of stage theories is that teachers might be able to use the model to identify what stage individual pupils are at, what sort of expectations they might have of these individuals, what targets they might set for these children and what teaching strategies they might usefully employ.

Criticisms of stage theory include the fact that they have a tendency to oversimplify the overall picture of how children develop their knowledge of spelling ability to become accurate spellers. Bourassa and Treiman (2010) point out, for example, the fact that phonological, orthographic and morphological knowledge are not homogenous. In this respect, Beard (1999) was accurate when he argued that each stage of Gentry's model represents complex patterns of thinking and behaviour. Aspects of several stages might be therefore evident in one piece of writing, meaning that children's spelling is not as one-dimensional as stage theories claim.

Linguistically based approaches to spelling development have superseded stage theories as a response to better support children in unravelling the sheer complexity of spelling and the English language. Linguistically based approaches focus more on a relational approach between phonological, orthographic and morphological knowledge and the resultant impact on the development of children's spelling knowledge and understanding (Herrington and Macken-Horank, 2015).

## Developing effective spelling pedagogy

Helping children to attend to the function of elements within words is a crucial aspect of spelling pedagogy (Herrington and Macken-Horank, 2015). Therefore, a linguistically based approach relies on strong teacher knowledge of how spelling works. This knowledge incorporates two strands. The first enables the teacher to teach children to look closely at the parts of words; the second enables them to look closely at and analyse children's errors of spelling approximations. An example of pedagogical practice in this way can be seen in Figure 18.1. Every week the children in Year 1 were encouraged to write a sentence, in response to a sentence starter, in their writing books.

Note the feedback the child has been given. There is nothing here about spelling errors. What the teacher did was to look at spelling across the class and identify common errors and patterns of mis-spellings, which she then used to guide her subsequent spelling input that week.

Phonic knowledge plays a significant role in the early stages of learning to spell. Children learn that segmenting words into their constituent phonemes ☞ for spelling is the reverse of blending phonemes into words for reading.

*Figure 18.1* Maisie's writing.
(Being kind means you can pass a pencil to your friend and if they lost their book you could help them find it and you can pass them their book and you can pass them a rubber)

They also need to learn to spell words accurately by combining the use of their grapheme–phoneme correspondence knowledge. As children's spelling understanding progresses, a linguistically based approach brings their attention to the role of morphological and etymological knowledge.

Consider what strategies you use when facing words you find hard to spell. Some general strategies that 'good' spellers use might be the following:

- sounding words out;
- drawing on a store of words they can spell, i.e. 'sight vocabulary';
- dividing words into syllables;
- knowing common prefixes and suffixes, for example, *pre-*, *dis-*, *-ed*, *-ing*;
- developing understanding of rules and their limitations;
- using dictionaries and spellcheckers;
- asking peers for help and/or confirmation.

O'Sullivan and Thomas (2000) found the following areas of spelling knowledge to be essential in the teaching and learning of spelling:

- extensive experience of written language;
- phonological awareness;
- letter names and alphabetic knowledge;
- known words which can form the basis for analogy making;
- visual awareness;
- awareness of common letter strings and word patterns, for example, *-at, -ad, -ee, -ing, -one, -ough*;
- knowledge of word structures and meanings, for example, prefixes, tenses, compound words, word roots and word origins;
- growing independence – where and how to get help; using dictionaries etc.
- making analogies and deducing rules.

We would add to this list the importance of extensive experience of spoken language, in line with recent research showing that children's early language experiences impact on later outcomes (Sylva *et al.*, 2010).

As children's spelling competence develops, teachers need to employ a variety of teaching approaches. These include:

- Visual strategies, where children are taught to recognise tricky or high frequency words, to use the 'Look–Cover–Say–Write–Check' procedure, to check critical features of the word such as shape and length and to ask questions such as 'Does it look right?'
- A morphemic and word analogy approach using known spellings as a basis for correctly spelling other words with similar patterns or related meanings and building words from awareness of the meaning or derivation of known words: here, there, everywhere; light, sight, bright.
- Knowing about and predicting the most likely order of letters in words – for example, what letter is likely to follow Q?
- Mnemonic approaches, which involve inventing and using personal mnemonic devices for remembering difficult words, such as 'two ships on the sea', 'one collar and two sleeves' – ne**cess**ary; 'there's an "e" for envelope in the middle of stationery'; or 'an "i" (eye) in the middle of "nose" makes a noise'.
- Thinking about and investigating spelling 'rules' and their exceptions.
- Games such as Scrabble, hangman, crosswords, word searches, Boggle and Countdown can all be useful in stimulating and motivating children's interest in words and spelling.

## Spelling tests and spelling homework

Weekly spelling tests are a common feature of primary classrooms. It is tempting to ask why this practice is so prevalent when there is very little evidence about its effectiveness. Historical tradition, and in particular the spelling test requirement in statutory tests, are the likely driving forces behind the practice. It is important to consider what the pros and cons of spelling tests are. A potentially positive feature of spelling tests is that lists of words are sent home to be learned. If this is carefully thought through, word lists can provide an activity that many parents feel confident to support their children with (although it is important to remember that some parents will not feel confident about this). Whilst these lists of words are often differentiated into ability levels, the National Curriculum now includes statutory lists of 100 words that must be learnt by children in each year of Key Stage 2. It is common for teachers to encourage children to put the words into a sentence context or even a paragraph context as part of the homework (see Figure 18.2 for a humorous response to using a particular set of words). Sometimes children are encouraged to identify problem words

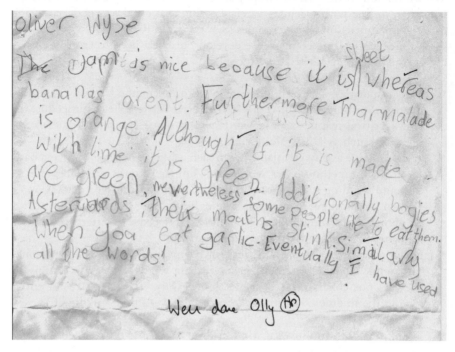

*Figure 18.2* Olly's writing.

from the ones they have been writing as part of their lessons or words that are of particular interest.

The claim that spelling tests are used as an assessment tool is questionable. The biggest problem is that the true test of learning a spelling is whether it is written correctly in the course of normal writing. The most useful assessment of children's spellings will, arguably, take place at the time that the children are actually engaged in the writing process in the classroom; every piece of writing that a child carries out gives teachers an opportunity to assess their spelling, and it should be remembered that the kind of writing that the child is doing will affect the nature of their spelling mistakes. A diary, for example, will create different challenges from a piece of scientific writing. In addition, however sensitively tests are handled, there will always be children who are poor spellers and whose self-esteem will be damaged each time they have to carry out a spelling test. It is important to remember that for children with dyslexia, learning about the writing system is slower in comparison with other children who follow a trajectory of typical spelling development. Research suggests, however, that children with dyslexia face the same stumbling blocks and make the same kinds of errors. Bourassa and Treiman (2010) therefore argue that whilst spelling instruction should be targeted at the same linguistic features for all children, it needs to be more intensive and more explicit for struggling spellers.

When word lists and spelling tests represent the main approach to spelling and when the lists are sent home *every* week, some children can become demotivated. There is no reason why spelling homework cannot feature a wide range of activities such as the ones covered in this chapter. Teachers need to create meaningful contexts for children's learning; for example, one activity is the simple but potentially valuable one of bringing home a draft of writing and asking for help with proofreading.

### Practice points

- In the early stages, the use of phonological understanding ☞ should be encouraged alongside visual strategies for conventional spelling.
- There needs to be a careful balance between encouraging invented spelling, which can aid composition, and standard spelling, which is the final goal.
- Every piece of writing provides an opportunity for the assessment of a child's spelling.

### Glossary

**Auditory strategies** – the use of sounds to help spelling of unknown words.

**Homographs** – words with the same spelling as another but different meaning: 'a lead pencil/the dog's lead'.

**Homophones** – words which sound the same but have different meanings or different spellings: 'read/reed'; 'right/write/rite'.

**Kinaesthetic strategies** – the use of the memory of physical actions to form words.

**Phoneme** – the smallest unit of sound in a spoken word, e.g. /b/ in 'bat'.

**Phonological understanding** – understanding the way that letters represent speech sounds.

**Visual strategies** – the use of visual memory of words including common patterns of letters.

### References

Beard, R. (1999). *National Literacy Strategy: Review of Research and Other Related Evidence.* London: DfEE Publications.

Bourassa, D. and Treiman, R. (2010). 'Linguistic foundations of spelling development', in D. Wyse, R. Andrews, and J. Hoffman (eds.), *The Routledge International Handbook of English, Language and Literacy Teaching.* London: Routledge.

Department for Education (DfE). (2013). *The National Curriculum in England: Framework Document. September 2014.* London: Department for Education.

Gentry, J. R. (1982). 'An analysis of developmental spelling in *Gnys at Wrk*', *Reading Teacher*, 36: 192–200.

Graham, S., Morphy, P., Harris, K., Fink-Chorzempa, B., Saddler, S. M. and Mason, L. (2008). 'Teaching spelling in the primary grades: A national survey of instructional practices and adaptations', *American Educational Research Journal*, 45(3): 796–825.

Herrington, M. H. and Macken-Horank, M. (2015). 'Linguistically informed teaching of spelling: Towards a relational approach', *Australian Journal of Language and Literacy*, 38(2): 61–71.

O'Sullivan, O. and Thomas, A. (2000). *Understanding Spelling.* London: CLPE.

Peters, M. L. (1985). *Spelling: Caught or Taught: A New Look.* London: Routledge and Kegan Paul.

Schlagel, B. (2007). 'Best practices in spelling and handwriting', in S. Graham, C. A. MacArthur and J. Fitzgerald (eds.), *Best Practices in Writing Instruction.* New York: The Guilford Press.

Sylva, K., Melhuish, E., Sammons, P., Siraj-Blatchford, I. and Taggart, B. (2010). *Early Childhood Matter: Evidence from the Effective Pre-School and Primary Education Project.* Oxford: Routledge.

### Annotated bibliography

Bourassa, D. C. and Treiman, R. (2010). 'Linguistic foundations of spelling development', in D. Wyse, R. Andrews and J. Hoffman (eds.), *The Routledge International Handbook of English, Language and Literacy Teaching.* London: Routledge.

An excellent overview of research on effective writing teaching and recommendations for research–informed practice.

L3 ★★★

Herrington, M. H. and Macken-Horank, M. (2015). 'Linguistically informed teaching of spelling: Towards a relational approach', *Australian Journal of Language and Literacy*, 38(2): 61–71.

An Australian study describing the importance of teacher knowledge in spelling. The study exemplifies in practical ways many of the key points raised in Bourassa and Treiman's (2010) study (see above).

L2 ★★

If you want to reflect on adults' spelling ability, and possibly your own spelling, try www.bbc.co.uk/skillswise/topic-group/spelling

L1 ★

# Chapter 19

# Handwriting

In England, the teaching of handwriting is still a statutory requirement of the National Curriculum (DfE, 2013). The development of a fluent, comfortable and legible handwriting style therefore helps children throughout their schooling, not least in exams. Basic principles of teaching handwriting are described in this chapter, along with suggestions for developing handwriting skills. We end with a section on handwriting problems.

Handwriting has always been taught; in England, for example, it has been a significant feature of the curriculum since the introduction of the 1870 Education Act. However, increasingly, children have access to means of producing print without having to hand-write. From 2014 onwards, it was claimed that in Finland there was no longer a requirement to teach cursive (joined up) handwriting in schools, instead teaching children to type using a keyboard (Hosie, 2017). In England, however, the teaching of handwriting is still an important part of the early years and primary curriculum. Children are taught handwriting skills from the age of 6 years old (DfE, 2013). The programmes of study for handwriting involve children learning how to form a fluent writing style through being taught effective ways to reproduce conventional letters. This is distinct from using writing as a means of producing meaningful communication with a clear audience in mind, although the ability to write fluently is an aspect of meaningful communication in terms of the reader's ability to read the writing.

## Considerations for teaching handwriting

Learning to form the individual letters of the alphabet and produce legible handwriting at a reasonable speed involves complex perceptuo-motor skills ☞. The National Curriculum (DfE) states that pupils' overall writing ability during Year 1 will generally develop at a slower pace than their reading. This is because they need to learn to encode the sounds they hear in words (spelling skills), develop the physical skill needed for handwriting and learn how to organise their ideas in writing. Not all pupils will have the spelling and hand-writing skills they need to write down everything that they can compose out loud. Table 19.1 summarises the statutory requirements for handwriting from Year 1 to Year 6.

*Table 19.1* Handwriting in England's national curriculum

| Year group | Statutory requirements for handwriting |
|---|---|
| **1 (Lower Key Stage 1)** | **Pupils should be taught to:**<br>• sit correctly at a table, holding a pencil comfortably and correctly<br>• begin to form lower-case letters in the correct direction, starting and finishing in the right place<br>• form capital letters<br>• form digits 0–9<br>• understand which letters belong to which handwriting 'families' (i.e. letters that are formed in similar ways) and to practise these. |
| **2 (Upper Key Stage 1)** | **Pupils should be taught to:**<br>• form lower-case letters of the correct size relative to one another<br>• start using some of the diagonal and horizontal strokes needed to join letters and understand which letters, when adjacent to one another, are best left unjoined<br>• write capital letters and digits of the correct size, orientation and relationship to one another and to lower case letters<br>• use spacing between words that reflects the size of the letters. |
| **3 and 4 (Lower Key Stage 2)** | **Pupils should be taught to:**<br>• use the diagonal and horizontal strokes that are needed to join letters and understand which letters, when adjacent to one another, are best left unjoined<br>• increase the legibility, consistency and quality of their handwriting [for example, by ensuring that the downstrokes of letters are parallel and equidistant; that lines of writing are spaced sufficiently so that the ascenders and descenders of letters do not touch]. |
| **5 and 6 (Upper Key Stage 2)** | **Pupils should be taught to:**<br>• write legibly, fluently and with increasing speed by:<br>• choosing which shape of a letter to use when given choices and deciding whether or not to join specific letters<br>• choosing the writing implement that is best suited for a task. |

Legibility will have different levels according to the purpose of the writing. Sassoon (1990) points out that children cannot be expected to produce their neatest handwriting all the time, so she advocates different levels of handwriting. While schools should have a handwriting policy in place which is shared with parents and which is adhered to when modelling writing and marking children's work, it is also important to acknowledge that handwriting needs to be fit for purpose. A calligraphic ☞ standard for special occasions, for example, might require a careful, deliberate approach which will be more time-consuming than a legible everyday hand. There will also be times when pupils are drafting text or making notes that they alone will read where a lower standard of legibility is appropriate. The importance of fluency is particularly important in tests and exams when time pressures are present.

A comfortable style requires an appropriate grip, appropriate seating and good posture. Posture and working space are important elements of handwriting. For right-handed pupils, it is suggested that the paper is tilted to about 45 degrees anti-clockwise. It is important that the writing implement is not gripped too hard, as this can also lead to muscle tension in the shoulder and pain in the wrist and hand. It is important that you are aware of all the left-handed children in your class from the beginning of the year. As far as handwriting is concerned, left-handed children have particular needs. Their writing moves inwardly – that is, in towards their bodies – thus tending to make it difficult for them to read what they have just written. Left-handers should be encouraged to turn their paper *clockwise* (to about 45 degrees), not to hold their pen or pencil too near to the actual point and to sit on a chair that is high enough to allow them good visibility. Left-handers can be helped by ensuring that they sit to the left of a right-hander so that their elbows are not competing for space. Whatever your own writing orientation, you will need to ensure that these issues are made explicit and discussed when you model handwriting.

## Basic handwriting concepts

Venerable Will played jazz sax 'til 3 o'clock in the morning before he quit.

The five boxing wizards jump quickly.

(Jarman, 1989: 101)

These examples, which contain all the letters of the alphabet, serve to remind us that quick brown foxes are not the only subjects for such sentences. They are, of course, quite a handy way to cover the formation of all the letters of the alphabet in one meaningful sentence.

Sassoon (1990) puts forward the concepts behind our writing system. Direction, movement and height are all crucial: left to right and top to bottom; the fact that letters have prescribed flowing movements with specific starting and exit points; the necessity to ensure that letters have particular height differences. In addition, the variance between upper and lower case must be recognised and correct spacing consistently applied. She also stresses the importance of taking particular care when teaching certain letters that have mirror images of each other, such as *b-d, m-w, n-u* and *p-q* to avoid confusion for young learners. She suggests that speed – but not too much speed – is important, as this can lead to fluency and greater efficiency. Modern classroom situations do require pupils to think, work and write reasonably quickly.

There are a small number of technical terms that are useful when talking about handwriting. *Ascenders* are the vertical lines that rise above the mid-line (or x-line) on letters like 'd'; *descenders* are the vertical lines that hang below the baseline on letters like 'g'. Letters have an *entry stroke* where you start the letter and an *exit stroke*. The letter 't' is interesting in that its horizontal line is called a *crossbar* and the height of the letter should only be three-quarters. This means that the top of the letter finishes between the mid-line and the ascender line. There are four important horizontal lines: the *descender line*, the *baseline*, the *mid-line* and the *ascender line*. For adults, only the baseline is visible; for children, other lines have to be used carefully because there is a danger that they can measure the length of a stroke by the distance to the line, not by understanding the differences in letter size. Children need to understand these concepts if they are to have legible and fluent handwriting.

Jarman (1989), like Sassoon, suggests that letters can be taught in families that are related by their patterns of movement. There are slight differences between their approaches, but both underline the importance of the idea of letter families. Jarman links specific patterns – which he regards as beneficial – with the families of letters. He also suggests that there are two kinds of join or 'ligature': horizontal joins and vertical joins. He points out that it is sensible to leave some letters unjoined – such as b g j p q and y – when joined to most vowels. Sassoon links together the following letters:

1    i l t u y j
2    r n m h b p k
3    c a d g q o e
4    s f
5    v w x z

Whilst we are outlining concepts and approaches here, it is important to be cognisant of the particular handwriting scheme that your school follows.

## Handwriting and writing

Peters (1985) discussed perceptuo–motor ability and argued that carefulness in handwriting goes hand in hand with swift handwriting, which in turn influences spelling ability. It is thought that for younger children in particular the kinaesthetic movements that are part of forming letters help with visual memory of letter shapes. Children who can fluently write letter-strings such as -ing, -able, -est, -tion and -ous are more likely to remember how to spell words containing these strings. It was also Peters' view that the teaching of 'joined up' (or 'cursive') writing, should begin long before the junior school, that is to say at Key Stage 1 rather than 2. The main advantages of this are:

- The concept of 'a word' (and the spaces between words) is acquired from the outset as distinct from 'a letter' (and the spaces between letters).
- Correct letter formation with appropriate exit strokes is learned from the beginning.
- The movement of joined-up writing assists successful spelling and is quicker than printing.
- Children do not have to cope with changing from one to the other at 7 or 8 years of age.

Sassoon (1990) suggests that with sufficient preparation in the 'movement' of letters and the different exit strokes required, pupils can begin to join up the simple letters by the end of the reception year or Year 1 at least. In England, joined up handwriting should be the norm from lower Key Stage 2 onward.

More recent research has begun to show the importance of automaticity in handwriting in order to benefit composing processes. Medwell and Wray (2007) argue that handwriting has been taught in mainstream schooling based on the link between correct spelling and the use of fluent, joined-up handwriting, but with little emphasis placed on the connection between automaticity in handwriting and its impact on compositional skills. In Scheuer et al.'s (2006) study, children aged 9 gave reflective accounts of learning to write. Their views represented writing as a developmental process which moves from very early attempts at writing to producing and understanding conventional writing over a relatively extended period of time.

Such research suggests that when teaching handwriting, we should focus on developing children's fine motor skills so that they are first and foremost able to hold a writing implement comfortably, and hand-eye coordination to support the 'uniform' formation of letters. Development of these early skills will support later automaticity and fluency in handwriting, which will impact on their future success as writers. Some activities to develop early handwriting skills are:

- **Hand control:** jigsaws, gluing, cutting, threading, painting, using a variety of implements with which to write and draw.
- **Learning to form letters (large-scale):** 'skywriting' the letters in the air, using a large brush and bucket of water to 'paint' a wall or shapes on the ground, writing with a stick or finger in a sand tray or in shaving foam, writing letters with chalk on the playground, writing with big pens at an easel.

- **Learning to write letters on paper:** use blank paper, sometimes draw a line on the paper, encourage a comfortable grip. As the child becomes more competent, they can think about spacing letters and words, using consistent letter sizes.

As children progress through primary school, handwriting can be taught through whole class sessions, guided writing sessions and one-to-one support. It can be taught on a weekly basis and according to need. One way to keep children engaged and encourage them to persist with handwriting practice is for them to know that there are a considerable number of professional writers, particularly writers of narrative and poetry, who find handwriting a better way to express themselves than the keyboard. Michael Morpurgo, a hugely popular children's author, is one such writer. Perhaps this may be a source of inspiration for them as they approach their own writing.

## Diagnosis of problems

Pupils entering the later phase of primary schooling sometimes arrive with handwriting problems. These might have been caused by indifferent teaching or they might, according to Sassoon (1990), be symptomatic of pupils' particular condition or set of circumstances. They might be considered 'dyspraxic', that is, prone to particular motor coordination problems. For example, they might not be able to catch a large ball, or cope easily with gymnastics or play games using the simplest of equipment. For these children, neat, regulation handwriting might be impossible. Teachers need to acknowledge this, not continually chastise them, and help them to develop handwriting that is reasonably swift and legible.

Sassoon argues that handwriting can be regarded as a diagnostic tool in itself, indicating certain problems that pupils have and teachers need to address. Hesitancy and a lack of confidence in spelling will interrupt the flow of handwriting. If this is accompanied by frequent attempts at correcting mistakes, the result is 'messy' handwriting that many teachers find unacceptable. This may well bring to their attention that these children need help in spelling as well as handwriting. Occasionally, children have psychological problems to cope with too, such as bullying, bereavement, divorce and so on, which, in addition to a range of behavioural changes, can become evident in their handwriting. Often the writing becomes very variable and sometimes illegible, when previously it was conventional.

Some children with poor handwriting skills are found to have weak auditory, perceptual and memory skills and are not able to remember sequences of movements or sequences of verbal instructions. Difficulties with handwriting might indicate poor eyesight or a squint and perhaps the need for spectacles for an individual child. Children suffering from fatigue due to the after-effects of an illness, a physical disability or just insufficient sleep at night might also reflect this in their handwriting. Finally, poor handwriting might bring the teacher's attention to the poor posture of some children when writing or even generally. Thus alerted and made aware of any of these cognitive, psychological or physical problems, teachers will be able to take steps, possibly with support from other professionals, to remedy them and aim to improve not just children's handwriting, but other important features of their pupils' learning and development in school.

It is clear that children experiencing handwriting difficulties need to be identified and supported, as these difficulties are likely to impact upon their ability to compose written language. Medwell and Wray (2007) suggest that there is evidence that intervention to teach handwriting to these children can improve not only their handwriting, but also their writing more generally.

### Practice points

- Familiarise yourself with the school's handwriting policies and schemes and ensure that you and other adults working with you model this when working with children.
- Handwriting should be taught on a weekly basis through whole group and small group sessions and supplemented by individual support during the writing process.
- A balance needs to be found between emphasis on standard letter formation and encouraging legibility and fluency.

### Glossary

**Calligraphic** – description of a particularly skilled way of writing. Handwritten italic script is often seen as calligraphy.

**Perceptuo–motor skills** – skills that rely on use of the senses, the brain and learned physical movement.

### References

Department for Education (DfE). (2013). *The National Curriculum in England: Framework Document. September 2014*. London: Department for Education.

Hosie, E. (2017). 'The uncertain future of handwriting: Future'. *BBC Global News Ltd*. Retrieved from www.bbc.com/future/story/20171108-the-uncertain-future-of-handwriting

Jarman, C. (1989). *The Development of Handwriting Skills: A Resource Book for Teachers*. Oxford: Basil Blackwell.

Medwell, J. and Wray, D. (2007). 'Handwriting: What do we know and what do we need to know?', *Literacy*, 41(1): 10–15.

Peters, M. L. (1985). *Spelling: Caught or Taught: A New Look*. London: Routledge and Kegan Paul.

Sassoon, R. (1990). *Handwriting: A New Perspective*. Leckhampton: Stanley Thornes.

Scheuer, N., de la Cruz., M., Pozo, J. and Neira, S. (2006). 'Children's autobiographies of learning to write', *British Journal of Educational Psychology*, 76: 709–725.

## Annotated bibliography

https://www.theguardian.com/teacher-network/2018/mar/07/does-being-able-to-write-by-hand-still-matter-in-2018
Useful exploration of some modern concerns about handwriting. Arose out of a Guardian round table discussion.
**L1** ★

Medwell, J. and Wray, D. (2010). 'Handwriting and writing', in D. Wyse, R. Andrews and J. Hoffman (eds.), *The Routledge International Handbook of English, Language and Literacy Teaching*. London: Routledge.
A timely and very useful overview of research in an area that receives limited attention.
**L3** ★★★

Sassoon, R. (1999). *Handwriting of the Twentieth Century*. London: Routledge.
Rosemary Sassoon made an important contribution to our understanding of the practice of handwriting. This book gives a historical perspective on the teaching of handwriting. Her book *Handwriting: The Way to Teach It* (Sassoon, 1990b; revised edition published in 2003 by Sage) focuses more on classroom practice.
**L1** ★

# Part V

# General issues

# Chapter 20

# Assessing writing, reading and oral language

Assessment is one of the most challenging aspects of early years and primary education. It is also a fundamental part of any teacher's role. In this chapter, we explore principles and processes that underpin how to assess learning in English, language and literacy across the modes of spoken language, reading and writing. We then discuss practices in summative and statutory assessment.

Reliable and accurate assessment of children's learning lies at the core of every teacher's daily practice. A teacher's interactions with a pupil to support their learning is based on moment by moment assessments of the pupil's learning. These assessments are *formative* assessments ☞, and the process has been called *assessment for learning* (AfL). This is not the same as testing or summative assessment *of* learning. When the teacher plans the curriculum for their class, the selection of activities is based on their assessment of all the pupils' learning on an ongoing basis. This also requires the teacher to plan for all pupils at different stages in their learning, including those with particular needs at any given time. The terms 'assessment for learning' and 'formative assessment' are often used synonymously. Essentially, this is an approach to teaching in which teachers elicit evidence about children's learning and then interpret this to inform future teaching and learning. It requires teachers to 'start from where the learner is' (Harrison and Howard, 2009). Where used most effectively, it requires active engagement from learners about knowing where they are, where they need to go and how best to achieve this (Assessment Reform Group, 2002).

## Summative and statutory assessment in England

Summative assessment ☞ is typified by in-school tasks such as end-of-unit reading and writing assessments and spelling tests to provide snapshots of pupils' attainment. The most challenging aspect of assessment concerns statutory assessment, or what are commonly called SATs. There is nothing intrinsically wrong with assessments that apply to a whole country, for example the assessment of pupils at secondary school level when they are age 15 to 16 has been taking place in England and other countries for decades. But the main problems arise when the purposes for assessment are confounded. When the SATs for primary pupils were first introduced from 1988 onwards, an influential report made it clear that the purposes for assessment should not be confused (TGAT Report citation). Yet consistently since their inception the use of the SATs has confused the purposes of a) assessment of pupils' learning with b) holding teachers and schools to account by linking tests outcomes to things like inspection outcomes, school placement in league tables and increasingly even linked to teachers' pay through performance management. When assessment for learning is confounded with assessment for accountability, significant problems arise.

As you read in the first chapter of this book, statutory assessment in England has been a controversial issue since the Education Reform Act of 1988. We waited to write this chapter until last because another wave of changes had been proposed by government and were under consultation. As we wrote this chapter, some of these changes were clearer but others were yet to be decided, as Table 20.1 illustrates.

*Table 20.1* Summary of changes to statutory assessment in England

| Academic year | Change |
| --- | --- |
| 2017 to 2018 | New statutory teacher assessment frameworks for writing implemented at KS1 and KS2<br>KS1 Grammar Punctuation and Spelling Test to be non-statutory in this and future years. |
| 2018 to 2019 | Revised statutory teacher assessment frameworks for reading mad mathematics and science.<br>Requirement for statutory teacher assessment in reading and mathematics at end of KS2 removed. |
| 2019 to 2020 | Statutory requirement to administer multiplication tables check to pupils at end of Year 4 introduced. |
| 2020 to 2021 | Possible changes to the Early Learning Goals implemented.<br>New reception baseline assessment implemented. |
| 2022 to 2023 | Requirement to administer statutory assessments at the end of Key Stage 1 removed. |

As you will see, the statutory assessment of writing has proven to be contentious, as has the assessment of reading through the phonics screening check. Another area of concern has been assessment of the youngest children in formal education. The original Early Years Foundation Stage profile and its associated Early Learning Goals were demanding of educators' time but did have a number of positive elements: they required assessment of all areas of the curriculum, and the assessments were built on educators' observations of children's learning. Although some may question whether teachers' assessment judgements are as reliable as an externally marked test. there was clear evidence, from a research study funded by the Department for Education, that language assessments made in relation to the early years goals did correlate robustly with children's later development (Snowling *et al.*, 2011). In spite of this, government pushed ahead with the idea of a more instrumental and less holistic *baseline assessment* to be used as a way to measure progress from a child's entry to school up to the end of their primary schooling. One of the many criticisms of this approach to the assessment of the youngest children was summed up in the word *datafication*, suggesting that children and teachers were being seen simply as data rather than as people with complex learning trajectories (Roberts-Holmes, 2014).

## Assessing writing

In addition to the emphasis in the National Curriculum programmes of study, the national statutory tests for 11-year-old pupils in England included for the first time in 2011 a separate spelling, punctuation and grammar test where formal grammar was emphasised. In addition, the requirements for teacher assessment of writing included a strong emphasis on grammar as part of the assessment criteria. In 2016, these emphases were still in place. For example, the national statutory test for Spelling, Punctuation and Grammar included a strong emphasis on formal grammar, including questions that required knowledge of grammatical terminology (for example, '**27.** Underline the **subordinate clause** in each sentence below' [UK Government, 2016: 17, emphasis in original]). All questions in the paper attracted one mark each. Although the 2016 criteria for statutory teacher assessment of writing, produced by pupils in lessons, included aspects such as 'creating atmosphere' in their writing, there was a strong emphasis on usage according to areas of formal grammar, such as 'passive and modal verbs' and 'adverbs, preposition phrases and expanded noun phrases', etc. (Standards and Testing Agency, 2015).

The politics and policies that led to the emphasis on formal grammar in England's national curriculum implemented from 2014 onwards began with a government white paper in 2010 that included the commitment to 'Review and reform the National Curriculum so that it becomes a benchmark outlining the knowledge and concepts pupils should be expected to master to take their place as educated members of society' (Department for Education, 2010: 41).

The link between statutory assessment, the curriculum and school account-ability was also made clear: 'The National Curriculum will continue to inform the design and content of assessment at the end of key stage two, which will apply to every child and which will provide a guide to the performance of primary schools' (op. cit.: 42). After publication of the white paper, the gov-ernment commissioned a review of assessment in England led by Lord Bew. Bew's final report noted that

> there are some elements of writing – spelling, grammar, punctuation, vocabulary – where there are clear 'right' and 'wrong' answers, which lend themselves to externally-marked testing . . . Internationally a number of jurisdictions conduct externally-marked tests of spelling, punctuation and grammar . . . These are essential skills and **we recommend that externally-marked tests of spelling, punctuation, grammar and vocabulary should be developed**.
>
> (Bew, 2011: 60; bold font in original)

In 2017, the House of Commons Education Select Committee carried out an enquiry about assessment in primary schools. Wyse's oral and written evidence to the enquiry was quoted in the final report:

> 34. However, moving away from the 'secure fit' model will not remove the focus on technical aspects of writing, something that was raised in evidence to our inquiry. Professor Dominic Wyse, UCL Institute of Edu-cation, wrote:
>
> > The assessment of writing in statutory tests in England in 2016, and for some years previously, suffers from two major flaws: 1. the undue separation of the composition of writing from the transcription elements of grammar, 2. An undue emphasis on decontextualised grammatical knowledge. Both of these flawed features of assessment are contrary to research evidence. . . .[36]
>
> 74. Professor Dominic Wyse argued that the focus should be on improv-ing teacher assessment, and not on the introduction of a formal test.[82] The Minister told us he was 'open minded' as to whether the assessment was a formal test or an observational model. The consultation proposal states 'this assessment would need to be appropriately teacher mediated, given the age of the children.' However, it does not give detail about the nature of the test.
>
> (House of Commons Education Committee, 2017)

As a result of the enquiry, and of a public consultation, the government did make some welcome changes to assessment of writing (as can be seen in Table 20.1); however, it remained to be seen whether this would be enough to stop the distortion of the primary curriculum that had been the case prior to 2018.

Assessing pupils' writing involves consideration of the compositional and transcriptional elements of the writing process. Some of the most important assessment of children's writing is carried out orally during the process of writing when the teacher works alongside a child or group of children to elicit their understanding and give feedback. This kind of feedback is likely to be most effective if the teacher has a clear idea of how children's writing should develop (→ Chapter 14) and a clear idea about each child as an individual, taking into account motivation to write and understanding of genre. Choices about when and how to give oral feedback are key in supporting a child to make progress as a writer.

It is helpful to give oral feedback to children in the immediacy of the writing moment. Information should be given to the child about the successes of their writing and the next steps to be taken, derived from the teacher's knowledge of the child's current writing dispositions and ability. At its best, this feedback is personalised and powerful in meeting the immediate needs of the child. Teachers give oral feedback to children using a variety of strategies, sometimes offering specific prompts, but also using questions and engaging in dialogue about their work. At all times, key principles are that children's efforts should be valued and also that a level of challenge is offered to provide for new learning. Getting children to talk about their writing in response to this feedback will increase their achievement and motivation.

The choice of pupils' writing for detailed assessment and feedback depends on the requirements of the school/setting as well as the nature of the writing and the teacher's approach to assessment. One of the ways to select can be on the basis of range: if on the last occasion you marked a story, a piece of non-fiction might be appropriate the next time; if on the previous occasion you had commented on the presentation of a final draft, you might want to comment on an early draft. It is worth remembering that appropriate written comments can be collected together to form the basis of an assessment profile for a child, and these could and should include assessment of writing that builds towards a final piece of writing.

One of the challenging aspects of marking children's work relates to the range of choices that are available for your response. One possible way to think about these choices can be to have a system for responding:

1    A specific positive comment about the writing.
2    A specific point about improving something that is individual to the child's writing.
3    A specific point about improving something that relates to a more general target for writing.

Using marking techniques such as ticks, stars, smiley faces, scores out of 10, etc., need to be thought about carefully. These give no information to the child about what in particular they did well or specifically how they could improve next time. However, for young children the extrinsic rewards such as stickers, smiley faces or merit awards can be an important way of rewarding hard work. As a teacher you need to be clear about the pros and cons of the different strategies. A tick at the end of a piece of work should perhaps be used only to indicate that you have checked that the child completed the work and that you have read the piece. A reward such as a sticker could be used to indicate that the child has worked particularly hard and ideally should accompany more detailed feedback, which could be oral and/or written. Children should also be given opportunities to assess their writing by themselves, for example, in relation to the oral and written feedback they have been given, or the success criteria. They should also be given opportunities to work collaboratively with a peer, evaluating each other's writing and making constructive suggestions for improvements. However, there have been some welcome moves to recognise that written marking by teachers of every single piece of work done by all children in the class is not the best way to ensure optimal learning. In our view, written marking should be one of a range of feedback strategies.

## Assessing reading

As we showed in the first chapter of the book, another controversial change to statutory assessment was the introduction of the phonics screening check for all 6-year-old children in England. As with other forms of high stakes assessment, one of the most serious concerns is the potential distortion of the curriculum. Although learning the alphabetic code is one important aspect of the teaching of reading, there are many others, such as motivating children to read for pleasure, developing their comprehension, developing critical appreciation and supporting a wide range of reading to learn. If the teaching of reading is dominated by phonics teaching to the exclusion of other aspects, children will not learn at optimal rates of progress.

In 2018 at Key Stage 1, there was an 'English reading test' consisting of two papers. The first test was designed to last for about 30 minutes and the second for about 40 minutes. Both required responding to a range of questions about texts presented as part of the tests. At key stage two there were three tests: (1) an English grammar, punctuation and spelling – questions; (2) English grammar, punctuation and spelling – spelling test; (3) English reading.

Classroom assessment of children's reading is built on knowledge of children's likely reading development (→ Chapter 8) and the teacher's ability to observe carefully and interact appropriately. Children's reading can be assessed during one-to-one interaction, and through guided reading or small group reading. Shared reading in whole class situations also provides opportunities for the teacher to assess children's progress (see examples of these teaching methods in Chapter 13).

There are a variety of strategies that can be used to assess individual readers; we consider reading interviews, observations of children reading, reading with children on a one-to-one basis, observing individuals within a group and using running records and miscue analysis ☞.

### Reading interviews

Through conducting reading interviews, the teacher can find out about the child's perceptions of themselves as a reader, their reading habits and interests and general attitude towards reading. When conducting an interview, ensure your pupil is relaxed and comfortable and that you have space and time where you can both concentrate. This will help you to find out about their reading in general, for example, where and when they read; their likes and dislikes; reading at home and school; experience of a range of genres, including comics, media and online texts. The questions to ask might include: what do you like reading? Which books do you like to read/look at? What is your favourite book? What do you read at home? A helpful survey can be found at https://researchrichpedagogies.org/_downloads/RfP_Childrens_Reading_Survey.pdf.

### Observations of a child reading

Observations are an essential tool for gathering information about a child's current reading and for planning next steps. An individual child can be observed reading, or being read to individually; in a group; with the class (for example, during shared and guided reading time or when the teacher is reading to the class); reading texts in all curriculum areas; reading work that they have written.

Questions to ask when observing children's reading behaviour include:

- Do they actively seek out books to look at or to be read to them?
- What texts do they choose?
- Do they read intently or flick through texts to avoid engagement?
- Do they spend time in the reading area?
- Do they try to retell a favourite story?
- What strategies do they use to interpret the text?

It is important not to forget early reading behaviours, such as looking at a book the right way round, turning the pages one at a time and retelling a favourite story whilst 'pretending' to read the text.

In addition, consider the following points to help focus your observations.

- Is the child reading for enjoyment/information?
- When is the child engaged, confident?
- Can the child read independently or are they dependent on support? Do they prefer reading with an adult?

- Can the child talk with interest and understanding about what has been read?
- In a small group context, does the child take a lead or wait for others? Are they engaged with the text? Do they feel comfortable in a group situation? How do they engage with the group?
- In a whole class context, how does the individual child respond when they are being read to?

### Reading with a child one-to-one (→ Chapter 10)

Although it is hard to find time in a busy classroom schedule to read with children on a one-to-one basis, this is still an important strategy in assessing where a child is in their reading and planning next steps. Having selected a book that you think will engage the child's interest, or one that the child has chosen, consider the following:

- Note positive and negative strategies, specific difficulties, understanding of the text, pleasure, etc.
- Discuss the book with the child, noting interest, understanding, awareness of bibliographical features, genre, etc.
- If appropriate, focus on semantic, syntactic and graphophonic cueing strategies. Note how these are used and where they could be developed, explicitly or implicitly, throughout the child's day. When planning your work within English and other curriculum areas, find opportunities to support and develop these strategies.

### Running records and miscue analysis

Running records were a method of assessment devised by Marie Clay as part of her highly successful Reading Recovery programme (see the International Literacy Centre at the IOE for the most up to date information on Reading Recovery: www.ucl.ac.uk/international-literacy). They can also be used as a method to assess the reading of children in your class who are not decoding with fluency. Conducting a running record will enable you to discover with individual children:

- which reading strategies are established;
- which reading strategies need to be practised and consolidated;
- which reading strategies need demonstration and development.

A book should be selected that a child can decode at approximately 95 per cent accuracy. As the child reads aloud from the text, the teacher records the reading on a photocopy of the text using a specified coding system, as devised by Clay (1979: 18):

- Tick words read correctly.
- Write incorrect guesses above the word in the text.

- If there is more than one guess, record each of them even if only one letter.
- Write 'SC' (Self Correction) if child corrects previous error.
- If the child makes no response put a dash – above the word.
- If the child has to be told the word, write 'T' above it.
- If the child asks for help, write 'A'. The teacher says 'you try it' first and then enters 'T' for told if the child still can't work out the word.

Here is a short example of how this works in practice. Helen, aged 7, read the following text, 'There was soup for dinner. Chicken soup for all the children' as, 'There is soup for dinner. Children . . . Chicken soup for all the children'. The teacher duly coded this as ✓ 'is' ✓ ✓ ✓ SC ✓ ✓ ✓ ✓ ✓, indicating that the child read the piece accurately apart from substituting the word 'is' for 'was'. Also, when Helen substituted 'children' for 'chicken', she immediately corrected herself when she realised that 'children soup' was very unlikely (except perhaps in a Roald Dahl story!).

Once a child has finished reading for the running record, the teacher might discuss with them some of the miscues made, particularly the substitutions ☞, to determine which cueing strategies the child is currently employing and how they might be taken forward in their reading development.

The great advantage of a running record is that it is immediate, in the sense that special materials are unnecessary (although there are published schemes available with ready-made materials). Provided you remember the codings, you can complete one or more records during one-to-one reading, or at any other suitable times of the day. There are limitations, however. Instant coding by the teacher is likely to be inaccurate from time to time, because some utterances by children need greater reflection or at least need to be listened to more than once or twice. The important message to take forward is how you analyse children's miscues to inform future planning.

### Reading journals

Observations of children's reading completed during one-to-one reading sessions, or when noteworthy things occur at other times, are sometimes collected in what are called reading journals (or diaries or logs). Teachers usually write brief observational jottings, appropriately dated, noting significant features of the child's reading as they occur. Over time, patterns of development and areas for further support become evident.

Reading journals can be sent home to help parents support their children's reading. If parents are able to write comments, then this log can offer an even richer picture of the child's reading development. Children too can contribute or keep their own reading journey. Periodically teachers and other adults may participate in a written dialogue with the child about their reading, the books they like to read in and out of school and even problems they are having with their reading. Some of the comments by children taken from actual diaries read like this: 'I enjoyed this book because it was in English as well as Gujerati'

[sic]; 'I found the story quite difficult because there were some very long words and lots of names that were a bit hard'; and 'I wish I could read like you, Miss. I enjoy it when you read to us because it makes the stories even better'.

### Small group work including guided work

The aim of group or guided reading sessions is to encourage and extend independent reading skills. During these sessions, the starting point is to consider which reading strategies are established for the pupils, based on your prior assessment. These could include both decoding and comprehension strategies. Throughout the session you can make observations and assessments about reading strategies that need to be consolidated or further practised and those that need more demonstration and development. This will enable you to plan next steps for the group. The constitution of the group should be fluid as you respond to the needs of individual learners. There are a variety of methods to record progress throughout a session, but a helpful way to structure this is to evidence children's responses to:

- reading strategies;
- the text;
- how they reflect on what they have learnt.

This will build in to planning for follow up independent activities, which will enable you to assess how pupils have responded to learning in the group session and to then plan objectives for future small group sessions.

## Assessing spoken language

Ironically, reductions in the emphasis on spoken language in the National Curriculum 2014 occurred at the same time that research continued to show the importance of explicit support for the development of talk (e.g. see Mercer and Littleton, 2013). Assessing spoken language serves two primary functions. Firstly, it allows the teacher to make judgements about the development of talk itself. Secondly, it affords the teacher an opportunity to assess other forms of understanding which are communicated through talk. As Britton (1970: 164) stated, 'reading and writing float on a sea of talk'. Different forms of assessment are needed at different stages of pupils' development in order to assess progression in features of talk and provide children with vital oracy skills.

According to Grugeon *et al.* (2005: 137), in the early years, the assessment of talk usually takes the form of observable features, such as:

- Does the child initiate and carry on conversations?
- Does the child listen carefully?
- Can the child's talk be easily understood?

- Does the child describe experiences?
- Does the child give instructions?
- Does the child follow verbal instructions?
- Does the child ask questions?
- Can the child contribute to a working group?
- Does the child 'think aloud'?
- Does the child modify talk for different audiences?

The Early Years Foundation Stage in England recommends that children are likely to develop more complex language when they talk about what they are most interested in. This has implications for planning opportunities for talk and highlights the importance of the teacher knowing what children's interests are. At later stages, other features of talk will take on greater significance in the assessment process. Teachers will find that they begin to look for evidence where the child is seen to be hypothesising, imagining, directing, exploring, practising, recalling, developing critical responses, explaining and sustaining talk. It is important to understand that a good listener is essential for developing good talkers, so the teacher's role as a sensitive partner in sustained shared thinking about children's activities, ideas, approaches to problem-solving and questions is key.

Talk is very much dependent on the context of its occurrence. For example, when early years teachers observe children in play contexts, they often see a very different language user. Some children, freed of the pressures of performance in front of the teacher, begin to demonstrate skills as language users and a preparedness to explore and experiment which teachers would otherwise never witness. These kinds of contexts are equally applicable among older children who perceive themselves as poor language users in the classroom, yet seem to be perfectly articulate and imaginative when outside at play. The point here is that teachers who are concerned about a child's development in this area should look at that same child at play for further evidence of language use.

For pupils in the primary phase, the features outlined above are all still very relevant. The classroom practices for talk, reading and writing outlined in previous chapters all afford opportunities for ongoing assessment of spoken language. The crucial element of a teacher's pedagogy is to ensure that there are planned opportunities to observe and assess individual pupils and also group and whole class interactions on a daily basis. Methods of recording these (such as observation proformas, annotated planning and evaluations and audio/visual recording) will help support how you then analyse and inform future planning. An aspect to consider is how you record and analyse responses to particular talk activities, for example in:

- talk partners;
- group work;
- drama;
- role-play;

- debates;
- responses in whole class discussions;
- formal presentations such as class assemblies/performances;
- self-evaluation.

All of these opportunities can be provided across a range of contexts and particular genres of reading and writing.

Some practical strategies to use for recording, monitoring and tracking your formative assessment of spoken language and oracy in curriculum planning could include the following:

- *Focusing on two or three children each week* (to ensure systematic coverage of the whole class).
- *Using objectives for whole class monitoring* (developing whole class lists of which children meet specific teaching objectives).
- *Integrating speaking and listening assessment with other records* (possibly building a page-per-pupil record which incorporates talk).
- *Termly checks* (looking for patterns, omissions, etc.).
- *Annual review* (to provide feedback for children and to enable target-setting and future planning).

(Department for Education and Skills (DfES) and<br>Qualifications and Curriculum Authority (QCA), 2003: 31)

Finally, a main concern for assessment is to consider how well the talk suits the kind of event in which children are participating, for example retelling a story or participating in a debate. Criteria for assessment are likely to be different depending on whether children are talking in a group, making a presentation to the class, engaged in a drama-related activity, discussing ideas in a cross-curricular focus, etc. The relative formality/informality of the context and the intended audience needs to be taken into account. There are many aspects of evaluating children's talk where great sensitivity is needed. The ways people talk can be closely related to their identities, and teachers need to be careful about making evaluations of some aspects of a child's way of speaking such as their accent (→ Chapter 5). As we have previously argued, teachers need to embrace the diversity of language within their classroom.

### Practice points

- Maintain a balance of assessment techniques to ensure that the curriculum does not neglect important elements.
- Reflect carefully on the processes of writing, not just the products.
- Use your assessments to improve future lessons.

## Glossary

**Formative assessment** – activity that provides students with developmental feedback on their progress during the learning programme and informs the design of their next steps in learning.

**Miscue analysis** – analysis of children's guesses at unknown words when reading

**Substitution** – an alternative word that a child suggests when trying to read a word that they are unsure of.

**Summative assessment** – typically end-of-learning assessment tasks, such as examinations and tests, to measure and record the levels of learning achieved.

## References

Assessment Reform Group. (2002). www.clpe.org.uk/library-and-resources/reading-and-writing-scales

Bew, P. (2011). *Independent Review of Key Stage 2 Testing, Assessment and Accountability: Final Report.* London: Department of Education.

Britton, J. (1970). *Language and Learning.* London: Allen Lane.

Clay, M. (1979). *The Early Detection of Reading Difficulties*, 3rd edn. Auckland: New Zealand: Heinemann Education.

Department for Education. (2010). *The Importance of Teaching: The Schools White Paper 2010.* Norwich: The Stationery Office.

Department for Education and Skills (DfES) and Qualifications and Curriculum Authority (QCA). (2003). *Speaking, Listening, Learning: Working with Children in Key Stages 1 and 2: Handbook.* London: DfES Publications.

Grugeon, E., Hubbard, L., Smith, C. and Dawes, L. (2005). *Teaching Speaking and Listening in the Primary School*, 3rd edn. London: David Fulton.

Harrison, C. and Howard, S. (2009). *Inside the Primary Black Box.* London: GL Assessment.

House of Commons Education Committee. (2017). *Primary Assessment: Eleventh Report of Session 2016–17: Report, Together with Formal Minutes Relating to the Report.* London: House of Commons.

Mercer, N. and Littleton, K. (2013). *Interthinking: Putting Talk to Work.* Abingdon: Routledge.

Roberts-Holmes, G. (2014). 'The "datafication" of early years pedagogy: "If the teaching is good, the data should be good and if there's bad teaching, there is bad data"', *Journal of Education Policy*, 30(3): 302–315.

Snowling, M., Hulme, C., Bailey, A., Stothard, S. and Lindsay, G. (2011). *Better Communication Research Programme: Language and Literacy Attainment of Pupils During Early Years and through KS2: Does Teacher Assessment at Five Provide a Valid Measure of Children's Current and Future Educational Attainments?* London: Department for Education.

Standards and Testing Agency. (2015). *2016 National Curriculum Assessments: Interim Teacher Assessment Frameworks at the End of Key Stage 2*. London: Standards and Testing Agency.

UK Government. (2016). *2016 Key Stage 2 English Grammar, Punctuation and Spelling: Paper 1: Questions*. London: UK Government.

### Annotated bibliography

Bradbury, A. and Roberts-Holmes, G. (2017). *The Datafication of Primary and Early Years Education: Playing with Numbers*. Abingdon: Routledge.
For those interested how assessment data has been used by government, this book provides a critical analysis of policy in England.
L3 ★★★

Wyse, D., Hayward, L. and Pandya, J. (eds.). (2015). *The SAGE Handbook of Curriculum, Pedagogy and Assessment*. London: Sage.
One section of this two-volume set has a series of chapters about assessment around the world. The editors' introduction argues for the importance of understanding curriculum pedagogy and assessment in combination.
L3 ★★★

www.clpe.org.uk/library-and-resources/reading-and-writing-scales
The Centre for Language in Primary Education has a long history of excellent work supporting teaching and assessment.
L1 ★

# Chapter 21

# Language and literacy difficulties

The societal consequences of literacy difficulties are highlighted, followed by an introduction to some general features of language and literacy difficulties. Reading difficulties and Reading Recovery are an important focus for the chapter, which concludes with some of the main requirements of the National Curriculum.

The development of literacy is vital for full engagement with modern society, and necessary for progression through all phases of education. As part of the government-commissioned survey of adult skills in England, approximately 15 per cent of the people who were interviewed and tested were assessed as attaining below level 1, which means that they 'may not be able to read bus or train timetables or check the pay and deductions on a wage slip' (Department for Business Innovation & Skills, 2012). If this is extrapolated to the approximately 30 million working population in the whole of the UK (Office for National Statistics, 2015), this means that 4,500,000 people may not be able to read a timetable. At entry level 1 or below, there are 1,500,000 people who 'may not be able to describe a child's symptoms to a doctor or use a cash point to withdraw cash'. Further evidence of the serious consequences of literacy difficulties include, for example, a report in the UK that found evidence that general learning difficulties affect up to 10 per cent of children, and that between 4 per cent and 8 per cent of children are affected by specific learning disability or dyslexia ☞ (The Government Office for Science, 2008). The long-lasting effects of dyslexia contribute to individuals' reduced wellbeing, self-esteem and mental capital, and even produce a predicted reduction in lifetime earnings to the individual of more than £80,000 (op. cit.).

It has been argued that difficulties with reading and/or writing can be attributed to problems in oral language development. For example, there is evidence that children who develop dyslexia, and for whom dyslexia runs in the family, 'have relatively broad oral language weaknesses that affect vocabulary knowledge and naming skills as well as phonological oral language skills' (Hulme and Snowling, 2014: 3). We also know that the quality of inter-action that children experience in the home is a vital part of their language development. For this reason, it is paramount that when young children enter formal education in nursery and reception classes, the opportunities for inter-action with adults and peers are numerous, varied and of high quality.

For all children by the time they are age 6, it is important to decide if diag-nostic assessments to determine if they have specific literacy difficulties are necessary. If diagnostic assessments are not carried out by age 6, there is a risk that literacy difficulties will be exacerbated because they have not been identi-fied with the intention to support children. We deal with reading difficulties in the latter half of the chapter, but at this point we consider some general aspects of language and literacy difficulties and support, and then some brief reflections on strategies for supporting difficulties with writing.

Unfortunately, although the identification of special needs should result in teaching and interaction that is ideally suited to the particular needs of indi-vidual children, there is worrying evidence that too often this is not the case (Webster and Blatchford, 2014). For example, children with special education needs often face segregation from the normal activities and interaction of their class: they may have separate places to work. Teaching assistants (TAs), who are often designated to support such children, can interrupt the normal teacher–pupil interaction that happens in lessons. TAs often have much more responsi-bility for the planning and teaching of children with special needs than the class teacher. This brings into question why the teacher, who is trained to a much higher level, is not taking more responsibility for the children who need the most sophisticated and skilled interaction. Webster and Blatchford's research also found that the differences between teachers' involvement and TAs' involvement was evident in less effective pedagogy employed by teaching assistants as well – for example, the use of repetitive undemanding tasks and interaction. Common to both groups are also often gaps in the teachers' and the TAs' knowledge of how to meet the needs of children with special needs most effectively.

If you accept that oral language development and quality of interaction are key factors in literacy difficulties, then it is instructive to consider literacy and writing more specifically, in particular the kinds of strategies that can make a positive difference. A strategy that has been proven on the basis of multiple experimental trial evidence of both typically developing children and children with writing difficulties is Self-Regulated Strategy Development (SRSD). In SRSD, children are taught to evaluate against criteria for how to plan, revise and edit their writing. They are also taught to set and meet explicit goals for their writing. A simpler strategy that has been shown to be effective is that when children are encouraged to dictate their compositions into a recording device, or when someone 'scribes' their compositions for them, their writing improves. Another strategy that has the benefit of multiple sources of confir-matory evidence of its success and appropriateness is the process approach to writing (see → Chapter 14). The process approach to writing makes explicit the different stages of the writing process; it incorporates writing for authentic purposes and audiences; and it involves direct instruction of writing skills for individuals and groups as the need arises.

The general features of language difficulties are implicated in more specific literacy difficulties. As part of this, the baseline checks that need to be done include checking children's hearing and vision. Reading difficulties are seen as particularly problematic because children who do not learn to read are denied access to the curriculum more generally.

## Understanding reading difficulties, including dyslexia

A child with reading difficulties can be defined as one whose reading achievements are significantly below the expected reading development of most children. But the definition of dyslexia is more precise than this: 'Developmental dyslexia is defined as specific and significant impairment in reading abilities, unexplained by any kind of deficit in general intelligence, learning opportunity, general motivation or sensory acuity (Critchley, 1970; World Health Organization, 1993)' (Habib, 2000: 2374). The important aspect to remember here is the fact that a child who is dyslexic is not regarded as having problems of the general kinds listed, for example intelligence as measured by Intelligence Quotient (IQ), hence the more specific term of dyslexia (children who have low IQ *and* reading difficulties are given the unfortunate name 'garden variety' poor readers by psychologists).

Although children are tested and registered as dyslexic, research has still not determined the precise nature of dyslexia. Goswami (2010) provides a useful overview of current understandings. From very early in life, babies begin to learn the auditory patterns of speech such as pitch, duration, rhythm and dynamics. Using this auditory information and their articulatory practice through, for example, babbling, they begin to make sense of the ways that these acoustic patterns represent meaning. In other words, they develop 'phonological representations' of the language or languages that they will eventually speak. Later, phonological awareness – an awareness of the component sounds of words – develops. Once children learn to read an alphabetic script, phonological awareness is refined to the point that the phonemes of individual words can be identified. The quality of the phonological representations that children develop is correlated with later acquisition of literacy. So, both articulatory and auditory processing are important precursors to learning to read. Brain imaging studies have shown that children with phonological difficulties rely more on articulatory networks than other children, probably as a result of compensation for poorer perceptual networks. Goswami (op. cit.) suggests that activities that emphasise rhythm and rhyme, such as nursery rhymes and songs, are likely to help support the development of phonological awareness. She suggests that because appropriately efficient ways of testing for dyslexia (that genuinely account for current research knowledge) are not yet available, all children should experience enriched linguistic and phonological environments. Consequently, she does not recommend trying to identify individual children for special programmes. However, as you will see below, this recommendation could be at odds with the success that one-to-one programmes have had, although this success may be with struggling readers more generally as opposed to those with the particular problem of dyslexia.

## Supporting children with reading difficulties

As a result of Marie Clay's work as early as the 1980s, we have known for many years that sensitive identification of children, by the time they are age 6, who struggle with reading is one important element. Sensitivity is required not least because the damaging effects on self-esteem for children who are inappropriately labelled are also understood. Clay (1979) outlined a diagnostic survey that includes a range of assessments. One of these is the 'running record': it is important to remember that this is a specific strategy for recording children's ability to read words (→ Chapter 20, 'Assessing writing, reading and oral language') rather than the more general meaning suggested by the term. Once identification has taken place, the most effective way to support children is through additional support in one-to-one sessions with a knowledgeable and skilled educator. Research evidence of the vital importance of one-to-one support can be seen in the work by Camilli et al. (2003), whose reanalysis of a meta-analysis from the National Reading Panel (NRP) in the US also questioned the strength of support for systematic phonics found by the NRP. Further evidence in relation to the importance of skilled one-to-one support can be found in relation to the international success of Marie Clay's approach called 'Reading Recovery' ☞, an approach that has been led in the UK for many years by the International Literacy Centre at the UCL Institute of Education (www.ucl.ac.uk/international-literacy/reading-recovery). Robust evidence of Reading Recovery's effectiveness can also be seen in the meta-analysis by D'Agostino and Harmey (2016); see also Wyse and Goswami (2008), who found significant experimental trials that had included Reading Recovery as the experimental group in the comparison with control groups and other approaches to reading. Evidence from multiple studies is also summarised by the US Department of Education (2013).

Reading Recovery is an early intervention programme for children with reading difficulties, and it is important to point out that 'Most children (80–90%) do not require these detailed, meticulous and special reading recovery procedures or any modification of them. They will learn to read more pleasurably without them' (Clay, 1979: 47). The teaching procedures for Reading Recovery include a range of ideas for enhancing children's reading. As far as the use of text is concerned, although Clay is critical of the controlled vocabulary of reading schemes (she emphasises the importance of natural language), she does not particularly emphasise the significance of the particular texts that children read.

Clay's procedures include learning about direction of text and pages; 'locating responses' that support one-to-one correspondence ☞ (e.g. locating words and spaces by pointing or indicating); spatial layout; writing stories; hearing the sounds in words; cut-up stories (i.e. cutting up texts and reassembling them); reading books; learning to look at print; linking sound sequences with letter sequences; word analysis; phrasing and fluency; sequencing; avoiding overuse of one strategy; memory; and helping children who are hard to accelerate.

Another procedure that Clay emphasises is the importance of 'teaching for operations or strategies' (ibid.: 71). Within this is the idea that readers need to be able to monitor their own reading and solve their own problems. It is suggested that teachers should encourage children to explain how they monitor their own reading. The process of explanation helps to consolidate the skills. Clay offers useful examples of language that teachers might use:

*Teacher:* What was the new word you read?
*Child:* Bicycle.
*Teacher:* How did you know it was bicycle?
*Child:* It was a bike (semantics).
*Teacher:* What did you expect to see?
*Child:* A 'b'.
*Teacher:* What else?
*Child:* A little word, but it wasn't.
*Teacher:* So what did you do?
*Child:* I thought of bicycle.
*Teacher:* (reinforcing the checking) Good, I liked the way you worked at that all by yourself.

*Teacher:* You almost got that page right. There was something wrong with this line. See if you can find what was wrong.
*Child:* (child silently rereads, checking) I said Lizard but it's Lizard's.
*Teacher:* How did you know?
*Child:* 'Cause it's got an 's'.
*Teacher:* Is there any other way we could know? (search further)
*Child:* (child reruns in a whisper) It's funny to say 'Lizard dinner'! It has to be Lizard's dinner like Peter's dinner, doesn't it?
*Teacher:* (reinforcing the searching) Yes, that was good. You found two ways to check on that tricky new word.

(Clay, 1979: 73–74)

## Special needs and the National Curriculum in England

The chapter on inclusion at the start of this book outlined some of the principles underlying the inclusive approach to children with particular needs. As far as the curriculum is concerned, the SEND Code of Practice makes it clear that 'all pupils should have access to a broad and balanced curriculum'

(Department for Education, 2015: 94). This requires teachers to plan their curriculum in a way that all pupils can access the activities and tasks as a result of differentiated provision that is built on assessments of the needs of different pupils. The emphasis is on ensuring as much as possible that all children can access the standard curriculum while recognising that a small minority of children will require specialist equipment and/or different approaches.

Teachers must also take account of pupils for whom English is not their first language. An important consideration is that these pupils' abilities are often in advance of their communication skills in English. So, for example, if a teaching assistant spoke their first language, these children would be able to engage much more easily with the curriculum planned by the teacher. It is important then that teachers do not have low expectations of pupils who speak English as an additional language. Instead, strategies have to be developed to support these children as their English language skills develop. These strategies include 'buddying' children with peers who speak the same language and are able to support their peers. Use of physical resources and pictures as part of activities can also be helpful. And as already mentioned, wherever possible having teaching assistants and other adults who can speak the languages that children speak, in addition to English.

### Practice points

- Identify children who are struggling as early as possible.
- Improve the relationship with the child and try to understand and empathise with their particular problems.
- Decisions should be made in terms of time and resources for extra support, including the use of classroom assistants.

### Glossary

**Dyslexia** – a formally recognised condition that results in specific difficulties with reading and writing.

**One-to-one correspondence** – the understanding that one word on the page corresponds with one spoken word. Evidence comes from finger pointing at words and numbers.

**Reading Recovery** – a set of techniques developed by Marie Clay designed to eradicate children's reading problems.

### References

Camilli, G., Vargas, S. and Yurecko, M. (2003). 'Teaching children to read: The fragile link between science and federal education policy', *Education Policy Analysis Archives*, 11(15). Retrieved 1 March 2006, from http://epaa.asu. edu/epaa/v11n15/

Clay, M. M. (1979). *The Early Detection of Reading Difficulties*, 3rd edn. Auckland, New Zealand: Heinemann Education.

D'Agostino, J., and Harmey, S. (2016). 'An international meta-analysis of Reading Recovery', *Journal of Education for Students Placed at Risk (JESPAR)*, 21(1): 29–46.

Department for Business Innovation & Skills. (2012). *The 2011 Skills for Life Survey: A Survey of Literacy, Numeracy and ICT Levels in England*. London: Department for Business Innovation & Skills.

Department for Education and Department of Health. (2015). *Special Educational Needs and Disability Code of Practice: 0 to 25 Years: Statutory Guidance for Organisations Which Work with and Support Children and Young People Who Have Special Educational Needs or Disabilities*. London: Department for Education.

Goswami, U. (2010). 'Phonology, reading, and reading difficulties', in K. Hall, U. Goswami, C. Harrison, S. Ellis and J. Soler (eds.), *Interdisciplinary Perspectives on Learning to Read: Culture, Cognition and Pedagogy*. London: Routledge.

The Government Office for Science. (2008). *Foresight Mental Capital and Wellbeing Project (2008): Final Project Report*. London: The Government Office for Science.

Habib, M. (2000). 'The neurological basis of developmental dyslexia: An overview and working hypothesis', *Brain*, 123: 2373–2399.

Hulme, C. and Snowling, M. (2014). 'The interface between spoken and written language: Developmental disorders', *Philosophical Transactions of the Royal Society B*, 369(20120395). doi: http://dx.doi.org/10.1098/rstb.2012.0395

Office for National Statistics. (2015). *UK Labour Market, February 2015*. London: Office for National Statistics.

U.S. Department of Education, Institute of Education Sciences, What Works Clearinghouse. (2013, July). *Beginning Reading Intervention Report: Reading Recovery®*. Retrieved from http://whatworks.ed.gov

Webster, R. and Blatchford, P. (2014). 'Worlds apart? The nature and quality of the educational experiences of pupils with a statement for special educational needs in mainstream primary schools', *British Educational Research Journal*. doi: 10.1002/berj.3144

Wyse, D. and Goswami, U. (2008). 'Synthetic phonics and the teaching of reading', *British Educational Research Journal*, 34(6): 691–710.

## Annotated bibliography

Clay, M. M. (1979). *The Early Detection of Reading Difficulties*, 3rd edn. Auckland, New Zealand: Heinemann Education.

This gives a full account of how to implement the Reading Recovery approach. One of the many useful aspects includes information on what a typical tutoring session looks like.

**L2** ★★

D'Agostino, J. and Harmey, S. (2016). 'An international meta-analysis of reading recovery', *Journal of Education for Students Placed at Risk (JESPAR)*, 21(1): 29–46.

In-depth analysis of the impact of Reading Recovery. Carries out statistical analysis of effects of a large number of relevant studies.

**L3** ★★★

**Reading Recovery at the International Literacy Centre** www.ucl. ac.uk/international-literacy/reading-recovery

The home for Reading Recovery in the UK.

**L1** ★★

U.S. Department of Education, Institute of Education Sciences, What Works Clearinghouse. (2013, July). *Beginning Reading Intervention Report: Reading Recovery®*. Retrieved from http://whatworks.ed.gov

Full report of the research evidence that demonstrates the long-standing effectiveness of Reading Recovery.

**L3** ★★★

# Chapter 22

# Planning

How to plan creatively within the parameters of the whole curriculum is the main concern of this chapter. We start by outlining the three levels of planning, then consider the issues involved in planning individual lessons and units of work. The premise is that good planning is enabling and should not be seen as a constraint.

Planning is commonly described in three levels: long-term, medium-term and short-term. Long-term planning tends to mean planning for a year or more and provides a broad framework of curricular provision for each year of every primary school year to ensure progression, balance, coherence and continuity. Medium-term planning refers to planning for half-terms and units of work. Short-term planning is usually weekly or daily and includes detailed lesson plans, with learning objectives that inform whole class/group and individual activities, differentiation, resources and assessment considerations. Key questions to be asked include:

- Does the planning take account of prior knowledge of pupils?
- Are you clear about what you intend the children to learn?
- Is the planning underpinned by secure subject knowledge and subject pedagogy?
- Does the planning take account of potential barriers to learning and needs of all learners in the class?
- Does the planning encourage creative and innovative responses to the task?
- Are children likely to be motivated by the suggested activities?
- Across a sequence of lessons, does the planning demonstrate progression?
- To what extent does planning take account of children's choices and interests?

Teachers need to provide high-quality, responsive literacy experiences for the pupils in their class, taking into account cross-curricular links to provide holistic learning experiences for pupils. Planning for English, language and literacy is complex because of the many strands involved and the interrelationships between them. Therein lies the challenge – and the reward!

## Planning individual lessons

Long-term and medium-term planning leads to lesson planning. Trainee and new teachers often need to devise individual lesson plans. These provide structure and security but should not be seen as completely inflexible; it is important to be responsive to the children during the session. Elements of a lesson plan are likely to include:

- administrative details (date and length of session, number and age of children, supporting adults, etc.);
- links to other curriculum areas/topics and to appropriate documentation, e.g. the National Curriculum or published schemes of work);
- continuity of learning (where the lesson fits within a sequence)
- learning objectives;
- success criteria;
- assessment opportunities;
- key vocabulary;
- resources;
- phases of the lesson (introduction, activities, plenary);
- differentiation;
- roles of other supporting adults.

Lesson plans should be annotated after teaching with evaluations and comments about children's learning so that planning for the next session can be adapted to take account of pupil progress.

There are many ways to find ideas for activities. The best way is for you to think about what you want your children to learn and the learning outcome of the lesson. You can then create suitable activities which will lead to this. When you create activities and resources yourself, you go through a process of development which includes having teaching objectives that are closely matched with the needs of your class. This creative process can, of course, involve you using ideas from all kinds of sources: colleagues, the internet, published materials, etc. While it can be tempting to use published schemes and materials, these must always be tailored for the specific learning needs of your class.

## Planning a unit of work

Our suggested approach to planning is the model developed from the Raising Boys' Achievement in Writing research (UKLA, 2004 – Figure 22.1 is taken from this). The model was also adopted by the Primary National Strategy (PNS) Literacy Framework which was in common use in English schools from

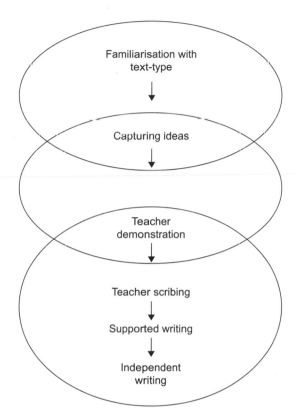

*Figure 22.1* Three circles model of planning a unit of work.

2006 to 2010. Its pedagogical strength is that it involves considering how a unit evolves from teaching reading into teaching writing, with speaking and listening objectives and activities integrated throughout. Most of this planning is in units lasting from around one to three weeks and is based around a particular genre. Every unit has a clear outcome and builds on children's prior learning. It is essential to consider the final outcome first, as all the teaching and activities you plan across the unit will lead to children being successful in achieving this.

The first part of the teaching process involves reading (including viewing where film texts are included), exploring and discussing the features of the text types covered by the unit. This is where knowledge of children's literature (see → Chapter 9) is key as it will enable you to provide high quality and motivating texts to share with the children. The purpose of this element of the teaching sequence is that pupils should enjoy exploring a range of texts and become confidently familiar with the features of the chosen text type and their language and organisational features. Planning should include time to 'play' with these features and experiment with grammatical aspects. Ideas for writing can then be captured and shaped in a variety of ways. This part of the process is where the class is likely to benefit from talk, drama or role-play activities to support their understanding of the whole text structure. Once there has been extensive experience of the text features, the process of teacher demonstrating/modelling and then scribing and guiding writing is used to support successful, sustained and independent writing.

This sequence can be planned on a piece of A4 paper using three interlocking circles as a starting point for generating ideas for creative approaches and activities that will lead to the children achieving successful outcomes. These ideas then need to be transposed onto medium-term and weekly planning with details across the week(s) of the unit for whole class, group and independent learning.

Planning on this proforma should also include grammar, punctuation, handwriting and spelling teaching and activities. It is usual for teachers to plan for guided reading separately, although, where possible, the texts used in guided reading should fit with the genre being taught in whole class sessions. Phonics sessions are also usually planned separately.

We include below a blank proforma for planning for a unit of work (Table 22.1). Planning proformas for phonics vary according the programme the school uses, so it is important to become familiar with these. For guided reading too, schools, have different expectations for planning, but a suggested sequence is provided in → Chapter 13.

*Table 22.1* Proforma for planning units and weeks of learning

| Session | Whole class teaching and learning | Link to objectives and success criteria | Activities/ independent learning | Plenary | Guided work |
|---------|---------|---------|---------|---------|---------|
| 1 | | | | | |
| 2 | | | | | |
| 3 | | | | | |
| Etc.... | | | | | |

Initial ideas for a three-week unit on narrative in Year 5 which covers traditional stories, fables, myths and legends might be as follows:

- Read wide range of myths, legends, fables and traditional stories. Discuss common themes. Identify features of these particular fiction genres.
- Read several different versions of same story, for example retellings from different times or countries, film versions. Draw out evidence of changing context and audience.
- Work in groups to explore and empathise with characters through drama activities. Children use a reading journal to record inferences and demonstrate understanding of characters by writing in the first person.

- Plan and tell stories orally. Show awareness of audience and use techniques such as humour or repetition. Performances could be recorded.
- Following teacher demonstration, plan and write a new version of a myth/legend/fable/traditional tale. Identify audience and adapt writing accordingly. Revise to produce polished version of at least one story.

These ideas could also be plotted on the three circles model. The teacher who uses such initial ideas flexibly will be able to follow children's interests and enthusiasms, adapting and innovating the suggested sequence in the light of their children's responses and the progress made.

### Practice points

- Consider your children's interests and enthusiasms as you think about planning.
- Carefully critique planning, including from published sources, and ensure that any used is adapted for the specific needs of your class.
- Familiarise yourself with the variety of unit planning, guided reading and phonics planning used in school.

### Reference

UKLA. (2004). *Raising Boys' Achievements in Writing*. Royston: UKLA.

### Annotated bibliography

Cremin, T., Reedy, D., Bearne, E. and Dombey, H. (2015). *Teaching English creatively*, 2nd edn. London: Routledge.
A useful text generally in planning to teach English creatively, but Chapter 12 is particularly helpful in providing principles and practical approaches to consider.
**L1 ★**
Brundrett, M. and Humphries, S. (2017). 'Planning', in D. Wyse and S. Rogers (eds.), *A Guide to Early Years and Primary Teaching*. London: Sage.
Good overview of planning based on practitioner author and academic author in collaboration. Includes podcasts from authors about key ideas.
**L1 ★★**
https://study.sagepub.com/wyseandrogers
Companion website to *A Guide to Early Years and Primary Teaching*. Includes a range of resources for trainee and practising teachers.

# Chapter 23

# Home–school links

> Research shows that children learn in different ways at home and at school. The importance of developing a partnership with parents/carers in order to keep communication between home and school environments open is addressed. We also explore issues surrounding the setting of homework and home-reading. The chapter includes some ideas from research on working collaboratively with families.

It is common practice in education settings to develop home–school links. By this we mean parents/carers acting as partners with educational professionals as part of their child's education. Having an understanding of learning at home and learning at school has the potential to enrich teaching and outcomes for children. Throughout the history of education in England there has been a tendency by some to think that children are like 'empty vessels' who know nothing until they are filled with knowledge by schools. This attitude resulted in parents/carers and children feeling powerless to engage in ways that would benefit their schools and their communities. However, we have now seen a growing concern within the early childhood education sector and beyond to empower parents to support the education of young children. Research now supports the notion of children learning from birth. It therefore follows that children will be learning in their home environments both before, and alongside, educational settings. This includes the development of language, reading and writing skills. The development of home–school links is particularly important in today's multicultural, multilingual society. As early as 1975, the Bullock Report signalled that a change in attitude was necessary (→ Chapter 1 for the historical sequence that led to this report):

> No child should be expected to cast off the language and culture of the home as he [sic] crosses the school threshold, nor to live and act as though home and school represent two totally separate and different cultures

which have to be kept firmly apart. The curriculum should reflect many elements of that part of his life which a child lives outside school.

(DES, 1975: S20.5)

Many schools work hard to involve parents/carers more in their children's education and determine to work collaboratively and inclusively within the immediate school community. Indeed, in England there is a current premise of parents as partners in their children's education. However, the ever increasing documentation that accompanies the school curriculum runs many risks, one of which is the alienation of parents/carers who may feel that they have nothing to offer in such a complex world. Nothing could be further from the truth. Today, parent workshops are commonplace in schools, for example workshops about phonics and learning to read. Parents' evenings can serve an additional purpose of introducing parents to the precepts of learning at school and how to support their child's language, reading, and writing skills.

## Learning at home and at school

Socio-cultural theories explain that children's differing perceptions and motivation to read and write could be attributed to different levels of exposure to, and experiences of, reading and writing in the cultural contexts of their home and community (Pellegrini, 2001; Compton-Lilly, 2006). To be aware of the kind of literacy activities that a child engages in at home can support an effective literacy curriculum in school. The 1980s in particular saw the rise of seminal research which offered evidence of the kinds of language and literacy learning that was taking place in children's homes. Tizard and Hughes's early work looked at the differences between talking and thinking at home and at school. They argued forcefully that the home was a 'very powerful learning environment' (1984: 249) and that school nurseries were not aware of this:

> Our observations of children at home showed them displaying a range of interests and linguistic skills which enabled them to be powerful learners. Yet observations of the same children at school showed a fundamental lack of awareness by the nursery staff of these skills and interests. There is no doubt that, in the world of the school, the child appears to be a much less active thinker than is the case at home. We do not believe that the schools can possibly be meeting their goals in the most efficient manner if they are unable to make use of so many of the children's skills.
>
> (ibid.: 264)

More recent research such as that of Shore (2015), which investigates the relationship between socio-economic background and school experiences, reveals how language levels still impact on children's learning. Whilst a small-scale study, the findings raise questions about the language children experience from an early age both in the home environment and at school and suggest that there is a significant part for schools to play in ensuring that they are not excluding some groups of children from participation.

As far as reading is concerned, there are two particularly powerful seminal studies that showed the positive influence of parents. Prior to 1966, Durkin carried out some research in the US and published it in a book called *Children Who Read Early* (Durkin, 1966). These were the things that were common to the children's experiences:

- Parents had read to their children frequently and also found time to talk with them and answer their questions.
- Real books were more commonly read by the children than typical school textbooks or reading scheme books.
- Whole-word learning had been used more than letter-sounds for reading, although letter-sounds had been used to help writing.
- Print in their everyday surroundings was of interest to the children.
- Many of the children had been interested in writing as well as reading.

A decade later, Clark (1976) published a very similar study in England. She said that although the child's natural abilities were important, 'the crucial role of the environment, the experiences which the child obtained, their relevance to his interest and the readiness of adults to encourage and to build upon these, should not be underestimated' (1976: 106).

And here are the factors that were common to the children in Clark's study:

- At least one parent, often the mother, had a deep involvement with their child and their progress.
- Parents welcomed the opportunity to talk to their children.
- Non-fiction and print in the home and local environment were mentioned as much as the reading of storybooks.
- Most of the children used the public library.
- Parents would happily break off from other activities and tell their child what words said if the child couldn't work them out independently from the context.
- Very few parents taught their children the letter-sounds. If they did, this was to help with writing more than reading. More children learned the letter names first.
- The children had a range of strategies for working out difficult words if their parent wasn't available to supply the word.
- Many of the children were interested in writing as well as reading.

There was very little evidence that systematic instruction in reading, including phonics, was part of these children's pre-school experiences, yet they all learned to read before they started school. Socio-economic differences were not a significant factor in the research, which showed that children from a range of family backgrounds learned to read. The implications for teachers are understanding that: (a) a range of approaches to reading is beneficial; (b) support for children should be based on understanding of home experiences; (c) other parents/carers may benefit from guidance in the kinds of approaches that the parents in the two studies used. Wyse (2007) offers guidance to parents/carers on supporting their children, not just in reading but also in writing because of the mutual benefits that occur from looking at both.

Research by those such as Marsh (2010) and Levy (2011) draw on Bhaba's (1994) 'third space theory', which reiterates the case for children's home literacy experiences to be valued. 'Third space theory' is a metaphor for the space in which cultures of home and school meet and explores how these diverse worlds can be brought together in educational settings. Millard (2003: 6) refers to this process as a 'transformative pedagogy of literacy fusion', suggesting that children's out-of-school interests can be fused with schooled literacy in classroom practice that pays attention to what happens when the two textual worlds collide.

## Home–school agreements

The Department for Education in England used to require governing bodies in educational settings to provide a home–school agreement explaining the school's aims and values, its responsibilities towards its pupils, the responsibilities of each pupil's parents and what the school expected of its pupils. Home–school agreements are no longer statutory, however. In England, the current Conservative government scrapped the requirement for such agreements in January 2016, arguing that in this way schools would benefit from a more fluid collaborative endeavour with parents/carers and pupils. This may see the development of more meaningful home–school agreements that reflect for example greater understanding of families' needs.

## Homework

Findings from the *Cambridge Primary Review* (Alexander, 2010) show that, in general, parents want schools to set homework and other outside-of-school activities for their children. The government recommendation is that children as young as 6 (Years 1 and 2) should do one hour a week of homework, ages 7–9 (Years 3 and 4) should do 1.5 hours a week, and those aged 10 and 11 (Years 5 and 6) should do 30 minutes a day.

As schools are required to set homework, it is important that they encourage children to engage in interesting activities. For example, one of our children's teachers suggested that they phone up a grandparent and ask them about the time when they were children. This activity inspired Esther, aged 7 years, to write one of her longest pieces of writing at home:

My grabad [grandad] and gramar didn't have a tely. they did hav a rabyo [radio]. they had a metul Ian [iron]. they had sum bens [beans] and vegtbuls. thee wa lots ov boms in the war. the shoos were brawn and blac Thay had long dresis. Thay had shun [short] trawsis and long socs.

There is always the danger that the pressures of time for teachers can result in photocopied homework sheets that are uninteresting and of questionable value. As is the case with many things in teaching, it is better to organise a limited number of really exciting homework tasks that are genuinely built on in the classroom than to set too many tasks where it is difficult for the teacher to monitor them all.

One of the most common strategies to support home–school links has been through book bags. Each child has a durable bag that contains books, often a reading scheme book and a free choice book, and the child takes this home on a regular basis. A reading journal or diary generally accompanies the books, and parents/carers are encouraged to note the date, title of the book and to make a comment about their child's reading. Table 23.1 shows an example of a parent's comments.

The idea of book bags was extended by the Basic Skills Agency (and subsequently build on by BookTrust), who set up a National Support Project to promote 'story sacks' throughout England and Wales. These have proved very popular and are widely used. Story sacks, containing a good children's book and supporting materials, are designed to stimulate reading activities. The sacks and the artefacts associated with the book can be made by parents/carers and other volunteers. This can be a good way to involve parents/carers in the school community. Other related items such as an audio recording, CD or animated video of the text, language game and other activities, can be used by parents/carers at home to bring reading to life and develop the child's receptive language skills.

Table 23.1 A parent's comments on a child's reading

| Date | Book | Comments |
| --- | --- | --- |
| 7/5 | Roll over | Well read. |
| 14/5 | Better than you | Fluent reading. |
| 17/5 | Big fish | Well read. Why no punctuation, i.e. question marks, speech marks? Esther commented on this. |
| | | [Teacher: I don't know. I will check.] |
| 21/5 | Sam's book | Well read. |
| 28/5 | Lion is ill | Well read. |

## Working with parents/carers in the classroom

Teachers have responsibility for managing other adults who work with them. As a teacher, you are likely to have the opportunity to work with parents/carers who have volunteered to help in the classroom. These parents/carers volunteer to support schools in their own time and are a precious resource.

One of the most important things to remember is that schools and teachers need to offer guidance to people who are supporting literacy in the classroom. Parental help is often invaluable in the group work section of the literacy sessions. They can also support struggling readers either individually or in groups, but it should be remembered that this is a skilled task and they will require the chance to discuss how things are going and how they can best help the children.

Knowsley Local Education Authority carried out a project that included the recruitment and training of large numbers of adult volunteers who helped primary pupils with their reading on a regular basis. An evaluation by Brooks *et al.* (1996: 3) concluded that the training for parents/carers and other volunteers was one of the most important components of the project, and 'it seemed to make the most significant difference to raising reading standards'. The idea of training parents/carers is one that the Basic Skills Agency has also been involved in. The main purpose of their family literacy initiatives was to raise the basic skills of both parents/carers and children together. For parents/carers, the emphasis was mainly on helping them to understand more about what happened in schools and how they could support this. The children's sessions involved hands-on motivational activities. Joint sessions were also held where parents/carers were encouraged to enjoy a natural interaction with their children during joint tasks. Historically, Brooks *et al.* (1999) found that these family literacy programmes – with some modifications – worked as well for ethnic minority families as for other families.

*Practice points*

- Develop strong home–school links through genuinely seek information from parents/carers about their children.
- Support engagement through thinking about how to incorporate children's learning outside the classroom within the school curriculum.
- Involve and support parents/carers who work in your classroom as much as possible.

*References*

Alexander, R. (ed.). (2010). *Children, Their World, Their Education: Final Report and Recommendations of the Cambridge Primary Review.* London: Routledge.

Bhaba, H. K. (1994). *The Location of Culture.* New York: Routledge.

Brooks, G., Cato, V., Fernandes, C. and Tregenza, A. (1996). *The Knowsley Reading Project: Using Trained Reading Helpers Effectively.* Slough: The National Foundation for Educational Research (NFER).

Brooks, G., Harman, J., Hutchison, D., Kendall, S. and Wilkin, A. (1999). *Family Literacy for New Groups.* London: The Basic Skills Agency.

Clark, M. M. (1976). *Young Fluent Readers.* London: Heinemann Educational Books Ltd.

Compton-Lilly, C. (2006). 'Identity, childhood culture, and literacy learning: A case study', *Journal of Early Childhood Literacy*, 6(1): 57–76.

DES (Department of Education and Science). (1975). *A Language for Life (The Bullock Report).* London: HMSO.

Durkin, D. (1966). *Children Who Read Early.* New York: Teachers College Press.

Levy, R. (2011). *Young Children Reading at Home and at School.* London: Sage.

Marsh, J. (2010). *Childhood, Culture and Creativity: A Literature Review.* Newcastle: Creativity, Culture and Education.

Millard, E. (2003). 'Transformative pedagogy: Towards a literacy of fusion', *Reading, Literacy and Language*, 37(1): 3–9.

Pellegrini, A. D. (2001). 'Some theoretical and methodological considerations in studying literacy in social context', in S. B. Neuman and D. K. Dickinson (eds.), *Handbook of Early Literacy Research.* New York: The Guildford Press.

Shore, L. M. (2015). 'Talking in class: A study of socio-economic difference in the primary school classroom', *Literacy*, 49(2): 98–104.

Tizard, B. and Hughes, M. (1984). *Young Children Learning: Talking and Thinking at Home and at School.* London: Fontana Press.

Wyse, D. (2007). *How to Help Your Child Read and Write.* London: Pearson and BBC Active.

### Annotated bibliography

Bonci, A. (2008). *A Research Review: The Importance of Families and the Home Environment.* London: National Literacy Trust (revised 2010 and March 2011). Thorough exploration of what the research says in relation to the development of literacy skills in the home environment.
L3 ★★★

Maher, D. and Twining, P. (2017). 'Bring your own device – a snapshot of two Australian primary schools', *Journal of Educational Research*, 59(1): 73–88. Using mobile devices from home such as tablets to support educational practice. Includes a comparison of English schools and discussion around the use of such a pedagogical approach to support home–school links.
L2 ★★★

The Skills for Life network includes resource aimed at adults who struggle with various skills, including literacy: www.skillsforlifenetwork.com/article/online-maths-english-courses/5178.
L1 ★

# Index

Note: Page numbers in *italics* refer to figures; **bold** numbers refer to tables.

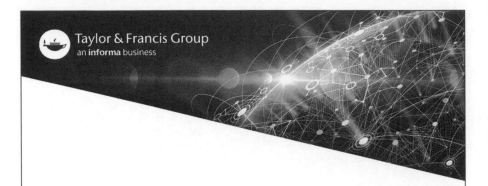